Paediatric Biomechanics and Motor Control

Paediatric Biomechanics and Motor Control: Theory and application brings together the very latest developmental research using biomechanical measurement and analysis techniques and is the first book to focus on biomechanical aspects of child development. The book is divided into four main sections: the biological changes in children; developmental changes in muscular force production; developmental changes in the biomechanics of postural control and fundamental motor skills; and, finally, the applications of research into paediatric biomechanics and motor control in selected clinical populations.

Written by a team of leading experts in paediatric exercise science, biomechanics and motor control from the UK, the US, Australia and Europe, the book is designed to highlight the key implications of this work for scientists, educators and clinicians. Each chapter is preceded by a short overview of the relevant theoretical concepts and concludes with a summary of the practical and clinical applications in relation to the existing literature on the topic. This book is important reading for any sport or exercise scientist, motor developmental specialist, health scientist, physical therapist, sports coach or clinician with an interest in child development or health.

Mark De Ste Croix PhD, is a Reader in Paediatric Sport and Exercise Physiology, Department of Sport and Exercise Science, University of Gloucestershire, UK.

Thomas Korff PhD, is a Senior Lecturer in Biomechanics, Centre for Sports Medicine and Human Performance, Brunel University, UK.

Routledge Research in Sport and Exercise Science

The *Routledge Research in Sport and Exercise Science* series is a showcase for cutting-edge research from across the sport and exercise sciences, including physiology, psychology, biomechanics, motor control, physical activity and health, and every core sub-discipline. Featuring the work of established and emerging scientists and practitioners from around the world, and covering the theoretical, investigative and applied dimensions of sport and exercise, this series is an important channel for new and ground-breaking research in the human movement sciences.

Also available in this series:

Mental Toughness in Sport
Developments in theory and research
Daniel Gucciardi and Sandy Gordon

Paediatric Biomechanics and Motor Control
Theory and application
Edited by Mark De Ste Croix and Thomas Korff

Attachment in Sport, Exercise and Wellness
Sam Carr

Paediatric Biomechanics and Motor Control
Theory and application

Edited by Mark De Ste Croix
and Thomas Korff

LONDON AND NEW YORK

First published 2012
by Routledge
2 Park Square, Milton Park, Abingdon, Oxon OX14 4RN

Simultaneously published in the USA and Canada
by Routledge
711 Third Avenue, New York, NY 10017

Routledge is an imprint of the Taylor & Francis Group, an informa business

© 2012 for selection and editorial material Mark De Ste Croix and
Thomas Korff; individual chapters, the contributors

British Library Cataloguing in Publication Data
A catalogue record for this book is available from the British Library

Library of Congress Cataloging-in-Publication Data
Paediatric biomechanics and motor control : theory and application /
edited by Mark De Ste Croix and Thomas Korff.
p. ; cm. – (Routledge research in sport and exercise science)
Includes bibliographical references and index.
1. Motor ability in children. 2. Biomechanics. 3. Child development.
I. De Ste Croix, Mark. II. Korff, Thomas. III. Series: Routledge research
in sport and exercise science.
[DNLM: 1. Motor Skills–physiology. 2. Adolescent. 3. Biomechanics. 4. Child Development. 5.
Child. 6. Infant. WE 103]
RJ133.P335 2012
618.92–dc23
2011016303

ISBN: 978–0–415–58018–2 (hbk)
ISBN: 978–0–203–85121–0 (ebk)

Typeset in Goudy
by Keystroke, Station Road, Codsall, Wolverhampton.

Printed and bound in Great Britain by
CPI Antony Rowe, Chippenham, Wiltshire

Mark: Martine, Lexie and Kina. Thanks for all your love and
  support.

Thomas: To my wife, Gemma.

Contents

Illustrations

Figures

Tables

Contributors

Anthony Blazevich PhD, Associate Professor in Biomechanics, School of Exercise, Biomedical and Health Sciences, Edith Cowan University, Australia.

Jane Clark PhD, Professor in Kinesiology, Department of Kinesiology, College of Health and Human Performance, University of Maryland, College Park, USA.

Diane Damiano PhD, Chief of the Functional and Applied Biomechanics Section, Rehabilitation Medicine Department, National Institute of Health Clinical Center, Maryland, USA.

Mark De Ste Croix PhD, Reader in Paediatric Sport and Exercise Physiology, Department of Sport and Exercise Science, University of Gloucestershire, UK.

Martine Deighan PhD, Senior Lecturer in Biomechanics, Department of Sport and Exercise Science, University of Gloucestershire, UK.

Florian Fath BSc, Research Student in Paediatric Biomechanics, Centre for Sports Medicine and Human Performance, School of Sport and Education, Brunel University, UK.

Caroline F. Finch PhD, NHMRC Principal Research Fellow and Research Professor, Australian Centre for Research into Injury in Sport and its Prevention, Monash Injury Research Institute, Monash University, Australia.

Vassilia Hatzitaki PhD, Assistant Professor in Motor Control, Laboratory of Motor Control and Learning, Department of Physical Education and Sports Sciences, Aristotle University of Thessaloniki, Greece.

Jody Jensen PhD, Professor in Kinesiology, Department of Kinesiology and Health Education, The University of Texas at Austin, USA.

Eleftherios Kellis PhD, Associate Professor in Sport Kinesiology, Laboratory of Neuromechanics, Department of Physical Education and Sports Sciences, Aristotle University of Thessaloniki, Greece.

Thomas Korff PhD, Senior Lecturer in Biomechanics, Centre for Sports Medicine and Human Performance, School of Sport and Education, Brunel University, UK.

Masayoshi Kubo ScD, Professor, Department of Physical Therapy, School of Medical Technology, Niigata University of Health and Welfare, Japan.

Stephen Langendorfer PhD, Professor in Kinesiology, School of Human Movement, Sport, and Leisure Studies, Bowling Green State University, Ohio, USA.

Jan Piek PhD, Professor of Developmental Psychology, School of Psychology and Speech Pathology, Curtin University, Australia.

Laura Prosser PhD, Postdoctoral Fellow, Functional and Applied Biomechanics Section, Rehabilitation Medicine Department, National Institute of Health Clinical Center, Maryland, USA.

Mary Ann Roberton PhD, Professor Emerita in Kinesiology, School of Human Movement, Sport, and Leisure Studies, Bowling Green State University, Ohio, USA.

David Stodden PhD, Associate Professor, Department of Health, Exercise and Sports Sciences, Texas Tech University, USA.

Dara Twomey PhD, Lecturer in Biomechanics, School of Human Movement and Sport Sciences, University of Ballarat, Australia.

Beverly Ulrich PhD, Professor and Director of Developmental Neuromotor Control Laboratory, School of Kinesiology, University of Michigan, USA.

Renate van Zandwijk MSc, Research Student in Movement Science, Department of Kinesiology and Health Education, The University of Texas at Austin, USA.

Charlie Waugh PhD, Post-doctoral Research Fellow, Centre for Sports Medicine and Human Performance, School of Sport and Education, Brunel University, UK.

Jill Whitall PhD, Professor in Neuromotor Control and Rehabilitation, Department of Physical Therapy and Rehabilitation, School of Medicine, University of Maryland, Baltimore, USA.

Craig A. Williams PhD, Associate Professor in Paediatric Physiology, Children's Health and Exercise Research Centre, School of Sport and Health Sciences, Exeter University, UK.

Louise Wood PhD, Senior Lecturer in Sports Biomechanics, University of Portsmouth, UK.

Foreword

Young people are not mini adults. Children grow and mature under the influence of individual biological clocks. Their psychological, physiological and biomechanical responses to exercise of different intensity, duration and frequency vary as they move from childhood through adolescence into young adulthood. The study of paediatric exercise science has emerged as a major component of sport and exercise science and medicine, and over the last two decades there has been a dramatic increase in published research focusing on the exercising child and adolescent. The publication of original research has led to several excellent books critically reviewing the literature in paediatric exercise physiology and psychology but, to date, no single text has adequately addressed developmental biomechanics and motor control. *Paediatric Biomechanics and Motor Control: Theory and application* fills this important omission in the extant literature.

The book consists of 13 chapters which not only synthesize and analyse relevant research findings, but also discuss the implications of the research for scientists, educators and clinicians. There is a common structure across all chapters, including sections on practical and/or clinical applications of the material presented. Students are well served by being introduced to the topics through a tutorial at the beginning of each chapter, and having the content summarized and reinforced through a comprehensive list of key points at the end of each chapter. The four sections of the book cover the key areas of *biological changes during motor development; motor development and force production; biomechanical aspects of the development of postural control and selected fundamental motor skills;* and *selected clinical applications*.

Editors Mark De Ste Croix and Thomas Korff are well known and respected as researchers and teachers in developmental muscle physiology and biomechanics. They have recruited a well-balanced international list of established authors and contributors who have recently emerged as leading researchers in the fields covered by their chapters. The book benefits greatly by the authors being able to enrich their contributions by drawing upon their own research to inform and challenge a wide spectrum of readers.

The publication of *Paediatric Biomechanics and Motor Control: Theory and application* will be welcomed by students, researchers, educators, scientists and clinicians working in paediatric exercise science and medicine. The editors have done a splendid job in bringing together so much material and disseminating it in such an accessible format.

Professor Neil Armstrong PhD, DSc
Exeter, 2011

Preface

Motor development is a multi-factorial process. When children learn how to sit, stand, walk, jump or throw, many factors contribute to how these tasks are performed and to how the execution of these tasks changes as children develop. These factors include the development of the central nervous system, neuro-musculoskeletal system, the accumulation of experience and repeated exposure as well as the environment in which the task is performed.

To fully understand how children become more skilled at performing motor tasks, we need to take all of these factors into consideration. What complicates this process (but makes it more exciting at the same time) is the fact that all these changes do not occur simultaneously and follow their own developmental trajectories. Therefore, understanding the complex interaction of factors that contribute towards motor control during growth and development is challenging.

Biomechanics is the application of mechanical principles to biological systems. It is a tool that allows us to gain unique insights into the mechanical make-up of human movement. With the advance of technology, it provides a seemingly endless number of opportunities to explore the mechanics of human movement. Historically, our knowledge may have been restricted by access to technology that allows us to explore the mechanisms associated with motor development. The use of magnetic resonance imaging (MRI), 2-D and 3-D video analysis, ultrasonography and electromyography have all helped to advance our understanding of the underlying biomechanical mechanisms that contribute towards the development of motor control.

It comes as no surprise that many researchers make use of biomechanical tools with the goal of unravelling the mysteries of motor development. The use of biomechanics has allowed us to dramatically increase our understanding of the mechanisms underlying motor skill acquisition in children with implications for teachers, coaches and clinicians. Therefore, the primary objective of this book is to synthesize current findings in which biomechanical principles have been used to increase our understanding of motor development and to point out applications and implications of this research for scientists, educators and clinicians. This may include the development of training programmes to enhance motor performance, measures to reduce relative risks of injury, the identification of atypically developing children or the implementation of appropriate clinical interventions.

The book is aimed at striking a balance between being scientific and educational, and every chapter begins with a description of the relevant biomechanical concepts and concludes with clinical and practical applications and implications. Chapters are self-contained but are cross-referenced to direct the reader to other parts of the book where specific issues or topics may be discussed in more detail. Each chapter ends with a list of key points so that the reader can clearly identify the most important 'take home messages' within a given research area.

Although developmental biomechanics is a rapidly growing discipline, and a significant amount of research has been published in this area over the last decade, our knowledge of paediatric biomechanics and motor control is still in its infancy. To this end, areas for future research are highlighted throughout the book. Therefore, the content of this book is a snapshot of our current knowledge in the area of developmental biomechanics and motor control. If it stimulates further debate in the area, it will have served its purpose.

Mark De Ste Croix
Thomas Korff
April, 2011

Part I

Biological changes during motor development

1 Growth and maturation during childhood

Craig A. Williams, Louise Wood and Mark De Ste Croix

Introduction

Understanding movement patterns and motor control during childhood is challenging, and the complex interaction of growth and maturation contributes towards this challenge. The monitoring of children's growth and maturation is not a simple task, and in part both ethical and methodological constraints have hindered our understanding of the growing child. Although growth and maturation are related concepts, it is important to acknowledge that although they are related, they are harmonized by differing time-scales and are probably controlled under separate biological regulation (Armstrong and Welsman 1997). The term 'development' has also been used in relation to growth and maturation but this term really refers to broader concepts that include behavioural and psychological as well as biological domains. Children are often placed into chronological age groups but it is well recognized that chronological age is a poor marker of biological maturity. The purpose of this chapter is to explore the key mechanisms of growth in preparation for later chapters where its impact upon motor development and motor control are discussed. Therefore, this chapter will primarily examine age- and sex-associated changes in stature, body mass, limb length and muscle size. Clinical, injury and performance applications will also be discussed.

Growth

Growth refers to an increase in the size of the body or any of its specific parts. Growth is a cyclic process where tissues and organs are in a constant process of growth, death and regeneration. Also, individual segments of the body do not grow at the same rate, and therefore the relative size and shape of the tissues and organs change throughout the life cycle. This has specific implications for motor control and will be discussed in more detail in relation to changes in limb length later in this chapter. The most common measurements of growth include stature and body mass, although other size measurements, including breadths, lengths, widths and girths are available. It is widely recognized that changes in body composition (fat-free mass, cross-sectional area, percentage body fat) provide paediatric researchers with more meaningful data regarding the relative proportions of tissue rather than relying simply on body mass per se.

Maturation

Maturation refers to the tempo and timing of progress towards the mature biological state. Maturation differs from growth in that it occurs at varying rates for individuals, but all individuals reach the same endpoint (e.g. fully mature). In its simplest terms, growth focuses on size and maturation on the progress of attaining size. It is well recognized that the duration of puberty embraces all of the physiological and morphological alterations that lead to the development of structure and function characteristics of adults. Maturation is a difficult concept to define, and techniques of assessing maturation vary depending on the biological system to be assessed. Commonly used techniques for determining maturation include the assessment of secondary sex characteristics, the age corresponding to peak growth in height, skeletal maturation and measurement of circulating hormones. Difficulties arise in that even these biological processes do not progress at a similar rate. Sexual maturation is highly related to the overall process of physiological maturation and is therefore a useful indicator of biological maturation. However, one method for the estimation of maturity status should not be used to predict another maturity status, e.g. using sexual maturity to predict skeletal maturity (Bielicki 1975; Bielicki *et al.* 1984). Assessment of secondary sex characteristics involves examination of breast, genital and pubic hair development (Tanner 1962), and as such there are considerable ethical implications associated with the technique. Subsequently, the development of non-invasive estimates of biological maturation using anthropometric measurements has become increasingly popular in the paediatric literature.

During puberty, the majority of individuals experience a growth spurt, which is most evident as rapid height gain (discussed later in this chapter). The age at which the growth spurt is most pronounced (i.e. the age at peak height velocity (PHV)) can be used as an indicator of maturity, with a child undergoing their growth spurt earlier than their peers being more mature. Mirwald *et al.* (2002) presented equations for boys and girls which included chronological age, gender, stature, sitting height and body mass to predict the offset from PHV. A further maturity assessment using X-ray scans assesses the degree of epiphyseal closure. The three most common methods – the Greulich-Pyle, Tanner-Whitehouse and Fels – all estimate the skeletal age of the bones in the wrist and hand relative to chronological age. It should be noted that the determined skeletal age by each of these methods may be different for the same child. Therefore, caution must be taken when comparing skeletal age across studies which have used different skeletal maturity methods. Other studies have examined the hormonal sex steroids as a marker of sexual maturation (Beunen *et al.* 2006; Falgairette *et al.* 1990). These include the determination of testosterone and oestradiol, but they often require a blood sample and some authors suggest information using these methods are limited due to large diurnal fluctuations and inter-individual differences (Matchock *et al.* 2007).

Despite the range of techniques available to estimate maturity status, a child who is classified as early, average or delayed in maturity by one method is likely to be classified similarly by other methods (Faulkner 1996). Although there appears to be no single system that will provide a complete description of an individual child's tempo of growth and maturation (Malina *et al.* 2004), it is important to understand that individual differences in this timing influence motor control and motor development. Very few studies investigating force development or motor control/development have directly taken into account the effect of biological maturation.

Basic theoretical concepts

The growth of bone and muscle play important roles in force development, musculoskeletal loading and motor control during childhood. The age and sex-associated changes in stature, limb length, body mass and muscle size are discussed in further detail later in this chapter. However, to understand these changes, the mechanisms that contribute to the physical growth of these tissues must be appreciated.

Bone growth

The skeleton accounts for around 15 per cent of body mass at birth and rises to about 17 per cent in young adults (Malina *et al.* 2004). Skeletal maturation and the growth of long bones have been reported as the most reliable method for assessing biological age, which can span the entire growth period (Faulkner 1996). Growth of the long bones at the epiphyseal plates results in increases in limb length and subsequently stature. Trapped osteoblasts in the bone tissue become osteocytes and begin to collect phosphate and calcium. This process continues until the plates become ossified. Therefore, a more mature child has more bone mineral and less cartilage than a less mature child. Growth of a long bone presents a unique problem, as the bone needs to grow in both length and width while maintaining its shape at the same time. During the early years of life growth of long bones is rapid, which is evident in the increase in stature during the first three years of life. Radiographs have shown that the length of the femur increases by about 58 per cent in the first year of life and then by 27 per cent in the second year. Growth of a long bone in length is influenced by both hormonal and nutritional factors. It is important to acknowledge that bone growth rates vary, which causes changes in body proportions with growth and maturation. Differential growth rates and proportionality changes play an important role in motor control and locomotion, and influences muscular loading and force development (see Chapters 4, 6 and 8).

Data also suggest that, on average, ossification of the secondary centres of the major long bones begins earlier and is completed earlier in girls than boys (Malina *et al.* 2004). Weight-bearing activities are important for the progress of this ossification, and there are a number of studies that have shown that increased physical activity during childhood helps to promote the development of bone mineral density (Debar *et al.* 2006; Heinonen *et al.* 1999; Mirwald *et al.* 1999).

This is important as in the growing child bone deposition occurs at a greater rate than bone resorption. However, by ~25 years of age, the drive for bone formation and resorption are similar. Bone mineral content increases in a linear manner with age, and there appear to be no sex differences during childhood. Girls tend to have on average a slightly greater bone mineral content than boys in early adolescence, which is probably reflective of their generally earlier growth spurt. As boys' growth spurts are generally later than girls', sex differences in bone mineral accruement are established in late adolescence. Total body bone mineral continues to increase into the 20s in boys but appears to reach a plateau at about 15–16 years of age in girls. This has important implications, especially in girls, where poor future bone health has been shown to be a predisposition for bone disorders such as osteoporosis. Recent data have also shown that excessive adiposity during childhood may be detrimental to development of bone strength parameters and bone mass accrual during growth (Mughal and Khadikar 2011). This chapter primarily focuses on longitudinal bone growth as opposed to bone mineral density as longitudinal growth has implications for biomechanical movement. The magnitude and rate of growth of the skeleton as a whole (stature) and differential bone growth are described during growth and maturation. Finally, the implications of these changes are discussed.

Muscle growth

Muscle accounts for the largest tissue mass in the human body and is the most metabolically active organ. Muscle growth begins early in foetal life, yet in these early stages there are substantial structural differences compared to adult muscle. Primarily, muscle fibres are both small in number and size and are widely separated by extracellular material. However, the size of the extracellular component decreases throughout foetal life due to an increase in the number of muscle fibres (Malina *et al.* 2004). The origin of the diversity in muscularity in adults occurs early in foetal development, at approximately the fifth week of gestation when some mesodermal cells differentiate into myoblasts. Most of the myoblasts fuse to form myotubes containing multiple nuclei that attach to the developing skeleton to form primordial muscles. The primordia of most muscle groups are well defined by the end of the ninth week of gestation. The others stay as mono-nucleate cells that become the satellite cells of more mature muscle, responsible for muscle cell regeneration. Within the myotubes of primordial muscles a chain of central nuclei form and soon after the contractile proteins actin and myosin with their characteristic striations are synthesized.

The muscularity of an individual is reliant on the size and number of muscle fibres. From 11 to 18 weeks hypertrophy of the muscles occurs due to both the multiplication of myofibrils and the addition of sarcomeres onto the ends of the muscle. By 23 weeks the nuclei of mature myotubes have moved to the edges of the muscle cell. At about 10 weeks of gestation outgrowths from the spinal motor neurones begin to innervate the developing muscle fibres. What initially begins with multiple synapses ends up with only one neuromuscular junction, usually in the centre of the muscle fibre. Muscle fibre type or the contractile and metabolic characteristics are determined at this early stage since muscle is a slave to its innervation or electrical frequency of stimulation. Generally, about half of the developing fibres express slow myosin isoforms and the other half, fast isoforms. There is some debate as to when the number of muscle fibres become fixed, with some authors suggesting that this happens as early as the fourth or fifth month of foetal life (Sinclair 1975), while others suggest that fibre numbers double from the 32nd week of gestation to around 4 months of age (Malina 1986). Some authors have suggested that an increase in muscle size is in part due to muscle fibres splitting longitudinally to form two daughter fibres (Edgerton *et al.* 1986; Sinclair 1975). The number of muscle fibres is due to the number of foetal myoblasts, which in part is genetically driven. The question of whether increases in muscle size during growth are due to hypertrophy or hyperplasia has been difficult to answer in situ due to the ethics of muscle biopsy and also limitations of non-invasive imaging techniques. However, it is clear from a cross-sectional study of whole autopsied vastus lateralis muscles from 22 males aged 5 to 37 years that, despite wide inter-individual variation, the average total number of fibres remains stable across age groups (Lexell *et al.* 1992). Therefore, during normal growth and development, increases in muscle cross-sectional area are generally agreed to be due to increased fibre size or hypertrophy rather than cellular hyperplasia.

Muscle fibre types appear to be undifferentiated prior to 30 weeks of gestation. Type I fibres appear around this time and represent around 40 per cent of the muscle at birth. These include an unusual type I 'large' fibre that decreases in number during gestation and their actual presence at birth is rare. Likewise, type II a and b fibres appear around 30 weeks of gestation and represent approximately 45 per cent of the muscle with the remaining 15 per cent consisting of undifferentiated type II c fibres, which change after birth. Within the quadriceps muscles, the muscle fibre distribution at the time of birth reflects an even distribution of type I and II fibres and fibres destined to become either type I or II contain both fast and slow enzymes at birth (Baldwin 1984). Unsurprisingly, the ethical constraints of obtaining muscle biopsies have limited the availability of data on muscle fibres types in children.

Examining muscle development in children differing in the stages of growth and maturation aids understanding of age and sex differences in strength and motor skill aptitude. As well as direct and indirect assessment of muscle size, body mass if considered alongside body composition (fat and fat-free mass) has been deemed to reflect muscle mass. In addition, body mass represents a load that must be supported by lower extremity muscle groups and therefore provides a stimulus for muscle development. Therefore, changes in stature, skeletal lengths and body mass during childhood and adolescence are important to describe and understand.

Age- and sex-associated changes in stature, body proportions, body mass and muscle size

Changes in stature with age in boys and girls

Clinically, the use of charts documenting the typical development of stature and body mass with age have been useful in monitoring the growth of healthy children and assisting in the identification of atypical development (Cole 1997; Freeman *et al.* 1995; Tanner and Whitehouse 1976; Tanner *et al.* 1966). When successive measurements of stature and body mass are used to calculate growth velocities (cm·y^{-1}) a valuable insight into the challenges to motor control and injury profiles of children can be gained.

To calculate growth velocities, stature and body mass have typically been measured every six months, with more frequent measurements (every three months)

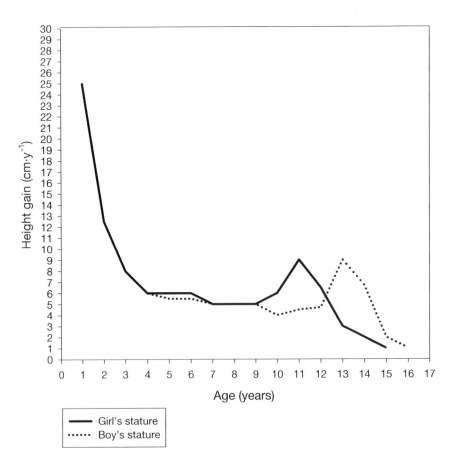

Figure 1.1 An individual velocity curve for stature for one boy and one girl (illustrative purposes only).

throughout adolescence (Tanner *et al.* 1966). When these growth velocities are plotted against age (Figure 1.1), the majority of children demonstrate an adolescent growth spurt, where the velocity of growth markedly increases. A smaller growth spurt has also been reported between 4 and 6 years of age (Westcott *et al.* 1997). Although there is considerable variation in the timing and magnitude of growth during adolescence, the average age at peak height velocity (PHV) is approximately 12 years of age for girls (range: 9.5–14.5 years) and 14 years of age for boys (range 10.5–17.5) (Baxter-Jones *et al.* 2005). Boys therefore have approximately an additional two years of growth prior to their growth spurt. The guidelines for classifying a child as an early or late maturer refer to the age at PHV as being two standard deviations below or above the average for boys and girls, respectively (Tanner and Whitehouse 1976). The studies by Faust (1977) and Tanner *et al.* (1966) are in agreement concerning the mean peak growth velocity reached during this adolescent growth period for boys and girls: 10.4 ± 1.6 cm·y^{-1} (boys) and 9.0 ± 1.4 cm·y^{-1} (girls) (Faust 1977) and 10.3 ± 0.22 cm·y^{-1} (boys) and 9.0 ± 0.16 cm·y^{-1} (girls) (Tanner *et al.* 1966). In some children, however, the growth velocity can reach 14 cm·y^{-1} (Faust 1977). To put these growth velocities into context, prior to the growth spurt, typical growth rates in both boys and girls are approximately 6 cm·y^{-1} (Tanner *et al.* 1966). The typical duration of the growth spurt is 2.81 years (SD = 0.53) for boys and 2.82 years (SD = 0.58) for girls with average stature gains of 21.1 ± 3.9 cm and 19.6 ± 4.7 cm, respectively over this period (Faust 1977). These growth data highlight that, for some children, growth rates will more than double, often within a year to a year and a half of the start of their growth spurt. Therefore, the magnitude and speed of growth in stature alone is important when considering implications for performance and injury. However, the unique challenges faced by children in their transition to adulthood are further emphasized by the differential growth of the body segments that collectively contribute to growth in stature.

Changes in body proportions with age in boys and girls

During the adolescent growth spurt, growth in leg length tends to precede growth in trunk length. Of the measurements presented for 67 boys and 94 girls by Faust (1977), the peak growth velocity of the legs preceded or was coincident with PHV in 77.6 per cent of the boys and 75.6 per cent of the girls. In contrast, the peak growth velocity of the trunk followed or was coincident with PHV in 83.5 per cent of boys and 71.3 per cent of girls. These differences in the timing of leg and trunk growth formed the rationale for their use as an indication of maturational status by Mirwald *et al.* (2002). The ratio of leg length to trunk length (i.e. sitting height) was observed to increase from four years prior to the age at PHV (mean 87.9 per cent in boys and 87.1 per cent in girls), reaching a maximum at the age at PHV (93.2 per cent in boys and 91.4 per cent in girls). The ratio then declined for the three years that were examined following the PHV (89.6 per cent in boys and 88.5 per cent in girls). The increase in the ratio therefore reflected the increased growth of the legs, and the later decrease indicates an increased growth of the trunk.

Although the pattern of growth in length of the legs and trunk is similar for boys and girls, shoulder and hip skeletal breadths show marked sex differences in their development. Prior to the onset of the pubertal growth spurt, girls (on average) have higher mean shoulder/hip width ratios than boys. However, at the onset of the growth spurt in stature, this ratio dramatically decreases in girls and increases in boys, resulting in significant sex differences at and following the age at PHV (Faust 1977). This reflects the greater development of hip pelvic width in females relative to the shoulder, and the greater shoulder width development of the males relative to the pelvis. The implications of the rapid growth of the skeleton and the different timing and rates of growth of its constituent bones will be discussed later in this chapter.

Changes in body mass and fat mass with age in boys and girls

A growth curve illustrating changes in body mass with age (Figure 1.2) shows similar patterns to the stature-age curve (see Figure 1.1). First, a rapid growth

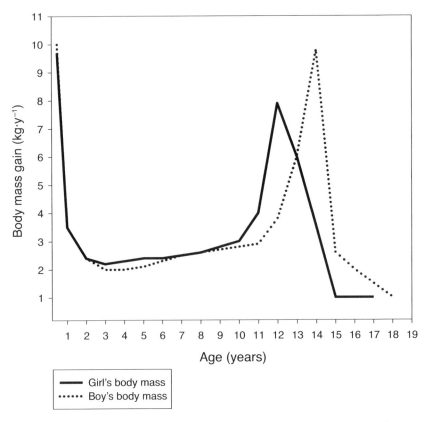

Figure 1.2 An individual velocity curve for body mass for one boy and one girl (illustrative purposes only).

period during infancy, followed by steady gains during childhood, then a rapid gain during adolescence, followed by a slower increase in adulthood are evident. As can be seen from Figure 1.2, body mass doubles by the end of year 1 and quadruples by the end of year 2. Typically, between the ages of 2–3 years there are minimal gains in body mass but thereafter the gains occur incrementally, albeit at a slower rate. During the onset of the adolescent growth spurt there are rapid gains in body mass, and typically the peak velocity for body mass occurs approximately one year after the age at PHV. Interestingly, this age roughly corresponds to the age of peak strength gains (see Chapter 4), suggesting that gains in body mass may partly reflect growth in muscle size alongside neuromuscular adaptation and coordination (Jones and Round 2000). Longitudinal bone growth and body mass (for lower extremity muscles) stimulate muscle growth. The existence and implications of musculoskeletal growth lags (delays between skeletal growth and muscle adaptation) will be explored later in this chapter.

It can also be seen from Figure 1.2 that for a time during the adolescent growth spurt, girls are heavier than boys. However, after boys begin their growth spurt they catch up and remain heavier. Overall, on average men tend to be taller and heavier than women due to their longer adolescence growth spans (Malina *et al.* 2004).

The presence of fat mass plays an important role during childhood growth. The estimated fat mass increases from ~15 per cent at birth to ~30 per cent at 1 year of age and thereafter shows minimal change until the age of 6 years. At this age, there is little sex difference. After this age, however, girls increase more rapidly than boys, and this is continued for girls throughout adolescence. Girls have been found to have a body fat of 14 per cent at 6 years, increasing to 25 per cent by 17 years. Female fat mass is on average 1.5 times that of males in late adolescence and adulthood. For boys, the increase in fat mass reaches a plateau near the time of the adolescent growth spurt. At 6 years of age, boys have ~11 per cent of body fat which increases to 15 per cent at 17 years of age. The key sex difference is that the increased body mass during adolescence for boys is attributed predominantly to an increase in muscle and skeletal mass, with relatively little change in fat mass. For girls, there is a continuous rise in fat mass with a lower increase in skeletal and muscle mass.

Relative differences in fat mass expressed as a percentage of body mass, show girls to be on average higher than boys throughout infancy, childhood, adolescence and adulthood. For adolescent girls, relative fatness approximates the same pattern as fat mass. In boys, however, unlike fat mass there is a decline in relative fat mass due to the rapid growth of fat-free mass (FFM). This is the major difference between boys and girls – that although there may be differences in muscle tissue during infancy and childhood, the sex differences become profoundly significant during adolescence.

Stature and body mass are easy and convenient measures to record to track growth and maturation, but both exhibit diurnal variations. Participants are taller and lighter in the morning and shorter and heavier later in the day. Stature is predominantly influenced by gravity and mass by diet, physical activity and the menstrual cycle. Therefore, standardized conditions should be ensured when recording height and body mass. Although it is easy to monitor the changes in body mass per se, increases in body mass with age and sex are achieved in differing ways. As body mass is a composite of both fat-free and fat tissue, increases in body mass can be achieved through changes in fat and/or FFM (as well as increases in body water with age).

Changes in muscle mass, fat-free mass and muscle size with age in boys and girls

Growth of muscle mass is similar to the increase of adipose tissue through child-hood. Although boys may have slightly greater amounts of muscle, it is during the adolescent growth period that the magnitude of change is significantly greater in boys compared to girls. At age 5 years, boys' muscle mass is estimated to be as high as 42 per cent of total body mass, increasing to 54 per cent by age 17 years. For girls, muscle mass increases from approximately 30 per cent before puberty to 40–45 per cent at age 17 years (Malina *et al.* 2004).

If FFM rather than muscle mass is considered, by their early 20s men have 50 per cent more lean body mass compared to women (Malina *et al.* 2004). When plotted, estimates of FFM show a growth curve similar to that of stature and body mass. Female adult values for FFM are usually attained earlier than males. On average, females reach adult FFM values about four years earlier than males. During late adolescence when adulthood is emerging, females have 1.5 times lower FFM than males (see Table 1.1). Eventually, adult females average FFM accounts for only 70 per cent of the mean value for adult males. This difference is a reflection of the growth spurt seen in male muscle mass and the sex difference in stature.

Measurement of muscle mass, fat-free mass and muscle size

There is no exact direct method to determine total muscle mass *in vivo*. Direct methods include measurements of either creatinine excretion or potassium concentration. The former method measures the creatinine excreted in urine and

Table 1.1 Estimated changes in densitometric estimates of body composition between 10 and 18 years of age

	Males	Females
FFM*	32.5kg	17.3kg
FFM	3.2 kg	7.1kg
% Fat	–2.7 %	+5.0 %

* Fat-free mass

Source: Reproduced from Malina *et al.* (2004).

assumes the volume excreted is a function of the muscle mass, because creatinine is a by-product of muscle metabolic pathways. However, the amount of creatinine produced in the urine can be influenced by other factors (e.g. nutrition, exercise, stress, the menstrual cycle and some diseases). Given the ease of collecting urine samples compared to other more invasive methods, its analysis can provide useful estimates of muscle mass. Measurements of potassium within muscle (50–70 per cent of potassium found within the body is located in muscle) is complicated because of the need to use radiation and therefore has ethical implications for scanning with children. More commonly found in the literature are instruments which have assessed the regional development of muscle tissue such as X-rays, Computed Tomography (CT), Dual X-ray Absorptiometry (DEXA), ultrasound or magnetic resonance imaging (MRI). The use of X-rays has had a long history of imaging bone, muscle and other organs but is limited as a research instrument because of the radiation dose. This limitation is also applicable to CT and DEXA. More recent studies have used ultrasound and MRI to provide information on muscle cross-sectional area, volume and the changes over time due to age and sex.

Measurement of muscle size in children

The size of a muscle (and therefore its strength capability) is often represented by its maximum cross-sectional area (CSA). In the paediatric exercise science literature, the majority of studies have measured CSA perpendicular to the long axis of the muscle (anatomical CSA or ACSA) or estimated muscle size indirectly using anthropometric methods. ACSA provides an inaccurate representation of the force-producing capability of muscles with a pennate muscle fibre arrangement. In these muscles physiological CSA (PCSA) – muscle CSA taken so that it bisects the parallel arrangement of muscle fibres – is more appropriate. However, this requires knowledge of the pennation angle (θ), muscle fibre length (l_{fibre}) and muscle volume ($volume_m$) if measured *in vivo*: PCSA = ($volume_m \cdot \cos\theta$) ÷ l_{fibre} (Fukunaga *et al.* 1996). Studies have shown that the pennation angle changes during growth and maturation (Binzoni *et al.* 2001) with sex differences apparent by adulthood (Abe *et al.* 1998; Chow *et al.* 2000). Pennation angle increases monotonically from birth and reaches a stable value after the adolescent growth spurt (Binzoni *et al.* 2001). The work of Binzoni *et al.* (2001) emphasizes the need to directly determine PCSA in the paediatric population and the procedures involved aid understanding of why ACSA has been most commonly estimated/measured. For more detailed information of the development of pennation angle, we refer the reader to Chapter 6.

Indirect measures of muscle size have previously utilized circumference and skinfold measurements (Housh *et al.* 1995; Roemmich and Sinning 1996). As the muscle comprises the major tissue compartment of the circumference, limb circumferences are often used to indicate muscular development. Roemmich and Sinning (1996) estimated arm muscle CSA using this method. The mid-arm circumference was adjusted for subcutaneous fat using biceps and triceps skinfolds.

This estimate of muscle plus bone CSA was then further adjusted to take into account the humerus CSA and the shape of the arm. Anthropometric cross-sections cannot be derived for individual muscles, and additional errors may be introduced by the assumptions used to predict the muscle CSA. However, due to the ease and convenience of this method, it is often utilized in nutritional field studies to assess components of under-nutrition. Conversely, this method would not be appropriate to use in participants who are obese, as it is unlikely that muscle would form the major constituent tissue in the limb. Muscle size has also been predicted on the basis of the principle of geometric similarity where ACSA is deemed to be proportional to the square of a length measure – typically height^2 or limb length^2 (Asmussen 1973; Parker *et al.* 1990). Although these measures might be useful in a field setting, they are not precise enough for the purposes of research and more direct measures should be sought.

The cost of the multiple scans required when using MRI to obtain an accurate estimate of maximal muscle ACSA and volume has resulted in the majority of studies measuring ACSA from a single scan at a fixed percentage of limb length (Davies *et al.* 1988; Kanehisa *et al.* 1994, 1997). This method does not acknowledge intra/inter-individual differences in the position of maximal muscle ACSA with growth and maturation, nor does it consider differences between the relative positions of maximal ACSA within a group of muscles (e.g. the muscles constituting the elbow flexor or knee extensor muscle groups) (Kawakami *et al.* 1994). When it is further considered that ACSA has commonly been estimated from anthropometric measurements or predicted on the basis of the principle of geometric similarity, the validity of these methods must be questioned. One of the key factors to remember is that anthropometry underestimates CSAs in comparison to a gold standard measure such as MRI (Housh *et al.* 1995; Knapik *et al.* 1996). Therefore, ACSA determination from a single site using anthropometric techniques should be avoided.

In addition to the direct measurement or indirect estimation of muscle CSA, muscle volume has also been assessed. Lower extremity muscles have been reported to develop in proportion to height cubed (volume), reflecting the stimulus of body weight in addition to the stretch imposed by the growing long bones. Also, since the volume of a muscle is proportional to its PCSA (Edgerton *et al.* 1986), it may more appropriately represent the force-producing capability of pennate muscles. Methods of estimating muscle volume invariably involve measurements of muscle/limb circumferences and lengths. For example, Kanehisa *et al.* (1995) estimated the volume of the plantar flexors and dorsiflexors in children by multiplying the respective muscle ACSA by lower leg length. However, the most well-known method is that of Jones and Pearson (1969) whereby measuring numerous circumferences of the upper and lower leg, lengths of the upper and lower leg and two skinfolds predicts the volume of the leg. These measurements were validated against water displacement and X-ray methods in order to produce the prediction equations. The correlation coefficients between the two methods ranged from r = 0.84 to 0.90, but anthropometry was always found to underestimate the quantity of adipose tissue. One of the assumptions of the technique by Jones and Pearson (1969) is that the limb is a cylinder and that the subcutaneous fat is evenly distributed. However, this is not necessarily the case in humans (Knapik *et al.* 1996; Malina 1986), and there are significant inter-individual differences in fat distribution which cannot be accommodated for by the limited and generalized regression equations. Winsley *et al.* (2003) measured leg volume of 16 prepubertal boys (9.9 ± 0.3 years) using the Jones and Pearson method and MRI leg scans. The range of potential underestimation by using the Jones and Pearson technique was 19–52 per cent for total thigh volume, 14–46 per cent for lean thigh volume and 5–98 per cent for fat thigh volume. Therefore, despite the simplicity of the Jones and Pearson method, the error of the method will only make it useful as an indicative measure rather than as a precise research tool.

Studies examining changes in muscle size or CSA have found that the muscle fibre CSA reaches a maximal or adult-like size by approximately 10 years in girls and 14 years in boys. While the maximal fibre CSA might have been attained by these ages, there is still considerable growth in the length of the limb. Whether assessments of muscle size are made by X-ray or MRI, differences in childhood tend to be negligible. For example, in the Harpenden growth study, boys' muscle widths, determined by X-ray, were greater than girls' but the differences were small and probably not meaningful. Using MRI, no significant sex differences were found in elbow and knee muscle CSA in 10–14-year-old boys and girls. It is not until the effects of the adolescence growth spurt occur that differences between boys and girls in muscle CSA manifest themselves. Deighan *et al.* (2003) observed in their MRI study of 9- and 24-year-old males and females an increase of 207 per cent, 210 per cent, 65 per cent and 78 per cent for elbow extensor and flexor muscle ACSA, respectively. In the most recent studies where PCSA has been calculated, as would be expected, the muscle size of adults is significantly greater than in children (Morse *et al.* 2008; O'Brien *et al.* 2009). In the latter study sex differences in PCSA were also found, but only in adult males compared to adult females and not between prepubertal boys and girls.

The use of different methods and measurements to represent muscle mass/size has influenced our understanding of age, sex and maturational differences in 'strength'. This will be discussed in more detail in Chapter 4.

Inter-relationship between skeletal and muscle growth

The previous sections have described age and sex differences in skeletal and muscle growth and development separately. However, the inter-relationship between the growth of these tissues may provide the most important insights into musculoskeletal injuries and motor control during childhood. As the skeleton grows in length and body proportions and shape change, the muscular system must also develop. The muscles must grow both in length (to normalize the tension caused by bone growth and the separation of the tendon origin(s) and insertion(s)); and also size/CSA to facilitate increased force production to support and move the larger and heavier skeleton. Although the changing hormonal milieu (e.g. growth hormone, insulin and testosterone) stimulates growth in muscle length and size, local mechanical factors also contribute to muscle adaptation (McComas 1996). Bone growth alters the distance separating the muscle attachments and the muscle fibres grow in length as a result of new sarcomeres being added in series at the level of the musculotendinous junction (Malina 1986; Sinclair and Dangerfield 1998; Williams and Goldspink, 1973). This adaptation is functionally important, as it ensures that for a given limb range of motion, the muscle will work at an optimum length or region of the length-tension curve (McComas 1996; Williams and Goldspink 1973). Since bone growth represents a local stimulus for growth in muscle length, a lag between growth in bone length and subsequent muscle length is suggested. This is supported by the concept of increased tissue preload during growth, which can be defined as the force to which a tissue is subjected when in a relaxed state (Hawkins and Metheny 2001). Increased tissue preload due to the disparity in bone and muscle-tendon growth rates may therefore be a contributory factor to the aetiology of traction apophyseal injuries in children (Mountjoy *et al.* 2008).

In addition to a lag in growth between bone and muscle-tendon length, there is also purported to be a delay between growth in muscle length and growth in muscle CSA (Kanehisa *et al.* 1995; Xu *et al.* 2009). This is anecdotally supported by the delay in peak strength gains relative to peak height velocity (described above). Muscle hypertrophy is required to enable greater muscular force to be generated. An increase in bone (lever) length and body mass without a simultaneous increase in muscle size imposes an overload on the muscle and requires a change in neuromuscular coordination to maintain equilibrium. Therefore, changes in anthropometry will alter gravitational forces and forces associated with motion, requiring muscular adaptations (Korff and Jensen 2008). A diagrammatic representation of musculoskeletal growth lags is presented in Figure 1.3.

In addition to longitudinal bone growth, the changes in mechanical loading due to increased body and muscle size will also cause changes in bone geometry

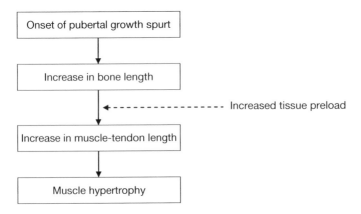

Figure 1.3 Musculoskeletal growth lags following the onset of a growth spurt.

and mass to increase bone strength (MacDonald *et al.* 2005, 2006; Ruff 2003). Changes in bone geometry during growth and maturation in boys and girls remain to be investigated. Studies examining differential growth rates of bone–ligament–bone and muscle–tendon–bone functional units relative to peak growth velocities will aid the development of injury prediction models in children (Hawkins and Metheny 2001).

Clinical and practical applications

There are three main areas where the combination of biomechanical and sport or health sciences has the potential to make a significant impact on our understanding of the paediatric participant. The first is from a clinical perspective when movement is compromised in the form of disease either related to musculoskeletal or nervous system disorders. The second is in relation to injury either as a consequence of participation in sport or recreation. The third is in the context of sports performance and the long-term athlete development plan, especially where technique and skill are being emphasized. The former two examples have a more extensive research base upon which to make recommendations, while the latter has yet to be exploited meaningfully.

From a clinical perspective, biomechanics has been traditionally applied to diseases where coordination and bodily movement is affected. Such examples include children with Developmental Coordination Disorder (DCD) (see Chapter 12) and neuromuscular disease (e.g. cerebral palsy – see Chapter 13), spastic diplegia or muscular dystrophy. The muscular dystrophies are a group of inherited diseases in which muscle degenerates and eventually is replaced with adipose and connective tissue. Duchenne Muscular Dystrophy (DMD) is the most common and severe of this group and results in the loss of skeletal muscle mass due to a deficiency of a protein, dystrophin, caused by a mutation to the DMD gene. The resulting progressive muscle weakness eventually leads to premature death, typically within the second decade of life (Lewis and Haller 1989). It currently affects 1 in 3,000 male births, presents clinically between the ages of 3–5 years, and there is no known cure (Emery and Mutoni 2003). The strategy for treating DMD is to alleviate the symptoms and delay the progression of the disease, thereby improving the quality of life for the young child. As the disease results in atrophy and weakness to the thighs, hips, back, shoulder, cardiac and respiratory muscles, tasks requiring considerable muscular effort such as rising from a chair, maintaining ambulation or climbing stairs become increasingly difficult. As the disease progresses patients usually become wheelchair bound by 10 years of age (Dallmeijer and Schnyder 2008). To date, many interventions (e.g. novel drug therapies, myoblast transfer or gene therapies) have been explored but they have not yet been successfully translated from promising animal work to humans. As DMD is characterized by a decreased muscle strength and endurance, endurance exercise and/or resistance training should be a useful therapy.

Biodynamic approaches to rehabilitative therapies have taken one of two approaches. The first aims to correct the abnormal kinematics in order to bring about a more normal gait (Holt et al. 1996). The other approaches the problem from a dynamics systems approach (see Chapter 2), which attempts to normalize the dynamic 'resource' (categorized according to the energy generation in the muscle, the conservation of the energy and the interchange of the kinetic and potential gravitational energy). Watkins (2007) argues that the dynamic resources approach, compared to the applied approach, has resulted in more successful interventions (Carmick 1995; Comeaux et al. 1998). Whichever approach is utilized, it is clear that this area is under-researched, and little is known about the trainability of children with neuromuscular diseases. Due to the importance of understanding movement and its relationships to other factors such as muscle force, strength and power, it is surprising that there is so little biomechanical research in these disease groups.

Implications for injury and performance

Considering the delays in the growth of muscle length and size that follow the rapid increases in bone length during the growth spurt, the potential for increased preload and muscular overload is clear. However, the injury/performance implications are truly emphasized by the fact that some muscles may span the trunk and lower extremity (both of which develop at different rates) and the sheer number of muscles that must adapt during growth. The likelihood of muscular imbalances occurring simply due to the adolescent growth spurt (let alone due to sports participation) is therefore high. Few studies have examined musculoskeletal malalignment and muscle imbalance in children encompassing the period of rapid growth. Fialka et al. (2001) performed such an examination in 27 top-level tennis players (8–15 years of age) participating in more than 10 national/international competitions per year and typically training in excess of 10 hours per week. Following musculoskeletal checks that were undertaken every six months for a total of two years, only three children were free from any musculoskeletal disorders at the start of the study. In seven children, pathological findings were attributed to a lack of muscular adaptation following a 'physiological growing episode'.

With regard to injury and performance risk, the rapid and asynchronous growth of the musculoskeletal system has also been shown to influence neuromuscular control. To maintain stability, the body's centre of mass must be maintained/controlled relative to the base of support (Westcott et al. 1997). The many systems that contribute to both dynamic and static stability (the somatosensory and vestibular senses, the motor system, which creates movement to maintain alignment, and the bone and joint skeletal frame that has to be moved/stabilized) change during puberty (Assaiante et al. 2005; Westcott et al. 1997). Indeed, in an earlier study, Assaiante (1998) described how adolescence may be a 'turning point' in postural control development, with teenagers moving from fixing body segment positions and movements using an 'en-bloc' strategy to simplify the movement solutions to an articulated strategy where body segments are controlled independently (Assaiante et al. 2005).

Coordination of neuromuscular control strategies can be altered by biomechanical malalignments, which may cause sensory information to be altered (Westcott *et al.* 1997). Winter *et al.* (1996) examined musculoskeletal control strategies used to maintain stability in different stance positions. They demonstrated that although there is considerable variation in the control strategies used, in bipedal stance (with feet together), the ankle plantar flexors and dorsiflexors primarily alter the centre of pressure (COP) anterior–posterior (a–p) position and the hip abductors/adductors alter medial–lateral (m–l) COP position. As the stance position changes to a tandem stance (with one foot in front of the other), the abductors/adductors now contribute to the maintenance of balance in the a–p direction and the ankle invertors/evertors to m–l balance. Thus, balance maintenance requires coordination of ankle and hip strategies, both of which may be affected (potentially at different times) during growth. Examination of neuromuscular control strategies during and immediately following a growth spurt will corroborate the suggestion of impaired neuromuscular control over this period.

Sports performance

The rapid growth of the skeleton and associated muscles during early childhood and adolescence presents many challenges to the young sports performer who must learn to move and control their longer and heavier body and limbs. Changes in moment of inertia with growth (Chester and Jensen 2005; Van Dam *et al.* 2009) will impact on the amount of muscular force required to rotate the limb. Studies by Jensen (1981) and Schneider and Zernicke (1992) have predicted changes in body segment moments of inertia with growth in children (see Chapter 6), and Lebiedowska and Polisiakiewicz (1997) demonstrated that moment of inertia was proportional to the fifth power of body height between the ages of 5 and 18 years. However, children were not tracked longitudinally in this study and therefore, based on the previous description of musculoskeletal growth lags, some divergence from this proportional relationship may occur during a rapid growth phase. Nevertheless, the findings are useful when trying to understand the challenge to the neuromuscular system in supporting and moving the growing skeleton, particularly since force output and range of motion have been shown to influence postural stability in children (Westcott *et al.* 1997).

Despite the recognized changes in growth that occur during puberty and the assumed links to impaired coordination and control (and hence injury and reductions in performance), there is limited empirical data to corroborate a cause–effect relationship (Fry and Hawkins 2001). This may reflect differences in the movements/skills examined. Impairment of motor coordination and control would be expected to be greater in complex multiple degree of freedom skills compared to those tasks that may be more dependent on strength (Hirtz and Starosta 2002). These authors reported that the growth spurt resulted in either an immediate impairment of coordination or a delayed impairment (approximately one year), which was more pronounced and protracted in tasks requiring greater rhythm and kinaesthetic differentiation. This delay corresponds to the timing of peak body mass and strength velocity. In 1947 Espenschade examined the gross motor coordination of 610 boys and girls. A marked adolescent lag was reported in the tests, which required dynamic balance. However, this was only apparent for the boys. Visser *et al.* (1998) undertook a three-year longitudinal study on boys aged 11.6 years at the start and 14 years at the end. For all boys for whom a growth spurt could be identified, height velocity was negatively correlated with the total score on the Movement ABC Assessment Battery. More recently, Hewett *et al.* (2004) observed a significant decrease in neuromuscular control of the muscles spanning the knee in girls from puberty onset extending into late puberty, which was not observed in the boys studied. The findings of these studies provide some support for altered coordination and potential 'adolescent awkwardness' following the adolescent growth spurt. However, further longitudinal studies are required on both boys and girls to aid understanding of links to performance and injury. Such studies will also contribute to knowledge of any sex differences in neuro-muscular control and coordination. It is important that coaches, teachers, parents and therapists are aware of when a child is undergoing or has undergone a growth spurt, so that children are given the opportunity to adapt to their new body shape and size. This will require children to formulate new movement solutions for a given skill as their neural and muscular systems adapt (McLester and St. Pierre 2008). Monitoring of musculoskeletal alignment and neuromuscular control and coordination during and for up to two years (at least) following the age of PHV should be a priority.

From a sports performance perspective, as puberty is such a significant event, there is great interest in monitoring and interpreting these changes in the context of training young athletes (Balyi and Williams 2009; Ford *et al.* 2011; Williams 2007). It has been well documented that there is a substantial increase in the muscle mass of boys compared to girls, and this greater muscle mass coupled with a lower body fat is considered to be an important advantage for boys' sports performance compared to girls. This is because the involvement of a greater amount of muscle mass will generally enhance performance, while the additional increase in fat mass for girls during weight-bearing activities (e.g. running) will decrease mechanical efficiency and reduce performance. It should be pointed out, however, that there is little empirical research on the quantification of growth relative to coordination and sports performance across the ages and between the sexes. Biomechanical modelling techniques have enabled us to gain some insights into the effects of growth and motor coordination (see Chapter 6 for more detail). Therefore, the measurement and modelling of the kinetics of coordination plus the interpretation of changes of force, torque and mechanical power output due to growth, maturation and sports practice makes this research area both inter-esting and complex.

Young male and female athletes who are biologically mature for their chrono-logical age are on average taller, heavier and have higher weight-to-height ratios than late maturers of the same chronological age. Also, early maturing boys gain advantages in increased shoulder breadth and upper body muscle mass. These body changes are associated with increases in strength and power (see Chapter 4) that also translate to increases in aerobic power, providing further enhanced capabilities in their sporting performance. Early female maturers, however, have broader hips and shorter legs, which are associated with increased body fatness and could be disadvantageous in some sports. Interestingly, late maturing girls exhibit more linear physiques, less weight-to-height ratio, less fatness, relatively longer legs and lower hip-to-shoulder ratios, factors that are predictors of success in sporting performances. These changes in growth have major implications for the selection of teams with typically larger, stronger and more powerful boys being selected ahead of smaller, less strong and less powerful boys. Despite the physiological differences associated with maturity being less pronounced in girls than boys, early maturing girls also dominate female sport. The selection of these boys and girls who are growing the fastest and are the most mature for their chronological age is known to most coaches and has its origins due to the 'birth date selection effect' – that is, the observations that those selected boys and girls in a squad predominantly have their birthdays in the first six months of a selection year. This observation means that those born earlier in the selection calendar have had an advantage of up to six months' extra growth period. Coaches therefore must look beyond the physique of the child and ensure that a changing body shape and its ensuing effect on performance does not cloud their judgement about retaining or releasing players from their squads.

Conclusions and future directions

Growth charts depicting changes in height and body mass with age for boys and girls are useful for monitoring growth, which allows for identifying atypical development and calculating growth velocities as indicators of maturity. Age, sex and maturational differences in muscle mass and size are less clear as a result of the use of different measurements and techniques. This limits the understanding of differences in strength and neuromuscular control as well as any subsequent implications for injury risk and/or performance. There is still considerable research to be undertaken about growth-related changes in children's movement patterns from a structural, physiological and biomechanical perspective. Currently, the higher metabolic costs of locomotion in children compared to adults can be attributed to numerous factors related to, but not solely limited to biomechanics (see Chapters 7 and 8). More research is required to increase the basic theoretical knowledge in this area and then how this is applied in the context of sport, injury and health.

Key points

- Growth and maturation measurements are well described and validated in paediatric studies and should be employed as a matter of course.
- Studies, which are able to describe changes in growth and maturation alongside quantifying the changes in such factors as force, torque, strength, gait and sport skill performance measures will have a greater impact on the discipline.

- Understanding injury mechanisms in children during growth and maturation requires longitudinal studies encompassing periods of rapid growth. The mechanical and morphological characteristics of muscle, bone, ligament and tendon, and the interactions between them during rapid growth phases remain to be determined.
- Despite the recognized changes in growth that occur during puberty and the assumed links to impaired coordination and control, there is limited empirical evidence to support a cause–effect relationship. There is some evidence to support this relationship in prepubertal children (see Chapter 6).

References

Abe, T., Brechue, W.F., Fujita, S. and Brown, J.B. (1998) 'Gender differences in FFM accumulation and architectural characteristics of muscle', *Medicine and Science in Sports and Exercise*, 30: 1066–70.

Armstrong, N. and Welsman, J.R. (1997) *Young People and Physical Activity*, Oxford: Oxford University Press.

Assaiante, C. (1998) 'Development of locomotor balance control in healthy children', *Neuroscience and Biobehavioural Reviews*, 22: 527–32.

Assaiante, C., Mallau, S., Viel, S., Jover, M. and Schmitz, C. (2005) 'Development of postural control in healthy children: a functional approach', *Neural Plasticity*, 12: 109–18.

Asmussen, E. (1973) 'Growth in muscular strength and power', in G. Rarick (ed.) *Physical Activity, Human Growth and Development*, London: Academic Press, pp. 60–80.

Baldwin, K. (1984) 'Muscle development: neonatal to adult', *Exercise and Sports Science Review*, 12: 1–19.

Balyi, I. and Williams, C.A. (2009) *Coaching the Young Developing Performer*, London: Coachwise.

Baxter-Jones, A.D.G., Eisenmann, J.C. and Sherer, L.B. (2005) 'Controlling for maturation in pediatric exercise science', *Pediatric Exercise Science*, 17: 18–30.

Bielicki, T. (1975) 'Interrelationships between various measures of maturation rate in girls during adolescence', *Studies in Physiological Anthropology*, 1: 51–64.

Bielicki, T., Koniarek, J. and Malina, R.M. (1984) 'Interrelationships among certain measures of growth and maturation rate in boys during adolescence', *Annals of Human Biology*, 11: 201–10.

Beunen, G.P., Rogol, A.D. and Malina, R.M. (2006) 'Indicators of biological maturation and secular changes in biological maturation', *Food and Nutrition Bulletin*, 27: S244–56.

Binzoni, T., Bianchi, S., Hanquinet, S., Kaelin, A., Sayegh, Y., Dumont, M. and Jéquier, S. (2001) 'Human gastrocnemius medialis pennation angle as a function of age: from newborn to the elderly', *Journal of Physiological Anthropology and Applied Human Science*, 20: 293–8.

Carmick, J. (1995) 'Managing equines in children with cerebral palsy: electrical stimulation to strengthen the triceps surae', *Developmental Medicine and Child Neurology*, 37: 965–75.

Chester, V.L. and Jensen, R.K. (2005) 'Changes in infant segment inertias during the first three months of independent walking', *Dynamic Medicine*, 4: 1–9.

Chow, R.S., Medri, M.K., Martin, D.C., Leekam, R.N., Agur, A.M. and McKee, N.H. (2000) 'Sonographic studies of human soleus and gastrocnemius muscle architecture: gender variability', *European Journal of Applied Physiology*, 82: 236–44.

Cole, T.J. (1997) 'Growth monitoring with the British 1990 growth reference', *Archives of Disease in Childhood*, 76: 47–9.

Comeaux, P., Patterson, N., Rubin, M. and Meiner, R. (1998) 'Effect of neuromuscular electrical stimulation during gait in children with cerebral palsy', *Pediatric Physical Therapy*, 9: 103–9.

Dallmeijer, A. and Schnyder, J. (2008) 'Exercise capacity and training in cerebral palsy and other neuromuscular diseases', in N. Armstrong and W. van Mechelen (eds) *Paediatric Exercise Science and Medicine*, 3rd edn, Oxford: Oxford University Press, pp. 467–76.

Davies, J., Parker, D.F., Rutherford, O.M. and Jones, D.A. (1988) 'Changes in strength and cross-sectional area of the elbow flexors as a result of isometric strength training', *European Journal of Applied Physiology and Occupational Physiology*, 57: 667–70.

Debar, L., Ritenbauch, C., Aickin, M., Orwoll, E., Elliot, E., Dickinson, D., Vuckovic, J., Stevens, M., Moe, E. and Irving, L.M. (2006) 'A health plan-based lifestyle intervention increases bone mineral density in adolescent girls', *Archives in Pediatric Adolescent Medicine*, 160: 1269–76.

Deighan, M., Armstrong, N., De Ste Croix, M., Welsman, J. and Barratt, V. (2003) 'Peak torque per MRU-determined cross-sectional area of knee extensors and flexors in children, teenagers and adults', *Journal of Sports Sciences*, 21 (4): 236 Abstract.

Edgerton, V.R., Roy, R.R. and Apor, P. (1986) 'Specific tension of human elbow flexor muscles', in B. Saltin (ed.) *Biochemistry of Exercise VI, International Series on Sport Sciences, 16*, Champaign, IL: Human Kinetics, pp. 487–500.

Emery, A.E.H. and Mutoni, F. (2003) *Duchenne Muscular Dystrophy*, 3rd edn, Oxford: Oxford University Press.

Espenschade, A. (1947) 'Motor development', *Review of Educational Research*, 17: 354–61.

Falgairette, G., Bedu, M., Fellmann, N., Van Praagh, E., Jarrige, J.F. and Coudert, J. (1990) 'Modifications of aerobic and anaerobic capacities in active boys during puberty', in G. Beunen, J. Ghesquiere, T. Reybrouck and A.L. Claessens (eds) *Children and Exercise XIV*, Stuttgart: Ferdinand Enke, pp. 42–9.

Faulkner, R.A. (1996) 'Maturation', in D. Docherty (ed.) *Measurement in Pediatric Exercise Science*, Champaign, IL: Human Kinetics, pp. 129–58.

Faust, M.S. (1977) 'Somatic development of adolescent girls', *Monographs of the Society of Research in Child Development*, 42: 1–90.

Fialka, C., Bockhorn, G., Weinstabl, R. and Bachl, N. (2001) 'Regular preventive musculo-skeletal check-up of top level juvenile tennis players', *Sport and Medicine Today*, Autumn: 38–9.

Ford, P., Oliver, J., Till, K., De Ste Croix, M., Lloyd, R., Meyers, R., Moosavi, M. and Williams, C.A. (2011) 'The Long-Term Athlete Development model – physiological evidence and application', *Journal of Sports Science*, 29: 389–402.

Freeman, J.V., Cole, T.J., Chinn, S., Jones, P.R.M., White, E.M. and Preece, M.A. (1995) 'Cross-sectional statue and weight reference curves for the UK, 1990', *Archives of Disease in Childhood*, 73: 17–24.

Fry, M. and Hawkins, D. (2001) 'Risk factors for overuse injuries in children: a modelling and movement simulation approach', paper presented at the American Society of Biomechanics, University of San Diego, August.

Fukunaga, T., Roy, R.R., Shellock, F.G., Hodgson, J.A. and Edgerton, V.R. (1996) 'Specific tension of human plantar flexors and dorsiflexors', *Journal of Applied Physiology*, 80: 158–65.

Hawkins, D. and Metheny, J. (2001) 'Overuse injuries in youth sports: biomechanical considerations', *Medicine and Science in Sports and Exercise*, 33: 1701–7.

Heinonen, A., Kannus, P. and Oja, P. (1999) 'Good maintenance of high-impact activity induced bone gain by voluntary, unsupervised exercises: an 8-month follow up of a randomised control trial', *Journal of Bone Mineral Research*, 14: 125–8.

Hewett, T.E., Myer, G.D. and Ford, K.R. (2004) 'Decrease in neuromuscular control about the knee with maturation in female athletes', *The Journal of Bone and Joint Surgery*, 86A: 1601–8.

Hirtz, P. and Starosta, W. (2002) 'Sensitive and critical periods of motor co-ordination development and its relation to motor learning', *Journal of Human Kinetics*, 7: 19–28.

Holt, K.G., Hamill, J. and Andres, P.O. (1996) 'Predicting the minimal energy costs of walking', *Medicine and Science in Sports and Exercise*, 23: 491–8.

Housh, D.J., Housh, T.J., Weir, J.P., Weir, L.L., Johnson, G.O. and Stout, J.R. (1995) 'Anthropometric estimation of thigh muscle cross-sectional area', *Medicine and Science in Sports and Exercise*, 27: 784–91.

Jensen, R.K. (1981) 'Effect of a 12-month growth period on the body moments of inertia of children', *Medicine and Science in Sports and Exercise*, 13 (4): 238–42.

Jones, P.R.M. and Pearson, J. (1969) 'Anthropometric determination of leg fat and muscle plus bone volumes in young male and female adults', *Journal of Physiology*, 240: 63–6.

Jones, D.A. and Round, J.M. (2000) 'Strength and muscle growth', in N. Armstrong and W. Van Mechelen (eds) *Paediatric Exercise Science and Medicine*, Oxford: Oxford University Press, pp. 133–42.

Kanehisa, H., Ikegawa, S. and Fukunaga, T. (1994) 'Comparison of muscle cross-sectional area and strength between untrained women and men', *European Journal of Applied Physiology and Occupational Physiology*, 68: 148–54.

Kanehisa, H., Ikegawa, S. and Fukunaga, T. (1997) 'Force-velocity relationships and fatiguability of strength and endurance-trained subjects', *International Journal of Sports Medicine*, 18: 106–12.

Kanehisa, H., Yata, H., Ikegawa, S. and Fukunaga, T. (1995) 'A cross-sectional study of the size and strength of the lower leg muscles during growth', *European Journal of Applied Physiology*, 72: 150–6.

Kawakami, Y., Nakazawa, K., Fujimoto, T., Nozaki, D., Miyashita, M. and Fukunaga, T. (1994) 'Specific tension of elbow flexor and extensor muscles based on magnetic resonance imaging', *European Journal of Applied Physiology and Occupational Physiology*, 68: 139–47.

Knapik, J.J., Staab, J.S. and Harman, E.A. (1996) 'Validity of an anthropometric estimate of thigh muscle cross sectional area', *Medicine and Science in Sports and Exercise*, 28: 1523–30.

Korff, T. and Jensen, J.L. (2008) 'Effect of relative changes in anthropometry during childhood on muscular power production in pedalling: a biomechanical simulation', *Pediatric Exercise Science*, 20: 292–304.

Lebiedowska, M.K. and Polisiakiewicz, A. (1997) 'Changes in the lower leg moment of inertia due to child's growth', *Journal of Biomechanics*, 30: 723–8.

Lewis, S.F. and Haller, R.G. (1989) 'Skeletal muscle disorders and associated factors that limit exercise performance', *Exercise and Sport Sciences Reviews*, 17: 67–113.

Lexell, J., Sjostrum, M., Nordlund, A. and Taylor, C.C. (1992) 'Growth and development of human muscle: morphological study of whole vastus lateralis from childhood to adult age', *Muscle & Nerve*, 15: 404–9.

MacDonald, H., Kontulainen, S., Petit, M., Janssen, P. and McKay, H. (2006) 'Bone strength and its determinants in pre- and early pubertal boys and girls', *Bone*, 39: 598–608.

MacDonald, H.M., Kontulainen, S.A., MacKelvie-O'Brien, K.J., Petit, M.A., Janssen, P., Khan, K.M. and McKay, H.A. (2005) 'Maturity- and sex-related changes in tibial bone geometry, strength and bone-muscle strength indices during growth: a 20-month pQCT study', *Bone*, 36: 1003–11.

Malina, R.M. (1986) 'Growth of muscle tissue and muscle mass', in F. Faulkner and J.M. Tanner (eds) *Human Growth, a Comprehensive Treatise (Vol. 2)*, 2nd edn, New York: Plenum Press, pp. 77–99.

Malina, R.M., Bouchard, C. and Bar-Or, O. (2004) *Growth, Maturation, and Physical Activity*, 2nd edn, Champaign, IL: Human Kinetics.

Matchock, R.L., Dorn, L.D. and Susman, E.J. (2007) 'Diurnal and seasonal cortisol, testosterone, and DHEA rhythms in boys and girls during puberty', *The Journal of Biological & Medical Rhythm Research*, 24: 969–91.

McComas, A.J. (1996) *Skeletal Muscle. Form and Function*, Champaign, IL: Human Kinetics.

McLester, J. and St. Pierre, P. (2008) *Applied Biomechanics, Concepts and Connections*, Belmont, CA: Thomson Wadsworth.

Mirwald, R.L., Baxter-Jones, A.D.G., Bailey, D.A. and Beunen, G.P. (2002) 'An assessment of maturity from anthropometric measurements', *Medicine and Science in Sports and Exercise*, 34: 689–94.

Mirwald, R.L., Bailey, D.A., McKay, H. and Crocker, P.E. (1999) 'Physical activity and bone mineral acquisition at the lumbar spine during the adolescent growth spurt', paper presented at the First International Conference on Children's Bone Health, Maastricht, May.

Morse, C.I., Tolfrey, K., Thom, J.M., Vassilopoulos, V., Maganaris, C.N. and Narici, M.V. (2008) 'Gastrocnemius muscle specific force in boys and men', *Journal of Applied Physiology*, 104: 469–74.

Mountjoy, M., Armstrong, N., Bizzini, L., Blimkie, C., Evans, J., Gerrard, D., Hangen, J., Knoll, K., Micheli, L., Sangenis, P. and Van Mechelen, W.V. (2008) 'IOC consensus statement: "training the elite child athlete"', *British Journal of Sports Medicine*, 42: 163–4.

Mughal M.Z and Khadikar A.V. (2011) 'The accrual of bone mass during childhood and puberty', *Current Opinion in Endocrinology, Diabetes and Obesity*, 18: 28–32.

O'Brien, T.D., Reeves, N.D., Baltzopoulos, V., Jones, D.A. and Maganaris, C.N. (2009) 'In vivo measurements of muscle specific tension in adults and children', *Experimental Physiology*, 95: 202–10.

Parker, D.F., Round, J.M., Sacco, P. and Jones, D.A. (1990) 'A cross-sectional survey of upper and lower limb strength in boys and girls during childhood and adolescence', *Annals of Human Biology*, 17: 199–211.

Roemmich, J.N. and Sinning, W.E. (1996) 'Sport-seasonal changes in body composition, growth, power and strength of adolescent wrestlers', *International Journal of Sports Medicine*, 17: 92–9.

Ruff, C. (2003) 'Growth in bone strength, body size, and muscle size in a juvenile longitudinal sample', *Bone*, 33: 317–29.

Schneider, K. and Zernicke, R.F. (1992) 'Mass, centre of mass and moment of inertia estimates for infant limb segments', *Journal of Biomechanics*, 25: 145–8.

Sinclair, D. (1975) *Human Growth after Birth*, Oxford: Oxford University Press.

Sinclair, D. and Dangerfield, P. (1998) *Human Growth after Birth*. Oxford: Oxford University Press.

Tanner, J.M. (1962) *Growth at Adolescence*, 2nd edn, Oxford: Blackwell Publishers.

Tanner, J.M. and Whitehouse, R.H. (1976) 'Clinical longitudinal standards for height, weight, height velocity, weight velocity, and stages of puberty', *Archives of Disease in Childhood*, 51: 170–9.

Tanner, J.M., Whitehouse, R.H. and Takaishi, M. (1966) 'Standards from birth to maturity for height, weight, height velocity, and weight velocity: British children, 1965, Part II', *Archives of Disease in Childhood*, 41: 613–35.

Van Dam, M., Hallemans, A. and Aerts, P. (2009) 'Growth of segment parameters and a morphological classification for children between 15 and 36 months', *Journal of Anatomy*, 214: 79–90.

Visser, J., Geuze, R.H. and Kalverboer, A.F. (1998) 'The relationship between physical growth, the level of activity and the development of motor skills in adolescence: differences between children with DCD and controls', *Human Movement Science*, 17: 573–608.

Watkins, J. (2007) *An Introduction to Biomechanics of Sport and Exercise*, Edinburgh: Churchill Livingstone.

Westcott, S.L., Lowes, L.P. and Richardson, P.K. (1997) 'Evaluation of postural stability in children: current theories and assessment tools', *Physical Therapy*, 77: 629–45.

Williams, C.A. (2007) 'Physiological changes of the young athlete and the effects on sports performance', *SportEX Medicine*: 6–11.

Williams, P.E. and Goldspink, G. (1973) 'The effect of immobilization on the longitudinal growth of striated muscle fibres', *Journal of Anatomy*, 116: 45–55.

Winsley, R., Armstrong, N. and Welsman, J. (2003) 'The validity of the Jones & Pearson anthropometric method to determine thigh volumes in young boys: a comparison with Magnetic Resonance Imaging', *Revista Portuguesa de Ciencias do Desporto*, 3: 94–5.

Winter, D.A., Prince, F., Frank, J.S., Powell, C. and Zabjek, K.F. (1996) 'Unified theory regarding A/P and M/L balance in quite stance', *Journal of Neurophysiology*, 75: 2334–43.

Xu, L., Nicholson, P., Wang, Q., Alén, M. and Cheng, S. (2009) 'Bone and muscle development during puberty in girls: a seven-year longitudinal study', *Journal of Bone and Mineral Research*, 24: 1693–8.

2 Sensory development and motor control in infants and children

Jan Piek

Introduction

The appropriate development of several sensory systems is essential for typical motor development during infancy and childhood. In particular, the visual and proprioceptive systems play a key role in motor control. In addition, the auditory system plays an important role, and an example is the sound of our steps as we are walking or running, which can provide valuable information on the type of surface we are traversing. As the sensory systems have been described extensively in many textbooks, the following tutorial provides only a brief overview of these sensory systems. In the second part of this chapter, the development of the sensory system as well as its relation to motor development is described.

Basic theoretical concepts: vision, proprioception and hearing

Vision

Vision is achieved through light-sensitive receptors located in the retina of the eye which detect light energy with wavelengths of 400–700 nanometers (Goldstein 2010). The retina contains two types of visual receptors: *rods*, which are more sensitive to light and important for night vision, and *cones*, which require higher levels of illumination to be stimulated and are essential for day vision and colour perception.

In order for the visual receptors to function, a complex set of muscle actions is required to regulate the incoming visual information. Vision is an excellent example of a sensory-motor system where muscles are responsible for the amount and quality of visual information received. This sensory information then provides feedback to the motor system on the quality of the information, and the cycle continues.

Muscles are responsible for a number of different visual processes. One such process is accommodation. Before the light stimulus reaches the retina it must pass through the cornea, the watery aqueous humor, the lens and then the gel-like vitreous humor (Figure 2.1a). This causes refraction of the light rays which can affect the quality of the visual image. In order to accommodate this change, the

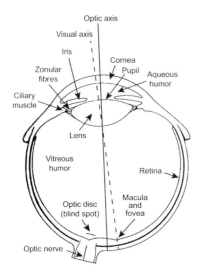

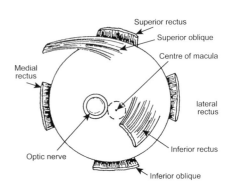

Figure 2.1a The major features of the human eyeball (adapted from Aslin (1987)).

Figure 2.1b The extraocular muscles with respect to the back of the eyeball (taken from Aslin (1987), Figure 10, p. 67).

ciliary muscle controls the shape of the lens. When this muscle relaxes it causes the lens to flatten by producing pressure on the zonular fibres. When the ciliary muscle is activated, it then releases the pressure on the zonular fibres, resulting in a thickening of the lens. This process ensures that the retinal image is in focus.

The eye also contains a set of six extraocular muscles (Figure 2.1b), which can move the eye vertically, horizontally and torsionally (Aslin 1987). When looking directly at an object, the visual image falls on the fovea, which is a small area of the retina containing only cones. Both head and body muscles can be used to foveate, but for rapid foveation the extraocular muscles are needed. They are also used for slow pursuit tracking of objects. In smooth pursuit tracking, the eyes can match the velocity of the moving stimulus. The extraocular muscles are also needed for binocular fixation, that is, moving both eyes to look directly at an object. The lack of binocular fixation can produce a double image (diplopia) or result in a loss of binocular depth perception or stereopsis (Aslin 1987).

Saccadic eye movements are very high velocity, ballistic movements, which are important for rapid changes in fixation. Once initiated, they cannot be modified or stopped, and, unlike other skeletal muscle systems, they do not use the antagonist muscles as a braking system. Rather, 'a saccade consists of an acceleration of the eyeball by a very brief neuromuscular pulse and deceleration by the friction or damping characteristics of the eyeball within the orbital tissues' (Aslin 1987, p. 68).

Proprioception

The term *proprioception* refers to the 'sensing of the body's own movements' (Prochazka 1996, p. 93) and includes the vestibular system as well as muscle, joint and cutaneous receptors (Schmidt and Lee 1999). An examination of the literature shows that definitions in relation to this sensory system vary considerably. For example, Goldstein (2010) describes the somatosensory system as including the cutaneous senses, proprioception (sense of body position) and kinaesthesis (sense of body movement), whereas Gandevia (1996) includes skin, joint and intramuscular sensations in his description of kinaesthesia. Others use the terms *proprioception* and *kinaesthesis* interchangeably. It is therefore important to have a clear definition from authors if they are referring to proprioception, kinaesthesis or the somatosensory system. In this chapter, kinaesthesis is described as information gained through sensors in the muscles, joints and tendons; the somatosensory systems includes kinaesthetic and tactile sensations; and proprioception refers to information gained from somatosensory and vestibular systems.

There are five receptor locations that provide information on the sense of movement: muscle receptors; tendon receptors; joint receptors; cutaneous receptors; and receptors in the vestibular apparatus.

1   *Muscle spindles*, located in the body of all skeletal muscles, are unique in that they contain muscle fibres called intrafusal fibres that are capable of movement. They are particularly important in providing information on the amount and rate of change of stretch in a muscle (Abernethy *et al.* 1996).
2   *Golgi tendon organs* are located in the tendons which attach the muscle to bone and provide information on the amount of tension produced in a tendon. Whereas the muscle spindles are most active during muscle stretching, the Golgi tendon organs are most active when the muscles contract, producing the greatest amount of tension on the tendon. These receptors appear to be very important in preventing damage at the muscle-tendon juncture which can result from a muscle contracting too forcefully.
3   *Raffini endings* and *Pacinian corpuscles* are joint receptors located in the joint capsule which holds lubricating fluid for the joint. Located within the ligaments binding the joint together is another receptor called the *Golgi organ*. Originally thought to be involved in detecting joint position, recent research has suggested that this may not be the case and these receptors may be important in signalling extreme ranges of motion and hence protecting the joint from injury (Abernethy *et al.* 1996).
4   Skin or cutaneous receptors are numerous as the skin is a very complex organ. Receptors such as *Pacinian corpuscles* (similar to those found in the joint) are located deep in the skin to sense deep deformation (e.g. from a heavy blow). Receptors that are close to the surface and monitor light pressure include the *Meissner corpuscles, Merkel's discs, Ruffini corpuscles* and *nerve endings*, either free or wrapped around hair follicles. One of the highest concentrations of skin receptors can be found in the fingertips.

5   An essential component of motor control is balance, and in order to maintain balance it is important to know how the body is oriented in space. This is achieved through the vestibular system, which consists of two different types of receptors. The *semicircular canals* consist of the superior, horizontal and posterior canals located in the inner ear. These canals contain a thick fluid that bends tiny hairs as the head changes position, signalling angular acceleration in the frontal, horizontal and sagittal planes of the body. The *otolith organs* comprise the *utricle* and *saccule* and are sensitive to linear acceleration.

Hearing

Sounds are produced by pressure changes in the air (or water) surrounding an object, which result in sound waves. When these waves fall between 20 Hz and 20,000 Hz, they can be detected by the human auditory system. Hearing for humans is most sensitive at frequencies of 2,000 to 4,000 Hz, the range that generally encompasses speech production (Goldstein 2010).

The ear contains three distinct sections. The outer ear consists of the most obvious part of the ear, the ear flap or pinna and the auditory canal, which direct the sound waves through to the ear drum or tympanic membrane of the middle ear. Vibrations on the tympanic membrane are transmitted to the ossicles of the middle ear, which are the three smallest bones in the body – the malleus, incus and stapes. Vibrations from these bones are then transmitted via another membrane on the oval window to the inner ear containing the complex structure, the cochlea, which is responsible for the conversion of the vibrations into electrical signals (Goldstein 2010).

An important function of hearing, particularly in relation to motor control, is locating the source of a sound. This relies on the time difference between sounds arriving at the two ears, the interaural time difference and the difference in the sound pressure level reaching the two ears, the interaural level difference (Goldstein 2010). Information gained using the combination of these two mechanisms allows us to locate objects via sound in three-dimensional space.

One of the more important uses of hearing for humans is speech perception and production. These rely on the identification of sounds and the discrimination of one sound from another based on differences in timing, intensity and frequency. Speech production involves the interaction of the auditory and motor systems, particularly the tongue, lips and vocal cords. This has led to much debate on how this sensory motor relationship develops, with recent studies suggesting that this link develops early in life and is dependent on experience (Imada *et al.* 2006).

Theoretical approaches to sensory-motor development

The relationship between sensory and motor development is one that has been of considerable interest over the last century. Glencross (1995) described the complex interplay between sensory and motor processes as 'the interaction and integration of sensory factors in the on-going organizational control of motor processes and hence movements and actions' (p. 3). How this integration occurs has been the subject of much debate as a result of several paradigm shifts over the last century. It is therefore important to begin with a brief description of these major theoretical approaches.

According to Clark and Whitall (1989), our understanding of processes associated with motor development began in the eighteenth century and can be divided into distinct stages. The first was the Precursor period, followed by the influential Maturational period in the first half of the twentieth century. Maturational theorists such as Myrtle McGraw and Arnold Gesell made a significant contribution in terms of mapping the motor milestones (Piek 2006). Following the Maturational period was the Normative/descriptive period from 1946 to 1970 (Clark and Whitall 1989) which focused on providing standardized quantitative measures of motor ability. These early stages were very important in laying the foundations for the current theoretical underpinnings which began in the second half of the twentieth century, described as the Process-oriented period (Clark and Whitall 1989). In this stage, the relationship between sensory and motor processes has been seriously considered. Various paradigms have emerged throughout this period that attempted to describe the mechanisms and processes responsible for motor development. Two approaches, the cognitive/information-processing approach and the ecological approach, resulted in intense debate in the early 1990s as a result of their very diverse philosophical viewpoints (Summers 1998). Both views remain influential in our understanding of the relationship between sensory and motor development. These views are summarized in the next two sections.

Cognitive/information-processing approach

Prior to the mid twentieth century, sensory systems were rarely linked or discussed in relation to motor systems. It was as if they functioned independently of one another. However, in the second half of the twentieth century there was a major shift in paradigm, primarily as a result of early cognitive theorists. One such theorist was Jean Piaget, one of the most influential developmental psychologists. Not only was he one of the first to discuss the integral relationship between sensory systems and motor control, but he also argued that we gain all knowledge as a result of our early motor actions.

Piaget (1953) proposed a stage theory of cognitive development, with the first stage termed the sensorimotor stage. He argued that through motor actions an individual acquires knowledge as a result of sensory feedback from that action. Piaget argued that the sensorimotor stage covers the first two years of life, the years of infancy, and that infants needed to actively explore their environment in order to understand the world around them. This was initially achieved through infant reflexes and then later as a result of developing voluntary movements. This is a very powerful proposal identifying the interrelationship between motor and sensory systems. Although Piaget's theories are now considered inaccurate and outdated, clearly his understanding of the close relationship between sensory and motor development was a major contribution to our understanding of motor development.

In the 1970s, the information-processing approach emerged as a key theoretical paradigm. In this approach, the central nervous system is considered a central processing unit, analogous to the hard drive in a computer. In its simplest form, the information-processing model consists of input from sensory systems, a processor (the central nervous system) and an output, which is the motor actions described by its biomechanical properties (Schmidt and Lee 1999). As depicted in Figure 2.2, the sensory receptors are responsible for feeding information into the central controlling systems. These sensory organs receive information from two sources, external and internal, from three main types of receptors. Intero- or enteroceptors provide information from within the body organs (e.g. sensing blood pressure, thirst, etc). Exteroceptors in organs such as the eye and ear pick up environmental information such as light and sound. Furthermore, the information on body motion and position is detected by proprioceptors located in the muscles and tendons (as discussed earlier in this chapter).

The sensory information that feeds into the system is required for both the detection and recognition of the relevant stimuli. Based on this input and the information stored in the memory component, response selection takes place. The memory component stores relevant information described by information-processing theorists as schemas, representations or motor programs (Glencross 1977). The motor plan is then executed as the final stage of the process when the motor or efferent commands are transmitted to the muscles.

As can be seen in Figure 2.2, several types of feedback loops have been proposed based on the information-processing approach. The movement itself results in sensory or afferent feedback, producing a closed loop feedback system. This sensory input provides knowledge of the results or outcome of the movement. However, given the time delay involved, it has been suggested that this type

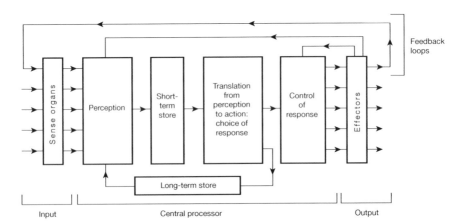

Figure 2.2 An early example of an information-processing model (adapted from Welford (1958)).

of feedback is only useful for movements longer than one third of a second (Abernethy *et al.* 1996). Hence, an open-loop model has been proposed where rapid movements are programmed and can be produced without the need for afferent feedback. An example of such a movement is the aforementioned ballistic saccadic eye movement.

In addition to afferent feedback, evidence has been provided for another feedback loop called *efference copy*, which is a replica of the original motor commands. This provides a reference for the afferent sensory information and provides a solution to the problem of determining whether errors are caused by changes in the environment or internally such as changes in body movement (Gandevia 1996).

In the 1970s and 1980s, concerns were raised that a prescriptive description of motor control, such as the information-processing approach, had many underlying problems, most related to the issue of information storage capacity – that is, there are an infinite number of solutions to any motor task, so how does the central nervous system store all of these exact specifications (Hopkins *et al.* 1993)? These concerns were initially raised by Nikolai Bernstein (1967) and elaborated in Whiting's (1984) reassessment of Bernstein's work. As a result of this debate, those who followed the prescriptive information-processing approach reassessed their description of the model and moved to a more abstract view of how motor programs are stored. Rather than a set of specific efferent commands, the notion of the generalized motor program was proposed (Schmidt 1975), which contained invariant features of the movements, but mapped in situation-specific parameters during the planning process.

Despite this re-evaluation of the information-processing approach, a new paradigm emerged based on Bernstein's work. Bernstein has been acknowledged as 'one of the premier minds of our century' (Latash 1998, p. v) and considered one of the influential researchers in the development of the ecological approach to motor control, which recognized the importance of understanding movement in terms of inertia and reactive forces as well as muscular force (Turvey *et al.* 1982).

Ecological approach

The ecological, natural-physical or action systems approach emerged in the late 1970s and is considered a result of the integration of views from Bernstein on movement coordination, Gibson's notion of direct perception and concepts of synergetics relating to pattern formation, as well as non-equilibrium thermodynamics relating to self-organizing systems (Summers 1998). Rather than movements being organized hierarchically via central control, this new theoretical paradigm argued for movements to emerge as a result of self-organization through natural laws.

A major influence in the development of the ecological approach was Gibson's (1979) perception-action coupling, where 'organisms do not simply perceive the physical properties of their environment', but 'perceive the physical properties of their environments *in relation to* their action capabilities' (Lockman 1990, p. 88)

– that is, perception and action are inseparable. Gibson and Gibson (1991) provided an 'active' theory of perception, arguing that organisms are designed to identify *invariants*, described as higher order relationships that detect constancies within the environment. Unlike the information-processing perspective, these constancies are not thought of as coded signals but as 'mathematical relations in a flowing array' (Gibson 1991, p. 506). As a result of actively seeking out and detecting these invariants, an organism gains an understanding of the environment by identifying *affordances*, which describe the experiences in functional terms. For example, a cave affords shelter from the rain, whereas rain affords getting wet.

Michaels and Beek (1996) suggested that three different perspectives emerged as part of the ecological shift: direct perception as described by Gibson (1979), non-equilibrium thermodynamics or kinematic theory, which examines how the many degrees of freedom in complex movements are constrained through muscle linkages or coordinative structures (Tuller *et al.* 1982), and the dynamical systems approach which investigates pattern formation through self-organization. Of the three, the dynamical systems approach is the one that has been most popular in the investigation of sensory motor development in infants and children.

Esther Thelen was considered one of the pioneers of research investigating motor development from a dynamical systems perspective. Thelen and Smith (1994) outline in detail how motor development can be interpreted as a self-organizing process where the continuous cycle of perception and action results in spontaneous pattern formation. The formation of new patterns or movements is dependent on many subsystems that mature at different rates, producing non-linear development. Hence, the system may be constrained by one or more of these collective variables (Ulrich and Ulrich 1993), which could include appropriate sensory development or postural control, muscle strength or cognitive factors such as a child's motivation or attention. The dynamical systems perspective continues to be influential in current research on sensory-motor development, as will be seen in the remainder of this chapter.

Development of the sensory systems

As a result of the nature–nurture debate by the early philosophers and theorists, the development of the sensory systems in infants has been extensively researched. Thomas (1996) describes the nature–nurture debate as 'the most fundamental concern of child development theorists' (p. 31). For example, John Locke, a seventeenth-century philosopher, was an environmentalist who had the extreme view that infants were born with an 'unmarked page or *tabula rasa*' (Thomas 1996, p. 31), and it was through experience that they developed an understanding of the world around them. In order to develop that understanding, infants were required to take in information, and this is achieved via the sensory systems. Hence, it was important to know how capable an infant's sensory system was at birth. Before the twentieth century, techniques had not been developed to investigate a newborn's sensory capabilities, but over the last century this has been extensively researched as a result of innovative research approaches. We now know that, not only do newborn infants have well-developed sensory systems at birth, but these also develop during gestation.

According to Gottlieb (1971), what is unique about humans (and other primates) is that they have a precocial sensory system where all of the senses can function prior to birth, but they have an altricial motor system where they have little mobility at birth and hence require nurturing from others. Given that the senses are available at birth, it is clear that these have developed *in utero*. Gottlieb (1971) attempted to predict the sequence of sensory development in the human foetus by examining the findings of animal studies. Similar to his earlier finding for chick embryos, he found that for mammals, the sequence was tactile, then vestibular, followed by auditory and lastly visual, although the specific time of onset varied between species. An obvious omission here is that of kinaesthesis, and this remains a difficult sensory system to investigate both *in utero* and in the infant.

Evidence for the presence of the tactile sense *in utero* comes from observations of the reactions of the foetus to clinical procedures. For example, aversive reactions to the needle have been noted when samples of amniotic fluid are taken (Lecanuet and Schaal 1996). Touch is highly developed in the newborn, with all of the specialized receptors for pain, pressure and temperature present at birth. Pain, for example, appears to be highly developed, with infants reacting to pain through responses such as crying or increased heart rate or blood pressure. These responses have been observed during procedures such as heel lancing to collect a blood sample or during circumcision, and they have been found in both full-term and preterm infants.

It appears that most sensory systems are functioning at an adult level at birth. These include vestibular, kinaesthetic, olfaction (the sense of smell) and gustation (the sense of taste). The auditory system has been extensively researched *in utero* and it has been known since early in the twentieth century that the foetus is sensitive to different sounds (Rosenblith 1992). However, the newborn is less sensitive to sounds than adults, partly due to gelatinous tissue that fills the inner ear which takes around a week to be reabsorbed (Haywood and Getchell 2009). In the first few months, infants are more sensitive to low-frequency sounds, allowing them to sense speech. Research has identified that the newborn may be biased towards listening to speech (Vouloumanos and Werker 2007), suggesting that language development begins at a very early age.

Gottlieb (1971) noted that vision was the sense that developed last, and this is supported by research on vision in the newborn. Newborn infants have difficulty with fixation, focusing and with the coordination of eye movements as the fovea and ocular muscles have not completely developed. One example of this limited capacity is the infant's reduced acuity at birth. It has been estimated that one-month-old infants have a visual acuity of between 20/400 and 20/600, meaning that an infant will see at 20 ft (6 m) what an adult will see at 400–600 ft (121–183 m) (Goldstein 2010), in contrast to normal adult 20/20 vision. This limited acuity allows the infant to clearly see only relatively large objects or features

(Salapatek and Banks 1978). However, Turkewitz and Kenny (1982) proposed that this limitation in vision at birth is important in limiting the amount and type of stimulation a newborn infant receives, therefore promoting subsequent perceptual organization.

From this very brief review, it is clear that the human infant is capable of taking in multiple sources of sensory information at birth. How the infant copes with these multiple sources is the topic of the next section.

Sensory integration

The individual sensory systems described earlier in this chapter are essential for appropriate motor control. However, few experiences are unimodal – that is, when an individual takes in sensory information from an object or an event, it is generally from several different sensory modalities. For example, the simple act of catching a ball includes visual, vestibular, kinaesthetic, tactile and can also include auditory information (e.g. if the ball is initially bounced or hits a bat). How do we make sense of this multiple array of sensations? In order to do this, the information needs to be integrated. This process of sensory-sensory integration or cross-modal integration is therefore an important process to consider in order to understand the relationship between sensory-motor integration.

As stated by Lewkowicz (2002), 'Perception of multimodally specified objects and events is generally more efficient and adaptive when the multiple, multimodal sources of information specifying these objects and events can be perceived in an integrated manner' (p. 41). When and how does this occur? This question has been pondered for centuries and has been an integral part of the nature–nurture debate. The early nature or nativist view was that the ability to integrate sensory information is innate and hence, readily available to the newborn infant. In contrast, theorists such as Piaget (1953) argued that infants are born with no ability to integrate sensory information and it is only through experience that they gradually acquire this skill during the first few months of life. Since then, there has been considerable research demonstrating that newborn infants are able to integrate sensory modalities, suggesting that Piaget and other nurture theorists may not necessarily be correct.

The seminal research of Meltzoff and Borton (1979) demonstrated that newborn infants between 26 and 33 days old could match two different modalities, vision and touch. In a simple but ingenious study, infants were given one of two different shaped pacifiers to orally explore without seeing them. They were then presented with large matching shapes made from styrofoam and the fixation time for each of the shapes was measured. It was found that a significantly larger proportion of infants looked longer at the pacifier they had tactually explored compared with the other shape – that is, they could translate the tactile information into a visual representation, strong evidence that infants have the ability to integrate sensory information at birth.

According to Gibson (1991), this finding would be expected based on their theory of direct perception. She argued that at birth, infants are amodal – that is, the information they receive is not specific for one modality. Although newborn infants cannot distinguish individual modalities from the stimulus array, they can perceive what that information affords them. It is only through experience that the infant then learns to differentiate the individual sensory modalities. For example, Gibson and Walker (1984) found that infants as young as one month old could identify the affordance of object rigidity (hard or soft). The infants initially mouthed either a hard plastic or soft sponge object without seeing them and then observed a rigid wooden or foam-rubber deforming object. Significantly more infants were found to look at the novel stimulus longer than the one they originally mouthed, suggesting that they can identify the affordance of hard versus soft. Importantly, this could be achieved intermodally, from tactile to vision.

Of concern in recent times is the number of studies that have failed to replicate the original work of Meltzhoff and Borton (1979). For example, Maurer *et al.* (1999) did an exact replication of the original study and failed to find any evidence of tactile–visual transfer. They carried out three experiments in all, and none provided evidence for cross-modal transfer between touch and vision in infants as young as one month. They do, however, point out that many studies have provided evidence of cross-modal integration in other sensory modalities. How can these discrepancies be understood? Lewkowicz (2002) argued that the earlier theories on sensory integration simplify what is a complex and dynamic developmental process. One important issue that has not been taken into account in these early views is that each of the sensory systems is at a different stage of development at birth as a result of different systems becoming functional at different times throughout gestation (Gottlieb 1971). According to Turkewitz and Kenny (1982), if different sensory systems are at different stages of development at birth, then it makes sense that this would influence the development of the integrative abilities of these senses – that is, 'the uneven rate of development and sequential onset of functioning of the sensory systems have consequences for the development of relationships between them' (p. 359). However, as Lickliter and Bahrick (2000) point out, it is very difficult to investigate these issues in humans, and they suggest that animal research is assisting in our understanding of the importance of prenatal and early experience on the emergence and development of intersensory perception. Just as animal research influenced our understanding of early developmental processes in the early twentieth century, it seems that it may again prove beneficial for our understanding of sensory integration in the early twenty-first century.

Sensory-motor development

Prenatal development

Just as sensory systems are known to develop throughout gestation, motor development also has its origins in this stage, and newborn infants already have a considerable movement repertoire, even though these movements are considered quite primitive.

Several techniques have been used to investigate motor development *in utero*. The normal gestational period for humans is 40 weeks, which is taken from the first day of the mother's last menstrual cycle. Human foetuses have been examined either through the body wall (e.g. using ultrasound) or by examining aborted foetuses, but these two approaches have led to very different findings. One of the most recognized early researchers in this area was Preyer who published *Die Spezielle Physiologie des Embryo* in 1885 (Provine 1993). By observing pregnant women and listening for palpitations using a stethoscope, Preyer was able to document foetal movements at 12–15 weeks gestational age, which is considerably earlier than those usually felt by the expectant mother. Preyer's early work on exteriorized animal foetuses provided some of the first evidence of early spontaneous motor activity, which has now been extensively documented using ultrasound. De Vries and colleagues (de Vries *et al.* 1982, 1985, 1988) discovered slow neck extensions at 7–7½ weeks, and at 8 weeks gestational age, startle and general movements were identified. From 8 weeks on the foetus develops a wide range of movements that were found to be continuous at birth.

The biomechanics of foetal movements have also been examined, initially in animal embryos, based on a procedure developed by Hamburger and Oppenheim (1967), where a small hole is made in the eggshell, which can then be resealed with paraffin or tape to allow the embryo to develop normally. Watson and Bekoff (1990) used this approach to take kinematic and electromyographic recordings of the movements in chick embryos and found that these were highly coordinated with the limb joints extending and flexing in synchrony. Similar synchronous joint movements have been identified by Heriza (1988a) in infants born preterm at 34 weeks gestational age.

Given that many of the sensory modalities are quite developed *in utero*, how then do these influence or function in relation to the movements of the foetus? One important function of these foetal movements is to prevent adhesions or local stasis of the circulation (Prechtl 1984). They may be important also in the appropriate development of the skeletal system. Early animal studies have found that if these movements are impeded using drugs such as curare that result in paralysis, there can be permanent joint malformations or muscle atrophy (Provine 1993).

The newborn infant

When an infant is born, the surrounding environment changes dramatically, with gravity being a key factor, influencing the infant's movements. However, despite this environmental change, the movements that have been identified *in utero* also appear once the infant is born, and de Vries *et al.* (1982) found no 'foetus-specific' patterns in their work using ultrasound to identify specific movements – that is, the infant can readily adapt the foetal movements to the very different sensory input provided at birth. Further evidence for this was provided by Heriza (1988a, 1988b) who used kinematic analyses to investigate leg flexion and extension in preterm infants born at 34 weeks gestational age. The highly organized synergies found between the hip, knee and ankle joints in these infants were no different from those of full-term newborn infants.

In addition to these spontaneous movements, other movement patterns called *primitive reflexes* can be found in the newborn infant. Some of these are found *in utero*, with others emerging following birth. These reflexes are initiated through the appropriate sensory stimulation. Many of these reflexes are essential for the infant's survival, such as the sucking reflex which is activated through touch and results in the jaw movements needed to express milk. These reflexes demonstrate the key role that sensory stimulation plays in the newborn infant.

Piaget (1953) considered that reflexes were the infant's first sensorimotor experiences and believed it was through these reflexes that infants gained knowledge. However, there is considerable evidence that newborns have coordinated sensorimotor experiences that are not related to reflexes. One key study that identified early intentional movements was that of von Hofsten (1982). He provided evidence that three-day-old infants extend their arm towards an object they fixate on, providing evidence for eye–hand coordination. Furthermore, hand-to-mouth movements have been identified *in utero* (de Vries *et al*. 1982) and continue at birth. Piaget would consider these movements unintentional and not involving sensory-motor coordination. However, a study by Butterworth and Hopkins (1988) using motion analysis found that newborns opened their mouth as the hand approached the face. This demonstrated intention, hence eye–hand coordination.

Sensory input and postural control

In the first few months after birth, an infant also demonstrates postural reflexes. Whereas touch plays a key role in initiating many of the primitive reflexes, postural reflexes are primarily influenced by the vestibular system. They are essential for maintaining the infant's body in the appropriate orientation to gravity. Apart from these reflexes, newborns appear to have little postural control. However, postural control is essential for stability, balance and orientation, all of which are required for voluntary movements.

Postural control is considered a perfect example of perception action coupling as the body continually moves to adapt to the changing sensory input in order to maintain balance and stability (Haywood and Getchell 2009). The sensory systems important for appropriate postural control include visual, vestibular and somatosensory systems, although there is some evidence that the auditory system may also play a role (Horak and MacPherson 1995, cited in Haywood and Getchell 2009). A key issue that has been investigated for over a quarter of a century is whether any one of these systems is more important than another. The results suggest that there may be a shift in the importance of different sensory systems with development.

Although the visual system is the most poorly developed of the sensory modalities at birth, it remains important for postural control in the newborn. For example, Jouen (1988) found that newborns responded to a light stimulus that was moved forward or away from them by making postural adjustments. These were assessed by measuring the infants' forward and backward head movements. However, it should be noted that the infants did not always respond in the appropriate direction.

Lee and Thomson (1982) developed the 'moving room' to investigate the influence of different sensory systems on balance control (Figure 2.3). Misleading visual information can be provided by moving the walls of the room while controlling for kinaesthetic, vestibular and tactile information – that is, the optical array is altered to suggest that there is a loss of balance even though there are no mechanical changes in the joints or muscles or the semicircular canals in the vestibular system to suggest such a shift in balance. The influence of the misleading visual information is determined by examining the sway produced as a result of this information. As a result of these studies, it was determined that vision was the dominant sense in adults in maintaining balance. Similar results have been found in sitting infants aged two months (Pope 1984, cited in Woollacott and Jensen 1996) and six to nine months (Barela *et al.* 2000). This is consistent with Bertenthal *et al.*'s (2000) suggestion that the infant's 'postural control system is not fundamentally different than the adult's system' (p. 313).

Using a similar paradigm, Barela *et al.* (2003) examined postural control in children aged 4, 6 and 8 years and adults. They employed a haptic moving room where the participant stands touching a metal touch plate with eyes closed. During the trials the touch plate was moved rhythmically at 0.2, 0.5 and 0.8 Hz, and even though no children and only one adult participant noticed it moving, there was a corresponding oscillatory postural sway to the oscillatory moving surface. Children and adults appeared to couple the moving surface and their postural sway equally. Children aged from 6 years had similar values of phase and gain to adults, suggesting a coupling of both position and velocity of the moving stimulus. This information comes from two different sensory sources, cutaneous

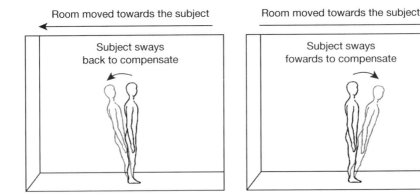

Figure 2.3 The 'moving room' designed by David Lee and colleagues to investigate the influence of different sensory systems on balance control (adapted from Ingle *et al.* (1982)).

stimulation at the fingertips and muscle spindle activity. However, 4-year-old children demonstrated a weaker coupling of position information, and it was suggested that they may have greater difficulty integrating the information from these two sources.

Children's postural sway also demonstrated higher variability compared with adults. However, the children's phase and gain plots were considered well-formed from 6 years old, suggesting their feedback process is equivalent to adults. As a result, it appeared that feed-forward mechanisms may be immature – that is, there may be an 'inadequate process of reweighting sensory information from different sources to generate an internal estimate of body orientation' (Barela *et al.*, 2003, p. 441). This was confirmed by Sparto *et al.* (2006) who found that, although children aged 7–12 years could use visual cues for postural control in a similar way to adults, they had greater difficulty than adults using somatosensory cues to stabilize posture when the visual cues conflict with the somatosensory cues.

This issue was extended further by Bair *et al.* (2007) who investigated multi-sensory reweighting during standing in a larger age range of children (4–10 years) in order to determine developmental changes in sensory integration. They proposed that as children become older they may be more able to reduce the importance or dominance of visual information by downweighting this in relation to other sensory inputs. This was tested using a paradigm that systematically varied the amplitude of both the visual scene and touch-bar oscillations simultaneously. Bair *et al.* (2007) found that children as young as 4 years of age were capable of reweighting in response to multisensory inputs, although the amount of reweighting increased with age. The young children could reweight intramodally, but only the older children could reweight intermodally.

In summary, the integration of information gained from several sensory systems is necessary to maintain balance. Depending on the environmental conditions, humans are able to reweight sensory information. The ability to reweight sensory information is acquired and refined during childhood. For a more comprehensive description of the biomechanics relating to the acquisition of postural control, the reader is referred to Chapter 7.

The aforementioned reweighting paradigm has been used to investigate the underlying deficits in children with motor disability, specifically those with developmental coordination disorder (DCD). Grove and Lazarus (2007) argued that previous research examining the underlying deficits in DCD had used functional performance measures to evaluate postural control and these 'alone do not discriminate between biomechanical, neuromuscular control, cognitive and/or sensory organization impairments' (p. 458). Using computerized platform posturography, Grove and Lazarus examined the ability of children with DCD to reweight somatosensory, visual and vestibular input under six different conditions of sensory feedback availability (normal or distorted). They found that children with DCD had significantly impaired postural control when they needed to rely primarily on vestibular information. This supported their hypothesis that children with DCD may have an underlying deficit in sensory reweighting, particularly when relying on vestibular information to maintain postural control. They suggested that interventions that focus on postural control training using vestibular feedback may assist in improving the motor performance of children with DCD. For more information relating to motor development and children with DCD, the reader is referred to Chapter 12.

The auditory system and motor development

There has been very little recent research that has examined the influence of the auditory system on motor development. However, research carried out in the late twentieth century focused on investigating motor proficiency in children with hearing impairments and generally found that hearing-impaired children performed more poorly on tasks measuring dynamic and visual-motor coordination (Wiegersma and Van Der Velde 1983). Such findings suggest that the auditory system is important for motor development. However, Crowe and Horak (1988) pointed out that although children with hearing impairments were known to have both vestibular and motor deficits, it was unclear whether the vestibular or hearing deficits were responsible for the motor impairment. Consequently, these authors examined a small sample of hearing-impaired children either with or without vestibular damage and concluded that the child's balance ability was related to the vestibular damage rather than the hearing loss. There was, however, no evidence that vestibular disorders were responsible for abnormal coordination, apart from balance activities that were dependent on vestibular function (Horak et al. 1988). The evidence suggested that motor impairment was associated with poor multimodal sensory integration rather than specific problems with the auditory or vestibular system.

The relationship between auditory and visual information was investigated further by Savelsbergh et al. (1991) in children who were congenitally deaf and children with no hearing loss. While no significant difference was found between the two groups for catching a ball when the ball approached from within the field of view, the children with no hearing loss were significantly better than the deaf children when the ball approached from the periphery outside the field of view. The authors argued that the additional auditory cues for the children with normal hearing resulted in faster reaction times, and they suggested that 'a lack of auditory stimulation during development can lead to deficiencies in the coordination of actions such as catching' (p. 489).

Clinical and practical applications

Problems with sensory development have long been associated with motor and other childhood disorders, and several clinical interventions have been developed based on the notion that underlying deficits in these disorders may be a result of problems with the functioning of the sensory systems.

One of the earliest approaches that viewed deficits in sensory development as a key factor in many childhood disorders was the Sensory Integration Theory of Ayres. Ayres (1972) proposed that some children have difficulty integrating sensory information and as a result of this may have problems with learning, motor control and social interaction. She developed Sensory Integration Therapy in which a child is presented with challenging sensory experiences involving the integration of a variety of sensations, primarily vestibular, kinaesthetic and tactile, which require an adaptive response. As a result of this exposure, it was argued that children increase their ability to integrate sensory information. This approach has been used primarily by occupational therapists on children with disorders such as autism, Asperger's syndrome, attention deficit hyperactivity disorder and learning disorders. However, despite its continued popularity over the past four decades, there is limited evidence that it is an effective treatment with little evidence supporting improvements in academic or motor performance, behavioural outcomes or perceptual skills. Based on a meta-analysis of 41 studies evaluating the Sensory Integration Therapy, Shaw (2002) concluded that there is no evidence of improvement in motor or psychoeducational skills as a result of the therapy, with any improvements identified being accounted for by maturational factors.

In the 1980s, research emerged which suggested that children with movement disorders such as DCD may have difficulties in processing kinaesthetic information (e.g. Smith and Glencross 1986). Laszlo and Bairstow (1985) developed the Kinaesthetic Sensitivity Test (KST) to measure kinaesthetic sensitivity and used it to map kinaesthetic developmental trends for children aged 5–12 years. The authors found that kinaesthetic perception was quite poorly developed in young children but could be improved with training. Laszlo and Sainsbury (1993) utilized this training technique as an intervention for children with DCD and found that both kinaesthetic sensitivity and motor ability improved as a result of training. However, this research has been criticized for a number of reasons. Sims and colleagues (Sims and Morton 1998; Sims *et al.* 1996) examined the effectiveness of the kinaesthetic training as a therapy for children with DCD. They confirmed that the training improved motor performance, but argued that the improvement was not a result of the specific kinaesthetic training as 'cognitive-affective' training produced similar improvements.

The KST has also been criticized for a number of reasons. Livesey and Coleman (1998) argued that this test is too difficult for young children and may have underestimated the kinaesthetic ability of the younger children as a result of this. Using a kinaesthetic acuity test (KAT) developed for preprimary children (Livesey and Parkes 1995), Livesey and Coleman (1998) found that children as young as 3 years of age had considerable kinaesthetic sensitivity, which rapidly improved at between 4 and 5 years of age. Also, children who performed poorly on the KAT had motor problems, which supported previous evidence that poor kinaesthetic acuity may be an underlying deficit of DCD.

The KST has also been criticized for its poor reliability and validity (Lord and Hulme 1987; Piek and Coleman-Carman 1995). For example, Hulme and colleagues examined visual, kinaesthetic and cross-modal judgements using a line length estimation paradigm. They found that children with poor motor ability had significant difficulties with visual matching but few difficulties with kinaesthetic or cross-modal matching (Hulme *et al.* 1982a, 1982b). Since then, there has been extensive research investigating visual-perceptual and visuo-motor ability in children with motor deficits.

Sigmundsson *et al.* (2003), for example, employed three tasks that measured both dynamic and static global visual processing and found that children with DCD were poorer at detecting both motion and form than their typically developing peers. They suggested that this was linked with cerebellar deficits, which have been previously linked with DCD. Other researchers have found mixed results and have argued that DCD is a heterogeneous disorder in terms of its underlying processes (e.g. Schoemaker *et al.* 2001; Tsai *et al.* 2008; Van Waelvelde *et al.* 2004). This notion is supported by results from the meta-analysis carried out by Wilson and McKenzie (1998). These authors analysed 50 studies investigating processing deficits in children with DCD. Overall, the meta-analysis indicated that visual-spatial, cross-modal and kinaesthetic deficits are all implicated in DCD.

Recent research has identified a relationship between visual movement imagery and kinaesthetic acuity (Livesey 2002). Furthermore, children with DCD have been identified with a motor imagery deficit – that is, they have problems producing an accurate visuospatial representation of the intended movement (Wilson *et al.* 2001). Consequently, the authors of this study proposed that children with DCD may have an inaccurate efference copy, leading to problems with motor sequencing, as rapid movements would rely on the efference copy as an error detection mechanism. This perspective has led to the proposal that motor imagery training may be useful for the treatment of children with DCD. Recent results using this approach have been promising, as motor imagery training appeared to lead to improvements in planning and coordinating movements that incorporated obstacles and objects within the environment (Wilson, 2005).

Clearly, an understanding of the relationship between sensory and motor development is essential for the understanding of motor disorders such as DCD. Different theoretical perspectives such as the dynamical systems and information-processing approach have provided insight into these issues. However, it is also clear that much work is still needed. The advances in areas such as biomechanics and neuroimaging will prove valuable tools to examine these issues in the future.

Key points

- Infants are born with all sensory systems functioning although the visual system is not as fully developed as the other systems.
- Sensory abilities appear to be amodal at birth, then discrimination of the senses occurs with age and experience.
- In order to use multiple senses there needs to be integration of the senses, and this process appears to develop in the first few months of life.
- Sensorimotor integration is present in some form at birth.
- The major sensory systems needed for motor coordination are vision, vestibular and somatosensory.

- In order to refine postural control, children acquire the ability to 'reweight' sensory information appropriately.
- Problems with sensory processing have been implicated in motor disorders such as developmental coordination disorder, resulting in new and innovative intervention approaches.

Acknowledgement

I wish to thank Daniela Rigoli for her assistance with this chapter and Dr Natalie Gasson for her valuable feedback.

References

Abernethy, B., Kippers, V., Mackinnon, L.T., Neal, R.J. and Hanrahan, S. (1996) 'Basic concepts of motor control: psychological perspectives', in B. Abernethy, V. Kippers, L.T. Mackinnon, R.J. Neal and S. Hanrahan (eds) *The Biophysical Foundations of Human Movement*, Melbourne: Macmillan Education Australia, pp. 295–311.

Aslin, R.N. (1987) 'Motor aspects of visual development in infancy', in P. Salapatek and L. Cohen (eds) *Handbook of Infant Perception*, New York: Academic Press, pp. 43–113.

Ayres, A.J. (1972) *Sensory Integration and Learning Disorders*, Los Angeles, CA: Western Psychological Services.

Bair, W.-N., Kiemel, T., Jeka, J.J. and Clark, J.E. (2007) 'Development of multisensory reweighting for posture control in children', *Experimental Brain Research*, 183: 435–46.

Barela, J.A., Godoi, D., Freitas, J. and Polastri, P.F. (2000) 'Visual information and body sway coupling in infants during sitting acquisition', *Infant Behavior and Development*, 23: 285–97.

Barela, J.A., Jeka, J.J. and Clark, J.E. (2003) 'Postural control in children: coupling to dynamic somatosensory information', *Experimental Brain Research*, 150: 434–42.

Bernstein, N. (1967) *The Co-ordination and Regulation of Movements*, Oxford: Pergamon Press.

Bertenthal, B.I., Boker, S.M. and Minquan, X. (2000) 'Analysis of the perception-action cycle for visually inducted postural sway in 9-month-old sitting infants', *Infant Behavior and Development*, 23: 299–315.

Butterworth, G. and Hopkins, B. (1988) 'Hand–mouth coordination in the new-born baby', *British Journal of Developmental Psychology*, 6: 303–14.

Clark, J.E. and Whitall, J. (1989) 'What is motor development? The lessons of history', *Quest*, 41: 183–202.

Crowe, T.K. and Horak, F.B. (1988) 'Motor proficiency associated with vestibular deficits in children with hearing impairments', *Physical Therapy*, 68: 1493–9.

de Vries, J.I.P., Visser, G.H.A. and Prechtl, H.F.R. (1982) 'The emergence of fetal behaviour. I. Qualitative aspects', *Early Human Development*, 7: 301–22.

de Vries, J.I.P., Visser, G.H.A. and Prechtl, H.F.R. (1985) 'The emergence of fetal behaviour. II. Quantitative aspects', *Early Human Development*, 16: 99–120.

de Vries, J.I.P., Visser, G.H.A. and Prechtl, H.F.R. (1988) 'The emergence of fetal behaviour. III. Individual differences and consistencies', *Early Human Development*, 16: 85–103.

Gandevia, S.C. (1996) 'Kinesthesia: roles for afferent signals and motor commands', in L.B. Rowell and T.J. Sheperd (eds) *Handbook of Physiology Exercise: Regulation and Integration of Multiple Systems*, New York: Oxford University Press, pp. 128–72.

Gibson, E.J. (1991) *An Odyssey in Learning and Perception*, Cambridge, MA: The MIT Press.

Gibson, E.J. and Gibson, J.J. (1991) 'The senses as information-seeking systems', in E.J. Gibson (ed.) *An Odyssey in Learning and Perception*, Cambridge, MA: The MIT Press, pp. 503–10.

Gibson, E.J. and Walker, A.S. (1984) 'Development of knowledge of visual-tactual affordances of substance', *Child Development*, 55: 453–60.

Gibson, J.J. (1979) *The Ecological Approach to Visual Perception*, Boston, MA: Houghton Mifflin.

Glencross, D.J. (1977) 'Control of skilled movements', *Psychological Bulletin*, 84: 14–29.

Glencross, D.J. (1995) 'Motor control and sensory-motor integration', in D.J. Glencross and J.P. Piek (eds) *Motor Control and Sensory Motor Integration*, Amsterdam: Elsevier, pp. 3–7.

Goldstein, E.B. (2010) *Sensation and Perception*, Belmont, CA: Wadsworth.

Gottlieb, G. (1971) 'Ontogenesis of sensory function in birds and mammals', in E. Tobach, L. Aronson and E. Shaw (eds) *The Biopsychology of Development*, New York: Academic Press, pp. 67–128.

Grove, C.R. and Lazarus, J.-A.C. (2007) 'Impaired re-weighting of sensory feedback for maintenance of postural control in children with developmental coordination disorder', *Human Movement Science*, 26: 457–76.

Hamburger, V. and Oppenheim, R. (1967) 'Prehatching motility and hatching behavior in the chick', *Journal of Experimental Zoology*, 166: 171–204.

Haywood, K.M. and Getchell, N. (2009) *Life Span Motor Development*, Champaign, IL: Human Kinetics.

Heriza, C.B. (1988a) 'Comparison of leg movements in preterm infants at term with healthy full-term infants', *Physical Therapy*, 68: 1687–93.

Heriza, C.B. (1988b) 'Organization of leg movements in preterm infants', *Physical Therapy*, 68: 1340–6.

Hopkins, B., Beek, P.J. and Kalverboer, A.F. (1993) 'Theoretical issues in the longitudinal study of motor development', in A.F. Kalverboer, B. Hopkins and R. Geuze (eds) *Motor Development in Early and Later Childhood: Longitudinal Approaches*, Cambridge: Cambridge University Press, pp. 343–71.

Horak, F.B., Shumway-Cook, A., Crowe, T.K. and Black, F.O. (1988) 'Vestibular function and motor proficiency of children with impaired hearing, or with learning disability and motor impairments', *Developmental Medicine and Child Neurology*, 30: 64–79.

Hulme, C., Smart, A. and Moran, G. (1982a) 'Visual perceptual deficits in clumsy children', *Neuropsychologia*, 20: 475–81.

Hulme, C., Biggerstaff, A., Moran, G. and McKinlay, I. (1982b) 'Visual, kinaesthetic and cross-modal judgements of length by normal and clumsy children', *Developmental Medicine and Child Neurology*, 24: 461–71.

Imada, T., Zhang, Y., Cheour, M., Taulu, S., Ahonen, A. and Kuhl, P.K. (2006) 'Infant speech perception activates Broca's area: a developmental magnetoencephalography study', *Brain Imaging*, 17: 957–62.

Ingle, D.J., Goodale, M.A. and Mansfield, R.J.W. (eds) (1982) *Analysis of Visual Behavior*, Cambidge, MA: The MIT Press.

Jouen, F. (1988) 'Visual-proprioceptive control of posture in newborn infants', in B. Amblard, A. Berthoz and F. Clarac (eds) *Posture and Gait: Development, Adaptation and Modulation*, Amsterdam: Elsevier, pp. 59–65.

Laszlo, J.I. and Bairstow, P.J. (1985) *Test of Kinaesthetic Sensitivity*, London: Holt, Rinehart & Winston.

Laszlo, J.I. and Sainsbury, K.M. (1993) 'Perceptual-motor development and prevention of clumsiness', *Psychological Research*, 55: 167–74.

Latash, M.L. (1998) *Progress in Motor Control Volume One: Bernstein's Traditions in Movement Studies*, Champaign, IL: Human Kinetics.

Lecanuet, J.-P. and Schaal, B. (1996) 'Fetal sensory competencies', *European Journal of Obstetrics and Gynecology and Reproductive Biology*, 68: 1–23.

Lee, D.N. and Thomson, J.A. (1982) 'Vision in action: the control of locomotion', in D.J. Ingle, M.A. Goodale and R.J.W. Mansfield (eds) *Analysis of Visual Behavior*, Cambridge, MA: The MIT Press, pp. 411–33.

Lewkowicz, D.J. (2002) 'Heterogeneity and heterochrony in the development of inter-sensory perception', *Cognitive Brain Research*, 14: 41–63.

Lickliter, R. and Bahrick, L.E. (2000) 'The development of infant intersensory perception: advantages of a comparative convergent-operations approach', *Psychological Bulletin*, 126: 260–80.

Livesey, D. (2002) 'Age differences in the relationship between visual movement imagery and performance on kinesthetic acuity tests', *Developmental Psychology*, 38: 279–87.

Livesey, D.J. and Coleman, R. (1998) 'The development of kinaesthesis and its rela-tionship to motor ability in pre-school children', in J.P. Piek (ed.) *Motor Behavior and Human Skill: A Multidisciplinary Approach*, Champaign, IL: Human Kinetics, pp. 253–69.

Livesey, D.J. and Parkes, N. (1995) 'Testing kinaesthetic acuity in pre-school children', *Australian Journal of Psychology*, 47: 160–3.

Lockman, J.J. (1990) 'Perceptuo-motor coordination in infancy', in C.A. Hauert (ed.) *Developmental Psychology: Cognitive, Perceptuo-Motor and Neuropsychological Perspectives*, Amsterdam: Elsevier, pp. 85–111.

Lord, R. and Hulme, C. (1987) 'Kinaesthetic sensitivity of normal and clumsy children', *Developmental Medicine and Child Neurology*, 29: 720–5.

Maurer, D., Stager, C.L. and Mondloch, C.J. (1999) 'Cross-modal transfer of shape is difficult to demonstrate in one-month-olds', *Child Development*, 70: 1047–57.

Meltzoff, A. and Borton, R.W. (1979) 'Intermodal matching by human neonates', *Nature*, 282: 403–4.

Michaels, C. and Beek, P. (1996) 'The state of ecological psychology', *Ecological Psychology*, 7: 259–78.

Piaget, J. (1953) *The Origin of the Intelligence in the Child*, London: Routledge & Kegan Paul.

Piek, J.P. (2006) *Infant Motor Development*, Champaign, IL: Human Kinetics.

Piek, J. and Coleman-Carman, R. (1995) 'Kinaesthetic sensitivity and motor performance of children with developmental co-ordination disorder', *Developmental Medicine and Child Neurology*, 37: 976–84.

Prechtl, H.F.R. (1984) 'Continuity and change in early neural development', in H.F.R. Prechtl (ed.) *Continuity of Neural Function from Prenatal to Postnatal Life*, Philadelphia, PA: J.B. Leffcott, pp. 1–15.

Prochazka, A. (1996) 'Proprioceptive feedback and movement regulation', in L.B. Rowell and J.T. Shepherd (eds) *Handbook of Physiology. Section 12: Exercise: Regulation and Integration of Multiple Systems*, New York: Oxford University Press, pp. 89–127.

Provine, R.R. (1993) 'Natural priorities for developmental study: neuroembryological perspectives of motor development', in A.F. Kalverboer, B. Hopkins and R. Geuze (eds) *Motor Development in Early and Later Childhood: Longitudinal Approaches*, Cambridge: Cambridge University Press, pp. 51–73.

Rosenblith, J.F. (1992) *In the Beginning: Development from Conception to Age Two*, Thousand Oaks, CA: Sage Publications.

Salapatek, P. and Banks, M.S. (1978) 'Infant sensory assessment: vision', in F.D. Minifie and L.L. Lloyd (eds) *Communicative and Cognitive Abilities: Early Behavioral Assessment*, Baltimore, MD: University Park Press, pp. 61–106.

Savelsbergh, G.J.P., Netelenbos, J.B. and Whiting, H.T.A. (1991) 'Auditory perception and the control of spatially coordinated action of deaf and hearing children', *Journal of Child Psychology and Psychiatry*, 32: 489–500.

Schmidt, R.A. (1975) 'A Schema theory of discrete motor skills learning', *Psychological Review*, 82: 225–60.

Schmidt, R.A. and Lee, T.D. (1999) *Motor Control and Learning: A Behavioral Emphasis*, Champaign, IL: Human Kinetics.

Schoemaker, M.M., van der Wees, M., Flapper, B., Verheij-Jansen, N., Scholten-Jaegers, S. and Geuze, R.H. (2001) 'Perceptual skills of children with developmental coordination disorder', *Human Movement Science*, 20: 111–13.

Shaw, S.R. (2002) 'A school psychologist investigates sensory integration therapies: promise, possiblity and the art of placebo', *National Association of School Psychologists Communique*, 31: 9–13.

Sigmundsson, H., Hansen, P. C., and Talcott, J. B. (2003) 'Do "clumsy" children have visual deficits', *Behavioural Brain Research*, 139: 123–9.

Sims, K. and Morton, J. (1998) 'Modelling the training effects of kinaesthesis acuity measurement in children', *Journal of Child Psychology and Psychiatry*, 39: 731–46.

Sims, K., Henderson, S.E., Hulme, C. and Morton, J. (1996) 'The remediation of clumsiness. I. An evaluation of Laszlo's kinaesthetic approach', *Developmental Medicine and Child Neurology*, 38: 976–87.

Smith, T.R. and Glencross, D.J. (1986) 'Information processing deficits in clumsy children', *Australian Journal of Psychology*, 38: 13–22.

Sparto, P.J., Redfern, M.S., Jasko, J.G., Casselbrant, M.L., Mandel, E.M. and Furman, J.M. (2006) 'The influence of dynamic visual cues for postural control in children aged 7–12 years', *Experimental Brain Research*, 168: 505–16.

Summers, J.J. (1998) 'Has ecological psychology delivered what it promised?', in J.P. Piek (ed.) *Motor Behavior and Human Skill: A Multidisciplinary Approach*, Champaign, IL: Human Kinetics, pp. 385–402.

Thelen, E. and Smith, L.B. (1994) *A Dynamic Systems Approach to the Development of Cognition and Action*, Cambridge, MA: The MIT Press.

Thomas, R.M. (1996) *Comparing Theories of Child Development*, Pacific Grove, CA: Brooks/Cole Publishing Company.

Tsai, C.-L., Wilson, P.H. and Wu, S.K. (2008) 'Role of visual-perceptual skills (non-motor) in children with developmental coordination disorder', *Human Movement Science*, 27: 649–64.

Tuller, B., Turvey, M. and Fitch, H. (1982) 'The Bernstein perspective. II. The concept of muscle linkage or coordinative structure', in J.A.S. Kelso (ed.) *Human Motor Behavior: An Introduction*, Hillsdale, NJ: Erlbaum, pp. 253–70.

Turkewitz, G. and Kenny, P.A. (1982) 'Limitations on input as a basis for neural organization and perceptual development: a preliminary theoretical statement', *Developmental Psychobiology*, 15: 357–68.

Turvey, M.T., Fitch, H.L. and Tuller, B. (1982) 'The Bernstein perspective. I. The problems of degrees of freedom and context-conditioned variability', in J.A.S. Kelso (ed.) *Human Motor Behavior: An Introduction*, Hillsdale, NJ: Erlbaum, pp. 239–52.

Ulrich, B.D. and Ulrich, D.A. (1993) 'Dynamic systems approach to understanding motor delay in infants with Down Syndrome', in G.J.P. Savelsbergh (ed.) *The Development of Coordination in Infancy*, Amsterdam: Elsevier, pp. 445–59.

Van Waelvelde, H., De Weerdt, W., De Cock, P. and Smits-Engelsman, B.C.M. (2004) 'Association between visual perceptual deficits and motor deficits in children with developmental coordination disorder', *Developmental Medicine & Child Neurology*, 46: 661–6.

von Hofsten, C. (1982) 'Eye–hand coordination in the newborn', *Developmental Psychology*, 18: 450–61.

Vouloumanos, A. and Werker, J.F. (2007) 'Listening to language at birth: evidence for a bias for speech in neonates', *Developmental Science*, 10: 159–71.

Watson, S. and Bekoff, A. (1990) 'A kinematic analysis of hindlimb motility in 9- and 10-day-old chick embryos', *Journal of Neurobiology*, 21: 651–61.

Welford, A.T. (1958) *Ageing and Human Skill*, London: Oxford University Press.

Whiting, H.T.A. (1984) *Human Motor Actions: Bernstein Reassessed*, Amsterdam: North-Holland.

Wiegersma, P.H. and Van Der Velde, A. (1983) 'Motor development of deaf children', *Journal of Child Psychology and Psychiatry*, 24: 103–11.

Wilson, P.H. (2005) 'Practitioner review: approaches to assessment and treatment of children with DCD: an evaluative review', *Journal of Child Psychology & Psychiatry*, 46: 806–23.

Wilson, P.H. and McKenzie, B.E. (1998) 'Information processing deficits associated with developmental coordination disorder: a meta-analysis of research findings', *Journal of Child Psychology & Psychiatry*, 39: 829–40.

Wilson, P.H., Maruff, P., Ives, S. and Currie, J. (2001) 'Abnormalities of motor and praxis imagery in children with DCD', *Human Movement Science*, 20: 135–59.

Woollacott, M. and Jensen, J.L. (1996) 'Posture and locomotion', in H. Heuer and S.W. Keele (eds) *Handbook of Perception and Action, Volume 2: Motor Skills*, San Diego, CA: Academic Press, pp. 333–403.

3 Development of neuromuscular coordination with implications in motor control

Eleftherios Kellis and Vassilia Hatzitaki

Introduction

During biological development the factors that contribute to physical fitness mature, and performance in exercise tests (strength, speed and endurance) as well as complex sport skills steadily improves. A major contributor to this improvement is neuromuscular development, which refers to the maturation of both neural and muscular systems and includes their integration. Selected issues of this integration with respect to growth and maturation are examined in this chapter. Particularly, the effects of growth on activation patterns of muscles either in isolation or in pairs during single joint maximum strength production are explored. This is followed by the presentation of neuromuscular development during force-control tasks. Next, we focus on the development of muscle synergies required for the production of purposeful multi-joint actions. Finally, we examine how the development of sensory and higher brain systems contributes to neuromuscular coordination during the performance of multi-joint daily life actions.

The most important factor which may explain the age-associated increases in muscle strength and, to some extent, improvements in task performance, is the change in muscle size with age (Kanehisa *et al.* 1995; Wood *et al.* 2004). However, there is an age effect for the increases in strength in children that cannot be accounted for by changes in body mass, fat-free mass (Belanger and McComas 1989), muscle–anatomical cross-sectional area (Kanehisa *et al.* 1994) or muscle volume (Kanehisa *et al.* 1995). Equally significant is the observation that strength is not proportional to the expected stature[2], but to stature[3,4] during the pubertal phase (Blimkie *et al.* 1990). This clearly indicates that, while muscle size and stature are related to increases in muscle strength, other factors mediate increases in strength during this phase (Kellis and Unnithan 1999). Among other factors, this age effect may be attributed to neural maturation that allows for a greater expression of muscular strength (Housh *et al.* 1996; Kellis and Unnithan 1999).

Not only do the contractile properties of muscles change during childhood, but so does the central nervous system and its control over the increasing force in the growing muscles. At birth, development of the spinal cord and lower brain systems is more advanced compared with higher brain structures. In addition, maturation of conduction velocities in the peripheral nervous system occurs earlier than in the central nervous system (Smits-Engelsman *et al.* 2003). In particular, both efferent and afferent (sensory) peripheral pathways (see Chapter 2) reach adult values around the age of 3 years (Muller *et al.* 1991; Smits-Engelsman *et al.* 2003). The rate at which signals are transmitted through nerve axons mainly depends on the development of myelin in the nerve cells as well as the size of axon diameters. Myelination in spinal root and peripheral nerves is complete by the age of 2 years and axon diameters attain adult values by 2–5 years (Connolly and Forssberg 1997; Rexed 1944).

The earlier development of the peripheral when compared to the central nervous system is supported by theories on child development which relate to the progressive appearance and disappearance of a variety of reflexes with the emergence and refinement of progressively more complex motor skills such as postural and locomotor skills, a process well described by the reflex-hierarchy theory (McGraw 1932). According to this theory, with further development of the cortex, more complex reflexes appear that are controlled by higher brain centres, which in turn inhibit lower level, more primitive reflexes (Assaiante and Amblard 1995). Studies using transcranial magnetic stimulation have shown that the time between the magnetic stimulus and the onset of the motor-evoked potential (latencies) decreases with age (Caramia *et al.* 1993; Connolly and Forssberg 1997; Eyre *et al.* 1991; Fietzek *et al.* 2000; Koh and Eyre 1988; Muller and Homberg 1992; Muller *et al.* 1991). These observations confirm that development of the corticospinal tract, which represents the efferent higher level centre of the brain, advances with age. The rate of this development varies in the literature as some studies have shown that these pathways reach maturity around 8–10 years of age (Muller and Homberg 1992; Muller *et al.* 1992) with possible extension up to 10–12 years of age (Caramia *et al.* 1993), while others have reported a constant (and high) corticospinal pathway for the first two years of life, which then falls exponentially with age and reaches adult values by age 16 years (Eyre *et al.* 1991). Other data suggest a steep increase in central corticospinal maturation during the first three years of life with less dramatic changes between the ages of 3 and 7 years (Fietzek *et al.* 2000). Methodological as well as differences in the type of test used to examine development of the corticospinal tract may account for these conflicting findings. In addition, the sensory (afferent) central pathways follow a somewhat faster maturational trend, reaching adult values by the age of 5–7 years (Muller *et al.* 1992).

According to a more recent view of neural development as described by the 'systems perspective', the emergence of multi-joint purposeful action is ascribed to complex interactions between (a) the developing nervous and musculoskeletal systems of the child, (b) the requirements of the task and (c) environmental constraints (Shumway-Cook and Woollacott 1985). Accordingly, the emergence and onset of a new skill in a child may be due to a number of contributing neural and musculoskeletal rate-limiting systems so that, until they mature sufficiently, the particular skill cannot emerge. Possible systems contributing to motor development include neural systems such as the different sensory modalities (visual, vestibular, somatosensory), motor systems (neuromuscular, musculoskeletal) and higher level adaptive-predictive systems. For example, the maturation of muscle response synergies or the development of sufficient muscle strength could be a rate-limiting factor in the development of independent stance and locomotion (Assaiante *et al.* 2000).

Basic theoretical concepts

Development of agonist muscle activation per unit of force/moment of force

Maximum isometric force (MVC) increases gradually in the preschool years, with a high increase occurring during the fourth year of life (Potter *et al.* 2006). Strength continues to increase gradually between 5 and 10 years of age, which is followed by a more evident increase after the age of 11 years (Smits-Engelsman *et al.* 2003). In turn, twitch peak moment of force (moment produced during electrical stimulation) increases gradually with age before undergoing a marked enhancement at puberty (Belanger and McComas 1989; Davies 1985; Grosset *et al.* 2005; Paasuke *et al.* 2000). Overall, twitch peak moment of force is lower in children compared with adults (Grosset *et al.* 2005; O'Brien *et al.* 2009b; Ramsay *et al.* 1990; Stackhouse *et al.* 2005). For example, Belanger and McComas (1989) found that activation level was 94 per cent for young boys (aged 6–13 years) compared to 99.4 per cent for older boys (aged 15–18 years). Recently, O'Brien *et al.* (2009b) reported that voluntary activation level in children was lower than adults, and this explained 7 per cent of the difference in MVC moment of males and 31 per cent of the difference in females. This value is higher in females as they may display lower motivation and effort during maximal strength tasks (O'Brien *et al.* 2009b). In addition, the increase in twitch peak moment of force/M-wave amplitude ratio (Grosset *et al.* 2005; Maffiuletti *et al.* 2001) with age indicates not only that contractile properties improve with growth and maturation, but also that as the child grows there is an improvement in neuromuscular efficiency.

The increase in twitch moment of force with age indicates that younger children cannot fully activate their muscles. The reasons for this observation are not fully understood. It has been suggested that incomplete activation may represent maturational differences in recruitment between children and adults (Tammik *et al.* 2007). This increase may be due to an increase in muscle fibre diameter with growth (Aherne *et al.* 1971) and an associated increase in the number of sacromeres in relation to muscle fibre size (Bowden and Goyer 1960). Furthermore, the activity of myosin ATPase, an important determinant of contractile activity, increases with growth, with higher levels reported in adults compared with children (Drachman and Johnston 1973).

Although children display lower maximal twitch moment of force than adults, there is evidence that the temporal characteristics of the twitch, i.e. the contraction time (elapsed time from the onset of force deflection following electrical stimulation to attainment of peak force) and the half-relaxation time (elapsed time from peak force to half-peak force during recovery), are not equally affected by growth (Davies 1985; McComas *et al.* 1993). Twitch contraction times of children aged 3 years were already within the adult range (McComas *et al.* 1993). Since the time course of a twitch is affected by muscle fibre type distribution (Grosset *et al.* 2005; Rice *et al.* 1988), this might explain the absence of age differences in time-related variables of the twitch. This is further enforced by the observation that muscle type distribution is acquired at the age of 3 years (Elder and Kakulas 1993). The above profile, however, may not apply to all children. For example, an age-related decrease of half-relaxation time was observed in males but not in females (Davies 1985). Further, a longer contraction time in older compared with younger children was found for the ankle dorsiflexors but not for the plantar flexors (Belanger and McComas 1989). These observations suggest that there is no single common development pattern of twitch characteristics, but specific patterns may exist depending on sex or the type of muscle examined.

Another important parameter of neuromuscular development is the electro-mechanical delay (EMD), which is defined as the time delay between the initial stimulus and mechanical response of the stimulated muscle (Grosset *et al.* 2005; Norman and Komi 1979; see Chapter 11). Studies have shown longer EMD of the elbow flexors (Asai and Aoki 1996) and triceps surae (Grosset *et al.* 2005) in children than for adults. The EMD is determined by the time taken for the contractile component to stretch the series elastic component of the muscle and it is affected by the propagation velocity of action potentials along the muscle-fibres membrane, the time course of the excitation–contraction coupling processes as well as the rate of shortening of the series elastic component of muscle (Norman and Komi 1979). An age-related decrease of EMD might suggest that children are less able to quickly activate their muscles until the first mechanical response or/and that children have a less stiff musculotendinous system than adults – i.e. a more compliant muscle–tendon system would need more time to produce a mechanical response, given the same stimulus. Since the time to shorten the series elastic component of muscle exceeds substantially the time for the activation of cross-bridges (Norman and Komi 1979), then the contribution of stretch properties of the musculotendinous unit to the observed age-related differences is likely to be higher than neuromuscular propagation (Grosset *et al.* 2005). This is further enforced by evidence that musculotendinous stiffness is higher in adults compared with children (Kubo *et al.* 2001; Lambertz *et al.* 2003).

The above studies have examined the activation of the muscles during maxi-mal isometric tests. Others, however, focused on the neuromuscular function in children during performance of dynamic single joint tasks by examining the electromyographic (EMG) activity of the muscles and the associated strength output during task performance. Cross-sectional as well as longitudinal studies have repeatedly found that moment of force: agonist EMG characteristics do not differ between children and adults (Kellis and Unnithan 1999; Seger and Thorstensson 1994, 2000). During isokinetic tests (see Figure 3.1), the eccentric moment of force: EMG ratio is significantly lower compared with the

Figure 3.1 Experimental set-up of an isokinetic testing and EMG recording protocol.

corresponding concentric action, indicating that the eccentric muscle actions are more efficient than concentric actions in children. Furthermore, at faster velocities the moment of force:EMG ratio decreases during concentric tests and increases or remains constant during eccentric tests. This difference in age-related observations between findings on isometric versus isokinetic tests is mainly related to methodological aspects, as studies which focused on isokinetic (dynamic) tests have used volitional moment of force and EMG measures, while studies which reported age differences used the twitch interpolation technique. It is clear that an enhanced understanding of neuromuscular activation patterns during dynamic efforts with respect to age, growth and maturation is needed.

Development of muscle coactivation during isometric and isokinetic single-joint force production tasks

Muscle activation is regulated by neural mechanisms of central and peripheral origin. With regard to peripheral neural mechanisms, it is known that in the typical adult, a sudden change in muscle stretch causes firing of Ia afferent fibres, which are excitatory on the agonist alpha motoneuron by way of a monosynaptic connection in the spinal cord and inhibitory on the antagonist alpha motoneuron through an inhibitory interneuron in the spinal cord (Myklebust *et al.* 1986). It is not known whether this chain of events occurs similarly in the developing child; in most cases, we assume that reflexes of this type hold true for children as well. The simultaneous facilitation of the agonist muscle and inhibition of the antagonist muscle is called reciprocal inhibition. From a mechanical point of view, the presence of high muscle coactivation from muscles that act in opposite directions (antagonistic muscles) around a joint will determine the moment of force exerted around the joint and, in turn, the characteristics of joint movement. The higher the coactivation of the agonists and antagonists, the less the resultant joint moment (Kellis 2003). Consequently, changes in muscle coactivation provide useful information regarding neuromuscular development in children, which is also essential for joint stability (see Chapter 11).

During isometric strength tests coactivation appears to differ between muscles and joint movement. For the hamstrings, the antagonist activity levels range from 3 per cent (Ikeda *et al.* 1998) to 22 per cent (Stackhouse *et al.* 2005), while quadriceps antagonist activity levels reach 22 per cent (Ikeda *et al.* 1998; Tedroff *et al.* 2008). For the ankle muscles, the tibialis anterior antagonist activation during isometric plantar flexion ranged from 11 per cent (Tedroff *et al.* 2008) to 25 per cent (Elder *et al.* 2003). The above antagonist activation values are similar to those reported for adults. However, none of the aforementioned studies examined age-related changes in coactivation and therefore no safe conclusions regarding developmental changes in coactivation at a single-joint level are possible.

Only a few studies have compared antagonist activation during isometric efforts between children and adults (Lambertz *et al.* 2003; O'Brien *et al.* 2009b) with somewhat conflicting findings. O'Brien *et al.* (2009b) reported higher antagonist hamstring coactivation during MVC in adult males than pre-pubertal boys, but no difference was found between females and girls. However, Lambertz *et al.* (2003) reported that tibialis anterior coactivation was significantly higher in younger (7 years) compared with older (10 years) children (boys and girls), while all children displayed higher coactivation compared with adults. This resulted in lower neuromuscular efficiency in the youngest children versus the older age group. Differences in the type of muscles and movements examined and methodological variations between studies prevent us from drawing definite conclusions regarding the progression of antagonist activation levels with age, at least during static actions.

Although the above evidence suggests an age-related progressive decrease in antagonist activation during isometric efforts, research studies examining dynamic (eccentric and concentric) strength tests have clearly shown that antagonist EMG does not differ between children (> 10 years) and adults, with values not exceeding 20 per cent of the maximum EMG of the same muscle when it works as agonist under the same testing conditions (Bassa *et al.* 2005; Kellis and Unnithan 1999). A higher antagonist EMG during concentric compared with eccentric efforts and an increase of antagonistic activation level with increasing velocity during concentric efforts has been reported in children (Bassa *et al.* 2005; Kellis and Unnithan 1999). Even when EMG is used to calculate the antagonist moment of force via EMG-moment models, pubertal children display similar antagonist moments to adults, with values ranging from 6 per cent (O'Brien *et al.*

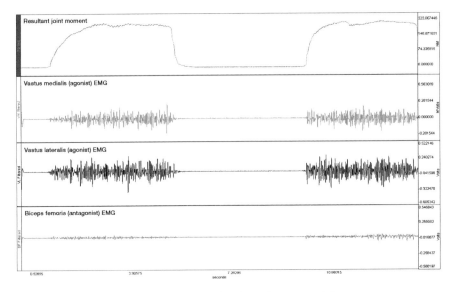

Figure 3.2 An example of electromyographic (EMG) raw signal during two consecutive
isokinetic knee extension efforts. The EMG is recorded from the vastus medialis
and vastus lateralis muscles (agonists) and the biceps femoris (antagonist). The
small magnitude of the antagonist EMG represents the amount of antagonist
activation, which is then normalized to a reference value (normalization) or
it can be divided by the agonist muscle EMG, providing the co-contraction
index.

2009b) to 27.6 per cent (Kellis 2003) of the moment recorded by the isokinetic
dynamometer (resultant joint moment) (Kellis 2003; O'Brien *et al.* 2009b). The
above clearly suggest that antagonist function during dynamic single-joint tests is
similar in pubertal children and adults.

There are several factors responsible for the absence of age-related differences
in antagonist coactivation during maximal voluntary efforts. First, as early as
infancy a single tendon tap causes high levels of muscle coactivation (Myklebust
et al. 1986). During development, this reflex activation is gradually reduced and
in the adult, only the stimulated muscle is activated by the stretch reflex (Tedroff
et al. 2008). In fact, the development of the ability to inhibit the recruitment of
motoneurons in the antagonistic muscles is significantly developed when children
reach 3 years of age (Gatev 1972). Further, it appears that, despite any differences
in joint and ligamentous properties between children and adults, the reflexive
excitation of the antagonist muscle does not differ in children compared with
adults. A possible explanation for this is that the relative level of joint forces
exerted may be similar or even lower compared with those observed in adults,
owing to a smaller relative strength capacity observed in children. In essence,
there is no apparent reason for a higher coactivation in children compared with
adults' as this would make movement more ineffective.

It is evident that more research is required to examine coactivation levels in children. We will mention three reasons: first, there is no systematic examination of age- and maturation-related development of antagonist coactivation. Most available data are the result of cross-sectional studies comparing children with cerebral palsy with healthy children (Elder *et al.* 2003; Stackhouse *et al.* 2005; Tedroff *et al.* 2008) or children with adults (Bassa *et al.* 2005; Kellis and Unnithan 1999; O'Brien *et al.* 2009a). The only study which examined younger children (< 10 years) reported an age-associated decrease in coactivation (Lambertz *et al.* 2003). Research examining other tasks has shown that differences in coactivation levels during gait between 7- and 10-year-olds were larger compared to the differences between 10- and 15-year-olds (Unnithan *et al.* 1996). Nevertheless, research on younger ages suggests that at about two months of walking experience, after the onset of autonomous walking, coactivation of tibialis anterior and lateral gastrocnemious during the stance phase of locomotion tends to decrease, giving way to more efficient reciprocal activation in successive step cycles (Assaiante *et al.* 1993). The quadriceps and hamstrings activity changes from a pattern of coactivation to reciprocal activation as well (Forssberg 1985). The exact time of this change is not, however, clear. Consequently, conclusions drawn for muscle coactivation differences between children versus adults may be specific to the age and/or maturational stage of the child (Unnithan *et al.* 1996). Second, information on age-related comparisons mainly refers to the knee extensors. However, lower limb muscles, such as the quadriceps and hamstrings, are actively involved in habitual activities and therefore they may be used more effectively compared with muscles of upper extremities (Ramsay *et al.* 1990). Finally, the absence of age-related differences in coactivation levels during well-controlled isometric/isokinetic tasks may be due to the way coactivation is assessed. Specifically, most studies have quantified coactivation in the amplitude and not in the time domain. Examination of the time sequence of muscle activity levels is a fundamental aspect of movement performance and it has not been evaluated during isometric or isokinetic tests in children.

The nature of the exercise examined in all the aforementioned studies refers to a known and well-controlled task, and therefore less uncertainty in movement (hence, less coactivation) is expected. Functional basic and athletic activities, such as walking and running, require the coordination of several muscle groups which are activated under different task and environmental constraints such as angular velocities, movement directions and types of muscle action. The age-related decrease in muscle coactivation observed during walking seems to support the idea that greater age-related differences in coactivation are expected to be seen in more complex daily life purposeful actions where experience also plays a role (Unnithan *et al.* 1996).

In conclusion, it appears that the age-associated relative strength differences at specific angular velocities cannot be attributed to higher activation levels of antagonist muscles in children, but rather to the lower efficiency of their neuromuscular system in effectively activating the agonist muscles.

Neuromuscular development during force-control single-joint tasks

Neuromuscular development affects not only the ability of the child to exert maximal strength, but also its ability to perform other functional tasks that require submaximal force exertion in order to control task precision or to move a segment towards a certain target while performing goal-directed actions.

The development of the ability to manually control force is affected by several mechanisms, including structural maturation of the neuromotor system and changes in the use of sensorimotor pathways. These pathways include efferent and afferent central routes and their development is crucial for performance of any task. The ability to lift an object is such a task in which the adaptation of grip and load force to the object's mass characteristics could be indicative of the level of sensorimotor development (Fietzek *et al.* 2000; Forssberg *et al.* 1991). Many studies have examined the motor development of the hand function using sub-maximal force control tasks (Potter *et al.* 2006). The increase of whole hand and finger grip force during the pre-school years has been attributed to an increase in strength due to muscle growth, an increase in motor unit activation of surrounding muscles and a better segmental coordination (Potter *et al.* 2006).

Important information on neural maturation can also be derived by examining the ability of children to execute a task at a constant force level. This is because movement patterns and performance outcomes in children are generally characterized by a large degree of variability in comparison to older children and adults. With increasing age throughout childhood, the degree of variability exhibited in the movements of children tends to decrease (Newell *et al.* 2000). This variability represents the amount of noise in the sensorimotor system. Lower noise may be a prerequisite for better performance (Newell *et al.* 2000). The oscillations of exerted force above and below the predetermined force level are reflected in the estimation of coefficient of variability (Jones *et al.* 2002). The ability to apply a grip force in order to match a visual target (force tracking ability) is low in 3–4-year-old children, while evidence for externally guided motor control by children is observed at the age of 4–5 years (Blank *et al.* 1999). By the age of 5 years, performance during force increase conditions improves significantly. Further neural maturation takes place after the age of 6 years and it is almost complete by 10 years of age despite further increases in maximum produced force until adulthood (Smits-Engelsman *et al.* 2003).

The gradual decline in force variability during development indicates a maturation of all neural mechanisms underlying successful motor output production. In general, maturation of the corticospinal tract parallels the skill of maintaining a constant force (Fietzek *et al.* 2000; Smits-Engelsman *et al.* 2003). However, the two events do not occur at the same time during normal development. This is supported by the observation that children show significant improvement in force variability with visual feedback between the ages of 6 and 10 years, but no difference in force variability across ages when vision is occluded (Deutsch and Newell 2001). This suggests that there is a gradual increase in the ability of children to use visual feedback to control performance output (Deutsch and

Newell 2001). Further information has also been obtained by examining the oscillations of force when executing a force task using power spectrum analysis. Power at low frequencies suggests that children use long loop visual and proprio-ceptive feedback to control force (Malmstrom and Lindstrom 1997), while a higher energy in the 10-Hz range is considered to indicate the use of feedforward strategies (Muller and Homberg 1992). Findings show that as the child grows up (especially after the age of 10 years), power at higher frequency bands (20–30 Hz) increases while power at lower frequency bands (< 10 Hz) decreases (Smits-Engelsman *et al.* 2003). This suggests a gradual change in strategy to control force from feedback-based control to feedforward control (Fietzek *et al.* 2000; Smits-Engelsman *et al.* 2003). It appears that motor performance improvements with age are due largely to the enhanced ability to organize appropriately the output of the sensorimotor system to meet the task demands (Deutsch and Newell 2001).

Another aspect of neural maturation during force control tasks is the time needed to exert force or to relax the muscle. Research findings on the develop-ment patterns in the time needed to contract the muscles differ, depending on the type of task examined and the methodology used. Potter *et al.* (2006) found that the rate of force rise for power and thumb–index finger precision grip did not significantly decrease with increasing age from 3 to 5 years. In contrast, other studies have shown that the time to establish target force levels decreases markedly between ages 3 and 6 years (Blank *et al.* 1999) and between ages 5 and 12 years (Smits-Engelsman *et al.* 2003). In fact, the time to relax the muscle decreases up to 7–8 years, indicating that faster muscular reactions are possible at that age, although this is evident in task performance by the age of 9–10 years (Smits-Engelsman *et al.* 2003).

In conclusion, as children grow their ability to control force output during a given task improves. Therefore, children gradually achieve higher levels of coordination through a reduction of noise in the sensorimotor system and a shift to feedforward strategies to control externally (visually) guided force exerted during single-joint tasks. On the other hand, the ability to coordinate the multiple degrees of freedom that are present during performance of a multi-joint motor task can be another rate-limiting element of development that is greatly affected by neuromuscular maturation.

Development of muscle synergies during multi-joint actions

The formation of multi-joint muscle synergies is a critical step in motor devel-opment since such synergies are a prerequisite for the execution of functional motor skills such as reaching, sitting, standing and walking. Moreover, these basic skills will form the basis for the acquisition of more complex, purposeful actions later in life.

The first signs of coordinated muscle activity appear in the neck and trunk muscles by the time the infant gradually develops the ability to sit independently (Woollacott *et al.* 1996). The nervous system first 'maps' relationships between sensory inputs and the neck muscles for controlling the upright position of the head and then extends these relationships to include the trunk muscles with the onset of independent sitting (Brauer *et al.* 2001).

The development of muscle synergies controlling independent standing occurs gradually, beginning at around 7–9 months, with infants gradually organizing the muscles within the leg and trunk into appropriate muscle synergies, with the ankle muscles being consistently activated first, followed by the upper leg and trunk muscles (Sveistrup and Woollacott 1996). Importantly, muscle activation patterns are highly variable and display greater co-contraction, which is necessary to maintain joint stability. The primary rate-limiting factor for the emergence of independent stance is the development of sufficient muscle force to support the body against gravity (Thelen and Fisher 1982). By 6 months of age, infants are producing forces well beyond their body weight (Roncesvalles *et al.* 2001). With the acquisition of upright posture, infants can control their body in relation to gravity, so that equilibrium control becomes global rather than segmental.

Independent walking appears later in life compared to independent standing because locomotion additionally requires higher levels of muscle strength and the ability to balance on one leg in a dynamically moving condition (Woollacott *et al.* 1996). At the onset of autonomous walking around the age of 3–4 years, lateral body stabilization is first initiated at the hip level in order to minimize upper body destabilization induced by leg movements and to prevent falling (Assaiante *et al.* 1993). At about two months of walking experience, shoulder stabilization in space improves followed by head stabilization. This suggests an ascending progression with age in the ability to control lateral balance during locomotion. The development of head–trunk coordination during locomotion under normal vision involves three periods in which blocking of the head on the trunk in order to reduce the degrees of freedom to be controlled is gradually reduced. In older children, after about 8 years of age, the head stabilization in space strategy is dominant (Assaiante and Amblard 1993). By the age of 8–11 years, young healthy children adopt a head stabilization in space strategy during walking by attenuating the acceleration going up from pelvis to head level, which is similar to adults (Mazza *et al.* 2010). Adaptation of the head stabilization in space strategy at the age of 7–8 years is illustrative of a descending temporal organization (Assaiante and Amblard 1993) that is associated with a selective control of the degrees of freedom at the neck level probably determined according to task constraints. Although maturation of the gait pattern is mainly complete by 3–4 years of age, the final development of locomotion requires many years to be refined with acceleration and braking capabilities developing only at 5–6 years (Clark *et al.* 1988).

The ability to develop the appropriate moment of force levels at the ankle and hip joints necessary to recover from a balance perturbation also changes with age. In contrast to older children, who rapidly generate large moments of force, younger children (9–23 months of age) use multiple moment of force adjustments before regaining balance (Roncesvalles *et al.* 2001). Shorter duration, faster and more stable muscle response synergies to visually or proprioceptively induced balance perturbations with increasing age confirm that postural response synergies do not mature until the age of at least 10 years (Forssberg and Nashner 1982). An unusual change is also noted in the muscle response characteristics of children aged 4–6 years, which show more variable and longer latency muscle responses when compared to children aged 2–3 years and 7–10 years (Shumway-Cook and Woollacott 1985). It has been hypothesized that this temporary regression in postural abilities is due to the fact that at this particular age children are beginning to expand their postural skill repertoire to include inter-sensory integration. By 7–10 years of age, muscle response synergies to balance perturbations of the support surface are essentially like those of the adult.

Multi-joint tasks are characterized by the presence of non-muscular, passive interactive forces developing between the linked segments within a limb. Intersegmental forces constitute a basic source of energy transfer for the lower limb and are responsible for the large increases in the distal segment speed during locomotion (Winter and Robertson 1978). The nervous system also exploits the intersegmental forces naturally arising between the segments to simplify the control of limb elevation over obstacles and minimize the energy cost during locomotion (Eng *et al.* 1997). As a result, muscle forces are always modulated in response to the passive reactions of the system so as to counteract or complement the motion-dependent forces generated by movement of the linked segments (Thelen *et al.* 1992). A possible rate-limiting factor for the development of appropriate multiple muscle synergies to control multi-joint action could be the slower development of the child's anthropometric characteristics (mass and inertia), leading to insufficient utilization of the passive interactive forces that are present in multi-joint tasks. Investigating this problem using a pedalling task, Brown and Jensen (2003) demonstrated that children adjusted the contribution of the proximal joint muscle moments of force to compensate for reduced contributions to the resultant pedal force by gravitational and inertial-interactive force components.

In summary, the development of muscle synergies to control multi-joint purposeful action appears to be organized in a cephalocaudal progression of control, starting with the head segment. The ability to apply force against gravity or develop the appropriate moment of force levels at a joint could be a rate-limiting factor in the completion of distinct motor milestones such as independent stance or locomotion. However, as the process of complex multi-joint behaviour involves the 'mapping' of sensory inputs to muscle activation patterns, it is also critical to examine the role of the sensory systems in the development of neuromuscular control.

The role of predictive mechanisms and higher brain centres in the control of multi-joint action

The development of complex multi-joint purposeful actions involves the capacity of the child to build up appropriate internal representations that reflect the rules for organizing sensory inputs and coordinating them with motor actions (Gurfinkel and Levik Yu 1979). For example, as a child gains experience in moving in a gravity environment, sensory-motor maps develop. These maps relate actions to incoming sensory inputs from vision, somatosensory and vestibular systems. In this way, rules for moving develop and are reflected in altered synaptic relationships. The path from sensation to motor actions proceeds via an internal representation structure or body schema that allows the emergence of predictive behaviour (Hirschfeld and Forssberg 1992).

A feedback-based system in the control of the motor output appears very early in life, as confirmed by experimental evidence showing that postural response synergies triggered by sensory perturbations are present as early as 15–31 months and have latencies comparable with those of adults (Shumway-Cook and Woollacott 1985). The first signs of the development of anticipatory control appear in 9-month-old infants during sitting, who activate the postural muscles of the trunk in advance of their reaching movement (Sveistrup and Woollacott 1996). In standing, children as young as 12–15 months are able to activate postural muscles in advance of arm movements (Forssberg and Nashner 1982). By 4–6 years, anticipatory postural adjustments preceding arm movements while standing are essentially mature (Nashner *et al.* 1983). Nevertheless, younger children until the age of 7 years rely predominantly on proprioceptive and visual feedback cues for maintaining postural control. This corresponds to an ascending organization of balance and posture from foot to head (in posture) and hip to head (in locomotion), which is more characteristic of children under the age of 7 years. A child's ability to apply feedforward control and activate the muscles in an anticipatory fashion in preparation for an upcoming perturbation greatly depends on the ability to control gravity and inertial forces (Berger *et al.* 1987; Grasso *et al.* 1998) and move the head independently of the trunk (Assaiante and Amblard 1995), skills that develop later, between 6 and 10 years of age. Balance control in this case is temporally organized in a descending order, from head to toe (Assaiante and Amblard 1995). By the age of 7–8 years, the complexity of the central nervous system strongly resembles that of an adult (Berger *et al.* 1987; Ledebt *et al.* 1998), enabling anticipatory, open-loop (feedforward) control of complex whole body movements, which continues through adulthood. In adults, the ascending, descending and mixed strategies all co-exist whereas the selection of one or another and their relevant contributions greatly depend on the particular task constraints (Grasso *et al.* 1998; Ledebt *et al.* 1998). In difficult balancing situations, for example, the descending strategy (from head to toe) is always dominant. Like adults, children aged 11–13 years have the ability to select between feedback and feedforward control strategies depending on the constraints of a particular task (Hatzitaki *et al.* 2002). Thus, dynamic actions seem to be pre-programmed in an open-loop fashion, pointing to the child's ability to plan in advance and apply feedforward control in order to reliably coordinate posture with the focal limb movement. Nevertheless, the development of motor prediction seems to depend on task complexity. Specifically, the development of postural adjustments during reaching from a sitting position turned out to have a non-linear and protracted course, which is not finished by the age of 11 years (van der Heide *et al.* 2003). Anticipatory postural muscle activity during this task, which was consistently present in adults, was virtually absent between 2 and 11 years of age. When the task requirements in motor planning increase even further, such as intercepting a partially occluded object while walking, even older children at the age of 10–13 years show important decrements in movement accuracy, confirming the decreased ability to plan their movement in advance (van Kampen *et al.* 2010). Moreover, different anticipatory control strategies have been shown in 10-year-old children as compared to adults when circumventing obstacles in the travel path while walking (Vallis and McFadyen 2005). The different head and trunk anticipatory segmental coordination suggests that children gather visual information differently when circumventing an obstacle in their travel path and are more dependent on visual input to guide their circumvention strategy.

To summarize, the child's capacity to build up internal representations that reflect the rules for organizing sensory inputs and coordinating them with muscle activation patterns to support goal-directed multi-joint action matures between 6 and 10 years of age. Nevertheless, this capacity is highly task dependent and shaped by environmental constraints depending on prior experience.

Clinical and practical applications

The ability to exert maximal or submaximal muscle force develops throughout childhood, reflecting the different rates of development of various neural mechanisms and systems. Assessment and evaluation of performance requiring maximum force development in single-joint controlled tasks should take into consideration that at an early age children are unable to recruit their entire motor unit pool. For example, at an early age, force-rate limiting capabilities in the child's ability to develop the appropriate levels of joint moment of force to control his/her body against gravity could delay the onset of independent standing.

Muscle coactivation of antagonistic muscle groups does not appear to be an important factor in explaining force differences between different age groups under well-controlled conditions. However, it is clear that experimental evidence is not adequate to make safe conclusions regarding the role of coactivation in motor performance development, especially during single-joint and well-controlled tasks. Therefore, we cannot confidently identify whether developmental differences in maximum strength output are due to altered magnitude of muscle coactivation.

Understanding the mechanisms of neural development is critical when a teacher/physical educator assesses performance improvements of his/her students, which in turn may influence the design of exercise or training programmes. It is clear that assessment of motor performance is based on the qualitative aspects of the motor performance, the chronological appearance of the successive motor milestones and the developmental sequence of functional motor behaviours. In addition, the variability and spontaneity of motor performance should also be assessed.

Physical educators, coaches and sports scientists require an understanding that children of different ages use different strategies to execute a given task. The ability to 'map' sensory input to muscle activation is critical for the development of multi-joint coordination synergies that support purposeful action and subsequently the performance of more complex sports skills. Not all senses, however, are mapped at the same stage of development, as vision is mapped first while sensory integration and organization abilities develop later in the child's life. Exercise regimes should therefore allow for the processing of complex environmental stimuli and exploit the child's exploration and adaptation capabilities to continuously and unpredictably changing environments.

Key points

- As children grow the ability to exert maximum forces increases. Although this increase is related to increases in muscle size, it is well recognized that older children and adults display an increased ability to recruit their muscles maximally compared with younger children.
- When children are requested to exert maximal force during single-joint tasks, the magnitude of coactivation seems to be similar to that observed for adults. However, there is not enough evidence to make safe conclusions regarding the development of coactivation under maximal strength conditions.
- When children are asked to perform constant force (submaximal) efforts, an age-associated decrease in force variability is observed. This is because younger children control force tasks by using long loop visual feedback, while older children (after 10 years) gradually learn to use feedforward strategies while they display faster responses to target force levels.
- Development of neuromuscular coordination at multiple joints relies on the ability of the nervous system to integrate sensory information to the corresponding muscle activation patterns.
- The mapping of individual sensory systems to muscle action precedes the mapping of multiple sensory systems. Sensory integration and organization abilities develop later in the child's life. This is also accompanied by the development of feedforward control and the increased ability of motor planning between 6 and 10 years of age.

References

Aherne, W., Ayyar, D.R., Clarke, P.A. and Walton, J.N. (1971) 'Muscle fibre size in normal infants, children and adolescents. An autopsy study', *Journal of Neurological Science*, 14: 171–82.

Asai, H. and Aoki, J. (1996) 'Force development of dynamic and static contractions in children and adults', *International Journal of Sports Medicine*, 17: 170–4.

Assaiante, C. and Amblard, B. (1993) 'Ontogenesis of head stabilization in space during locomotion in children: influence of visual cues', *Experimental Brain Research*, 93: 499–515.

Assaiante, C. and Amblard, B. (1995) 'An ontogenetic model for the sensorimotor organization of balance control in humans', *Human Movement Science*, 14: 13–43.

Assaiante, C., Thomachot, B. and Aurenty, R. (1993) 'Hip stabilization and lateral balance control in toddlers during the first four months of autonomous walking', *Neuroreport*, 4: 875–8.

Assaiante, C., Woollacott, M. and Amblard, B. (2000) 'Development of postural adjustment during gait initiation: kinematic and EMG analysis', *Journal of Motor Behavior*, 32: 211–26.

Bassa, E., Patikas, D. and Kotzamanidis, C. (2005) 'Activation of antagonist knee muscles during isokinetic efforts in prepubertal and adult males', *Pediatric Exercise Science*, 17: 171–81.

Belanger, A. and McComas, A. (1989) 'Contractile properties of human skeletal muscle in childhood and adolescence', *European Journal of Applied Physiology*, 58: 563–7.

Berger, W., Quintern, J. and Dietz, V. (1987) 'Afferent and efferent control of stance and gait: developmental changes in children', *Electroencephalography and Clinical Neurophysiology*, 66: 244–52.

Blank, R., Heizer, W. and von Voss, H. (1999) 'Externally guided control of static grip forces by visual feedback-age and task effects in 3–6-year-old children and in adults', *Neuroscience Letters*, 271: 41–4.

Blimkie, C., Sale, D. and Bar-Or, O. (1990) 'Voluntary strength, evoked twitch contractile properties and motor unit activation of knee extensors in obese and non-obese adolescent males', *European Journal of Applied Physiology*, 61: 313–18.

Bowden, D.H. and Goyer, R.A. (1960) 'The size of muscle fibers in infants and children', *Archives of Pathology*, 69: 188–9.

Brauer, S.G., Woollacott, M. and Shumway-Cook, A. (2001) 'The interacting effects of cognitive demand and recovery of postural stability in balance-impaired elderly persons', *Journal of Gerontology Series A*, 56: M489–96.

Brown, N.A. and Jensen, J.L. (2003) 'The development of contact force construction in the dynamic-contact task of cycling [corrected]', *Journal of Biomechanics*, 36: 1–8.

Caramia, M.D., Desiato, M.T., Cicinelli, P., Iani, C. and Rossini, P.M. (1993) 'Latency jump of "relaxed" versus "contracted" motor evoked potentials as a marker of corticospinal maturation', *Electroencephalography and Clinical Neurophysiology*, 89: 61–6.

Clark, J., Whitall, J. and Phillips, S. (1988) 'Human interlimb coordination: the first 6 months of independent walking', *Developmental Psychobiology*, 21: 445–56.

Connolly, K.L. and Forssberg, H. (1997) *Neurophysiology and Neuropsychology of Motor Development*, London: MacKeith Press.

Davies, C. (1985) 'Strength training and mechanical properties of muscle in children and young adults', *Scandinavian Journal of Sports Sciences*, 7: 11–15.

Deutsch, K.M. and Newell, K.M. (2001) 'Age differences in noise and variability of isometric force production', *Journal of Experimental and Child Psychology*, 80: 392–408.

Drachman, D.B. and Johnston, D.M. (1973) 'Development of a mammalian fast muscle: dynamic and biochemical properties correlated', *Journal of Physiology*, 234: 29–42.

Elder, G.C. and Kakulas, B.A. (1993) 'Histochemical and contractile property changes during human muscle development', *Muscle and Nerve*, 16: 1246–53.

Elder, G.C., Kirk, J., Stewart, G., Cook, K., Weir, D., Marshall, A. and Leahey, L. (2003) 'Contributing factors to muscle weakness in children with cerebral palsy', *Developmental Medicine and Child Neurology*, 45: 542–50.

Eng, J.J., Winter, D.A. and Patla, A.E. (1997) 'Intralimb dynamics simplify reactive control strategies during locomotion', *Journal of Biomechanics*, 30: 581–8.

Eyre, J.A., Miller, S. and Ramesh, V. (1991) 'Constancy of central conduction delays during development in man: investigation of motor and somatosensory pathways', *Journal of Physiology*, 434: 441–52.

Fietzek, U.M., Heinen, F., Berweck, S., Maute, S., Hufschmidt, A., Schulte-Monting, J., Lucking, C.H. and Korinthenberg, R. (2000) 'Development of the corticospinal system and hand motor function: central conduction times and motor performance tests', *Developmental Medicine and Child Neurology*, 42: 220–7.

Forssberg, H. (1985) 'Ontogeny of human locomotor control. I. Infant stepping, supported locomotion and transition to independent locomotion', *Experimental Brain Research*, 57: 480–93.

Forssberg, H., Eliasson, A.C., Kinoshita, H., Johansson, R.S. and Westling, G. (1991) 'Development of human precision grip. I: Basic coordination of force', *Experimental Brain Research*, 85: 451–7.

Forssberg, H. and Nashner, L.M. (1982) 'Ontogenetic development of postural control in man: adaptation to altered support and visual conditions during stance', *Journal of Neuroscience*, 2: 545–52.

Gatev, V. (1972) 'Role of inhibition in the development of motor co-ordination in early childhood', *Developmental Medicine and Child Neurology*, 14: 336–41.

Grasso, R., Assaiante, C., Prevost, P. and Berthoz, A. (1998) 'Development of anticipatory orienting strategies during locomotor tasks in children', *Neuroscience and Biobehavioral Reviews*, 22: 533–9.

Grosset, J-F., Mora, I., Lambertz, D. and Perot, C. (2005) 'Age-related changes in twitch properties of plantar flexor muscles in prepubertal children', *Pediatric Research*, 58: 966–70.

Gurfinkel, V.S. and Levik Yu, S. (1979) 'Sensory complexes and sensomotor integration', *Human Physiology*, 5: 269–81.

Hatzitaki, V., Zisi, V., Kollias, I. and Kioumourtzoglou, E. (2002) 'Perceptual-motor contributions to static and dynamic balance control in children', *Journal of Motor Behavior*, 34: 161–70.

Hirschfeld, H. and Forssberg, H. (1992) 'Development of anticipatory postural adjustments during locomotion in children', *Journal of Neurophysiology*, 68: 542–50.

Housh, T.J., Johnson, G.O., Housh, D., Stout, J., Weir, J., Weir, L. and Eckerson, J. (1996) 'Isokinetic peak torque in young wrestlers', *Pediatric Exercise Science*, 8: 143–55.

Ikeda, A.J., Abel, M.F., Granata, K.P. and Damiano, D.L. (1998) 'Quantification of cocontraction in spastic cerebral palsy', *Electromyography and Clinical Neurophysiology*, 38: 497–504.

Jones, K.E., Hamilton, A.F. and Wolpert, D.M. (2002) 'Sources of signal-dependent noise during isometric force production', *Journal of Neurophysiology*, 88: 1533–44.

Kanehisa, H., Ikegawa, S., Tsunoda, N. and Fukunaga, T. (1994) 'Strength and cross-sectional area of knee extensor muscles in children', *European Journal of Applied Physiology*, 68: 402–5.

Kanehisa, H., Ikegawa, S., Tsunoda, N. and Fukunaga, T. (1995) 'Strength and cross-sectional areas of reciprocal muscle groups in the upper arm and thigh during adolescence', *International Journal of Sports Medicine*, 16: 54–60.

Kellis, E. (2003) 'Antagonist moment of force during maximal knee extension in pubertal boys: effects of quadriceps fatigue', *European Journal of Applied Physiology*, 81: 71–80.

Kellis, E. and Unnithan, V. (1999) 'Co-activation of vastus lateralis and biceps femoris muscles in pubertal children and adults', *European Journal of Applied Physiology*, 79: 504–11.

Koh, T.H. and Eyre, J.A. (1988) 'Maturation of corticospinal tracts assessed by electromagnetic stimulation of the motor cortex', *Archives of Disease in Childhood*, 63: 1347–52.

Kubo, K., Kanehisa, H., Kawakami, Y. and Fukanaga, T. (2001) 'Growth changes in the elastic properties of human tendon structures', *International Journal of Sports Medicine*, 22: 138–43.

Lambertz, D., Mora, I., Grosset, J.F. and Perot, C. (2003) 'Evaluation of musculotendinous stiffness in prepubertal children and adults, taking into account muscle activity', *Journal of Applied Physiology*, 95: 64–72.

Ledebt, A., Bril, B. and Breniere, Y. (1998) 'The build-up of anticipatory behaviour. An analysis of the development of gait initiation in children', *Experimental Brain Research*, 120: 9–17.

Maffiuletti, N.A., Martin, A., Babault, N., Pensini, M., Lucas, B. and Schieppati, M. (2001) 'Electrical and mechanical H(max)-to-M(max) ratio in power- and endurance-trained athletes', *Journal of Applied Physiology*, 90: 3–9.

Malmstrom, J.E. and Lindstrom, L. (1997) 'Propagation velocity of muscle action potentials in the growing normal child', *Muscle and Nerve*, 20: 403–10.

Mazza, C., Zok, M. and Cappozzo, A. (2010) 'Head stabilization in children of both genders during level walking', *Gait & Posture*, 31: 429–32.

McComas, A.J., Galea, V. and deBruin, H. (1993) 'Motor unit populations in healthy and diseased muscles', *Physical Therapy*, 73: 868–77.

McGraw, M.B. (1932) 'From reflex to muscular control in the assumption of an erect posture and ambulation in the human infant', *Child Development*, 3: 291–7.

Muller, K. and Homberg, V. (1992) 'Development of speed of repetitive movements in children is determined by structural changes in corticospinal efferents', *Neuroscience Letters*, 144: 57–60.

Muller, K., Homberg, V. and Lenard, H.G. (1991) 'Magnetic stimulation of motor cortex and nerve roots in children. Maturation of cortico-motoneuronal projections', *Electroencephalography and Clinical Neurophysiology*, 81: 63–70.

Muller, K., Homberg, V., Coppenrath, P. and Lenard, H.G. (1992) 'Maturation of set-modulation of lower extremity EMG responses to postural perturbations', *Neuropediatrics*, 23: 82–91.

Myklebust, B.M., Gottlieb, G.L. and Agarwal, G.C. (1986) 'Stretch reflexes of the normal infant', *Developmental Medicine and Child Neurology*, 28: 440–9.

Nashner, L.M., Shumway-Cook, A. and Marin, O. (1983) 'Stance posture control in select groups of children with cerebral palsy: deficits in sensory organization and muscular coordination', *Experimental Brain Research*, 49: 393–409.

Newell, K.M., Deutsch, K.M. and Morrison, S. (2000) 'On learning to move randomly', *Journal of Motor Behavior*, 32: 314–20.

Norman, R.W. and Komi, P.V. (1979) 'Electromechanical delay in skeletal muscle under normal movement conditions', *Acta Physiologica Scandinavica*, 106: 241–8.

O'Brien, T.D., Reeves, N.D., Baltzopoulos, V., Jones, D.A. and Maganaris, C.N. (2009a) 'The effects of agonist and antagonist muscle activation on the knee extension moment-angle relationship in adults and children', *European Journal of Applied Physiology*, 106: 849–56.

O'Brien, T.D., Reeves, N.D., Baltzopoulos, V., Jones, D.A. and Maganaris, C.N. (2009b) 'In vivo measurements of muscle specific tension in adults and children', *Experimental Physiology*, 95: 202–10.

Paasuke, M., Ereline, J. and Gapeyeva, H. (2000) 'Twitch contraction properties of plantar flexor muscles in pre- and post-pubertal boys and men', *European Journal of Applied Physiology*, 82: 459–64.

Potter, N.L., Kent, R.D., Lindstrom, M.J. and Lazarus, J.A. (2006) 'Power and precision grip force control in three-to-five-year-old children: velocity control precedes amplitude control in development', *Experimental Brain Research*, 172: 246–60.

Ramsay, J.A., Blimkie, C.J., Smith, K., Garner, S., MacDougall, J.D. and Sale, D.G. (1990)

'Strength training effects in prepubescent boys', *Medicine and Science in Sports and Exercise*, 22: 605–14.

Rexed, B. (1944) 'Contribution to the knowledge of the postnatal development of the peripheral nervous system in man', *Acta Psychiatrica Neurologica*, 33: 1–205.

Rice, C.L., Cunningham, D.A., Taylor, A.W. and Paterson, D.H. (1988) 'Comparison of the histochemical and contractile properties of human triceps surae', *European Journal of Applied Physiology*, 58: 165–70.

Roncesvalles, M.N., Woollacott, M.H. and Jensen, J.L. (2001) 'Development of lower extremity kinetics for balance control in infants and young children', *Journal of Motor Behavior*, 33: 180–92.

Seger, J. and Thorstensson, A. (1994) 'Muscle strength and myoelectric activity in prepubertal and adult males and females', *European Journal of Applied Physiology*, 69: 81–7.

Seger, J.Y. and Thorstensson, A. (2000) 'Muscle strength and electromyogram in boys and girls followed through puberty', *European Journal of Applied Physiology*, 81: 54–61.

Shumway-Cook, A. and Woollacott, M.H. (1985) 'The growth of stability: postural control from a development perspective', *Journal of Motor Behavior*, 17: 131–47.

Smits-Engelsman, B.C., Westenberg, Y. and Duysens, J. (2003) 'Development of isometric force and force control in children', *Cognitive Brain Research*, 17: 68–74.

Stackhouse, S.K., Binder-Macleod, S.A. and Lee, S.C. (2005) 'Voluntary muscle activation, contractile properties, and fatigability in children with and without cerebral palsy', *Muscle & Nerve*, 31: 594–601.

Sveistrup, H. and Woollacott, M.H. (1996) 'Longitudinal development of the automatic postural response in infants', *Journal of Motor Behavior*, 28: 58–70.

Tammik, K., Matlep, M., Ereline, J., Gapeyeva, H. and Paasuke, M. (2007) 'Muscle contractile properties in children with spastic diplegia', *Brain Development*, 29: 553–8.

Tedroff, K., Knutson, L.M. and Soderberg, G.L. (2008) 'Co-activity during maximum voluntary contraction: a study of four lower-extremity muscles in children with and without cerebral palsy', *Developmental Medicine and Child Neurology*, 50: 377–81.

Thelen, E. and Fisher, D.M. (1982) 'Newborn stepping: an explanation for a "disappearing" reflex', *Developmental Psychology*, 18: 760–75.

Thelen, E., Zernicke, R., Schneider, K., Jensen, J., Kamm, K. and Corbetta, D. (1992) 'The role of intersegmental dynamics in infant neuromotor development', in G.E. Stelmach and J. Requin (eds) *Tutorials in Motor Behavior II*, Amsterdam: Elsevier, pp. 533–48.

Unnithan, V.B., Dowling, J., Frost, G., Ayub, B. and Bar-Or, O. (1996) 'Cocontraction and phasic activity during gait in children with cerebral palsy', *Electromyography and Clinical Neurophysiology*, 46: 487–94.

Vallis, L.A. and McFadyen, B.J. (2005) 'Children use different anticipatory control strategies than adults to circumvent an obstacle in the travel path', *Experimental Brain Research*, 167: 119–27.

van der Heide, J.C., Otten, B., van Eykern, L.A. and Hadders-Algra, M. (2003) 'Development of postural adjustments during reaching in sitting children', *Experimental Brain Research*, 151: 32–45.

van Kampen, P.M., Ledebt, A. and Savelsbergh, G.J. (2010) 'Planning of an interceptive movement in children', *Neuroscience Letters*, 473: 110–14.

Winter, D.A. and Robertson, D.G. (1978) 'Joint torque and energy patterns in normal gait', *Biological Cybernetics*, 29: 137–42.

Wood, L.E., Dixon, S., Grant, C. and Armstrong, N. (2004) 'Elbow flexion and extension strength relative to body or muscle size in children', *Medicine and Science in Sports and Exercise*, 36: 1977–84.

Woollacott, M.H., Assaiante, C. and Amblard, B. (1996) 'Development of balance and gait control', in M. Bronstein, T. Brandt and M.H. Woollacott (eds) *Clinical Disorders of Balance and Gait*, London: Arnold.

Part II

Motor development and force production

4 Development of strength during childhood

Louise Wood and Mark De Ste Croix

Introduction

The development of equipment, technology and an increased understanding of growth and maturation issues have recently provided new insights into paediatric strength development. Measurement of muscle force is certainly not a new concept, and the development of force production during childhood has been studied for decades. Despite this acknowledgement of the importance of strength to both physical performance and health and well-being, our understanding of the age- and sex-associated changes in strength is relatively limited compared to other physiological parameters. Nevertheless, those studies that are available describing the age- and sex-associated change in dynamic strength are relatively consistent, especially for the lower limbs. However, the complex interactions of factors that explain the differences in strength during childhood and adolescence are still poorly understood (De Ste Croix 2007). This may be due in part to the fact that there are few well-controlled longitudinal strength studies that have concurrently examined the influence of known explanatory variables using appropriate statistical techniques (Wood *et al.* 2004, 2006). There are a number of excellent reviews describing force production during childhood (e.g. De Ste Croix 2007, 2009), but few have focused predominantly on the measurement of muscle size and biomechanical changes during childhood. Therefore, the main purpose of this chapter is to explore the age- and sex-associated changes in muscle strength from a biomechanical position, focusing on physiological muscle cross-sectional area (CSA) and muscle moment arms. We will briefly describe the age- and sex-associated development in strength, but concentrate primarily on the biomechanical factors that may contribute towards this development. We will then derive clinical and practical applications based on biomechanical factors that influence age- and sex-associated changes in strength.

Age- and sex-related development of strength in children

Strength appears to increase in both boys and girls until the age of about 14 years, where it begins to plateau in girls and a spurt is evident in boys. By 18 years there are few overlaps in strength between boys and girls with force production being generally greater in males. The exact age at which sex differences become apparent is unclear, and the extent of any sex differences is both muscle-group and muscle-action specific (De Ste Croix 2007). For example, the male–female difference in strength is much greater in the trunk and upper extremity than in the lower extremity from a younger age (Beunen and Malina 1988). It has been suggested that male upper limb strength may reach almost double that of females (Jones and Round 2000). The differing magnitude of sex differences in strength according to the muscle groups examined may reflect the effects of adrenal and sex steroids, the growth of the long bones, body weight and ground-reaction forces (Parker *et al.* 1990). While the general pattern of strength changes with age in boys and girls can be summarized, descriptions of age and sex differences in strength must be interpreted in the context of the definition of 'strength'. This will influence the factors contributing to any differences observed.

In longitudinal studies, stature and limb length consistently appear to be key factors in strength development alongside muscle size, body mass and neuro-muscular coordination. The importance of stature and limb length may be attributed to changes in mechanical advantage, including muscle moment arm lengths and the stimulation of muscle growth as a consequence of long bone growth. Evidence suggests that although these variables have a significant role to play when examined individually, the picture is very different when they are analysed together and the contribution of some factors becomes non-significant. For example, most studies have shown that maturation does not exert a significant independent effect when stature and body mass are accounted for, particularly for the lower extremities (De Ste Croix *et al.* 2003). Current data also suggest that muscle size cannot fully account for age and sex differences in strength during childhood when examined with other known variables. This finding may be clouded by the majority of studies examining anatomical CSA (ACSA) rather than physiological CSA (PCSA) when interpreting differences in strength (see Chapter 1 for more detail). Whether PCSA is the most important determinant of strength development during childhood remains to be identified. Unfortunately, to our knowledge, no longitudinal studies appear to have analysed the relative contribution of neuromuscular factors, muscle moment arms and mechanical advantage together when exploring differences in strength development in children. Therefore, it is difficult to assess their contribution alongside other known variables, and longitudinal studies are required that employ multilevel modelling techniques to understand the contribution of the key determinants of strength.

The tutorial below summarizes the key biomechanical concepts, which contribute to our understanding of age- and sex-related differences in strength and include the measurement of muscle agonist/antagonist activation, muscle size, muscle architecture and muscle-tendon moment arms/mechanical advantage. It is important to note that measurement of these key parameters present many challenges to researchers, especially in the paediatric population due to the individual timing and tempo of growth and maturation (see Chapter 1). Therefore, despite historical comprehension of the importance of muscle design, function and lever systems, the development of musculoskeletal imaging techniques has only recently opened up new opportunities for muscle-tendon and fascicle visualization *in vivo* (De Ste Croix 2007). To understand the reasons for differences in strength development during growth and maturation, the measurement method must be considered. This is especially crucial when interpreting data collected on boys and girls differing in age and maturity.

Basic theoretical concepts

Definitions of force and torque

Up to this point, the term 'strength' rather than 'force' or 'torque' (moment of force) has been used. Force represents the tension generated during voluntary/involuntary muscle activation and is appropriate to use when strength is measured *in vivo* at the point of load application (Blimkie 1989; Blimkie and Macauley 2000). This includes direct measurement within muscle-tendons (Ravary *et al.* 2004) as well as forces measured using strain gauge transducers mounted in line with the applied force (Nevill *et al.* 1998; Parker *et al.* 1990; Round *et al.* 1999). If a load is applied distal to a joint axis/centre of rotation and measurement is made at the axis of rotation (for example in isokinetic dynamometry), the term 'torque' is appropriate. A torque can be defined as the product of the force and the perpendicular distance from the line of action of the force to the axis/centre of rotation. It represents the tendency of a force to cause body segment rotation about an axis/centre of rotation (Pandy 1999). In most instances, more than one force may be acting about the axis of rotation. Therefore, the measured torque represents the net torque (sum of all products of forces and their corresponding moment arms). The joint torque varies depending on joint angle and velocity of movement. For the purpose of this chapter, where we refer to the term 'strength' we will be referring to torque.

When using an isokinetic dynamometer, isometric torques, or gravity-corrected joint torques, may be recorded. Assuming sufficient body stabilization and joint isolation, these can be defined as the net torque resulting from both agonist and antagonist activation. Therefore, when measuring concentric isokinetic knee extension torque, greater activation of the hamstrings would reduce the measured net knee extension torque. Any estimation of maximal knee extensor muscle torque must therefore take into account muscle coactivation. This can be achieved by deriving the torque–EMG relationship for the antagonist muscles (hamstrings) and adding the estimated antagonist torque to the net knee extension torque (Aagaard *et al.* 2000; Kellis 2003; Kellis and Baltzopoulos 1997; O'Brien *et al.* 2010). Failure to consider antagonist coactivation would lead to an underestimation of knee extensor torque and consequently inaccuracies in any subsequent calculations and comparisons of strength development. The torque (adjusted for any contribution from the antagonists) represents the resultant torque from all of the agonist muscles (Equation 4.1). However, this assumes that the angular acceleration is zero (the measurement of joint torque occurs within the isokinetic window). For knee extension, the adjusted knee extension torque acting via the patella tendon can be considered to be a resultant torque generated by the individual quadriceps muscles.

$$T = \sum_{i=1}^{n} (r_i \times F_i) \tag{4.1}$$

where T is the resultant torque adjusted for any contribution from the antagonists, r_i is the moment arm for the corresponding force, F_i is force of the ith muscle and n is the number of muscles contributing to the torque (Herzog 2000). Angular acceleration is considered to be zero.

Estimation of muscle-tendon forces when torque is measured at the axis of rotation

If the agonist muscles can be considered to insert via a common tendon (for example, the plantar flexors and knee extensors), the muscle-tendon force (F) can be estimated from the resultant muscle torque (T) if the moment arm (r) is known (Equation 4.2):

$$F = \frac{T}{r} \tag{4.2}$$

If the agonist muscles do not insert via a common tendon, Equation 4.1 is often described as indeterminate, since there are too many unknown parameters (Tsirakos *et al.* 1997). For example, if only the prime movers for elbow flexion are considered, the elbow flexion torque would be a product of the forces generated by biceps brachii (bb), brachialis (b) and brachioradialis (br) and their corresponding moment arms (r_{bb}, r_b, r_{br}, respectively) (Equation 4.3). This equation assumes that the triceps brachii antagonist torque has been taken into account.

$$T = \sum (r_{bb} \times F_{bb}) + (r_b \times F_b) + (r_{br} \times F_{br}) \tag{4.3}$$

Equation 4.3 demonstrates that even if the moment arm length of each elbow flexor muscle was known, the force at each muscle-tendon still cannot be estimated. This would require knowledge of how the resultant torque is distributed between the muscles (Herzog 2000). Approaches to solve the problem of indeterminacy include reducing the redundancy until the number of equations and number of unknown parameters is equal or using optimization methods (Tsirakos *et al.* 1997). Unsurprisingly, most studies examining developmental changes in muscle-tendon force have focused on the knee extensors (O'Brien *et al.* 2009b, 2010) or the ankle plantar flexors (Morse *et al.* 2008) because these muscle groups insert via a common tendon (the patella tendon and Achilles tendon, respectively). Therefore, if the moment arm of the tendon is measured, muscle-tendon forces can be calculated for these muscle groups.

Estimation of muscle-tendon forces when torque is measured distal to the axis of rotation

If torque or force is measured distal to the joint centre/axis of rotation, the length of the skeletal lever arm relative to the muscle moment arm length(s) must be considered to estimate muscle-tendon force. The moment arm length/skeletal lever length ratio is described as the mechanical advantage of the system. The mechanical advantage (r/L) is illustrated in Figure 4.1 for isometric elbow flexion, assuming the linear and angular acceleration of the elbow is zero and that the biceps brachii is representative of the muscles contributing to the elbow flexion torque. The contribution from the elbow extensors is considered to be negligible.

In Figure 4.1 the net muscular flexion torque must balance the extension torque caused by the weight of the forearm acting through the forearm plus hand centre of mass (Equation 4.4):

$$l \times W = r \times F \tag{4.4}$$

where 'l' represents skeletal lever length and 'W' is the combined weight of the forearm and hand.

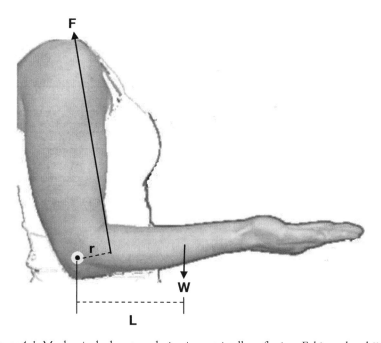

Figure 4.1 Mechanical advantage during isometric elbow flexion. F: biceps brachii muscle force; r: biceps brachii moment arm; W: mass of forearm plus hand multiplied by gravitational acceleration; L: distance from elbow joint axis of rotation to centre of mass of the forearm plus hand.

If the weight of the hand and forearm of a child is 10 N, and the length of the skeletal lever (from the elbow joint axis of rotation to the centre of mass of the forearm) is 0.11 m, the torque created by this force about the axis of rotation will be 1.1 Nm. To balance this extension torque the elbow flexor torque must be equal to 1.1 Nm in the opposite direction. The elbow flexor muscle-tendon forces must therefore be greater than 10 N, because the flexor moment arm about the elbow joint is smaller than the skeletal lever length. If it is assumed that the biceps brachii contributes approximately 57 per cent of the elbow flexion torque (Ikegawa *et al.* 1985) and the moment arm of the biceps brachii is 0.02 m, the biceps brachii tendon force would be 31.4 N (1.1 Nm × 0.57/ 0.02 m). Similar reasoning suggests that forces measured distal to the muscle attachment during strength testing will be lower than those measured directly in the muscle/tendon. Hence the musculoskeletal system is considered to act at a mechanical disadvantage, since measurements of force/torque are typically much smaller than actual tension within the muscle-tendon (Blimkie 1989). Estimates of muscle-tendon forces when the external force is measured distal to the centre/axis of rotation therefore requires the mechanical advantage to be measured.

When muscle-tendon forces have been measured or estimated, the reason for any differences observed between boys/girls differing in age and maturational status can be explored. For example, the muscle-tendon force may be normalized to muscle CSA to produce an estimate of 'specific tension'. Consequently, if despite differences in muscle-tendon force and CSA specific tension remains constant across individuals or within the same individual over time, it may be assumed that the potential to generate force is the same with the same level of muscle activation (there are no differences in muscle 'quality'). Differences in muscle-tendon forces may therefore be attributed to differences in the quantity of contractile muscle tissue available. In the paediatric literature, most studies have measured muscle ACSA as opposed to PCSA. As PCSA is a more accurate predictor of muscle strength, the use of ACSA can lead to inaccurate estimates of the potential of pennate muscles to generate tension (see Chapter 1) and subsequently limit the interpretation of age- and sex-related differences in strength.

Summary

These described biomechanical concepts emphasize that, unless force is directly measured in the muscle-tendon, *in vivo* measurements of strength (force at the site of application or joint torque) will be influenced by musculoskeletal lever systems. To seek the determinants of sex, age and maturational differences in strength, muscle-tendon forces should ideally be estimated. By nature of their calculation/ estimation, these consider agonist and antagonist coactivation and force moment arms/mechanical advantage. Differences due to muscle force-producing capability – muscle 'quality' (force/PCSA) or muscle 'quantity' (PCSA) – can then be determined.

It is difficult to evaluate the potential impact of studies which have concurrently measured the neuromusculoskeletal parameters required to estimate muscle-tendon or fascicle forces in children *in vivo*. Most of the available literature highlight pitfalls in the interpretation of differences in strength when any or all of the determinants of 'strength' (muscle size, muscle architecture, muscle tendon moment arms/mechanical advantage, agonist/antagonist activation) are not directly measured or assumptions are made regarding relations to anthropometric measurements and/or muscle activation.

Measurement of muscle moment arms/mechanical advantage in children

The measurement of muscle moment arms is not straightforward, and different approaches have been used to enable estimation of muscle-tendon forces *in vivo* (An *et al.* 1984). The most common procedures for estimating muscle moment arms can be classified under one of two general headings: the geometric method or the tendon excursion method. Detailed descriptions and evaluations of these methods can be found elsewhere in the literature (e.g. An *et al.* 1984; Fath *et al.* 2010; Maganaris 2000; Maganaris *et al.* 2000; Pandy 1999; Spoor and Van Leeuwen 1992). The geometric method conforms to the definition of a muscle moment arm when examined in two dimensions, requiring the joint centre/axis of rotation or contact point, and the muscle-tendon origin, insertion and line of action to be identified. The perpendicular distance from the rotation centre/axis to the force line (moment arm) can then be estimated.

The tendon excursion technique focuses on the principle that the excursion of a muscle-tendon during the rotation of a joint is determined by its moment arm (Murray *et al.* 2000). Muscle moment arms can therefore be estimated from the magnitude of tendon displacement that occurs for a given rotation. The relationship is derived in Spoor *et al.* (1990), using expressions of muscle and joint work. It is assumed that the work done by a force (F) over a given tendon displacement (ds), equals the work done by a torque about the axis of rotation during a given angular rotation ($d\theta$):

$$F \cdot ds = T \cdot d\theta \tag{4.5}$$

From Equation 4.1: $T = r \cdot F$

$$\frac{T}{F} = r \tag{4.6}$$

Substituting Equation 4.5 into Equation 4.6:

$$\frac{ds}{d\theta} = \frac{T}{F} = r \tag{4.7}$$

Moment arms calculated using the tendon excursion technique have typically (as the name suggests) measured the displacement of the tendon for a given joint rotation (Maganaris *et al.* 2001) or examined the displacement of pennated muscles by visualizing the attachment of one or more fascicles to the central tendon (aponeurosis) using ultrasound (Fukunaga *et al.* 1996a, 1996b; Ito *et al.* 2000; Kubo *et al.* 1999; Maganaris and Paul 1999). During changes in joint angle, the position of a marked fascicle attachment is tracked, enabling displacement to be estimated. Invariably, both of these methods require the use of imaging techniques (e.g. ultrasound or magnetic resonance imaging) that may be difficult in paediatric studies.

With growth and maturation, muscle moment arms would be expected to increase. This may reflect changes in, for example, bone structure, joint kinematics and the line of action of the muscle-tendon unit. Joint kinematics and the joint centre/axis of rotation will be affected by the shape of the articulating bones (Murray *et al.* 1995; O'Brien *et al.* 2009a). Therefore, bone development (see Chapter 1) and genetic influences on bone structure may influence moment arm lengths in children. Figure 4.2 illustrates differences in the humerus and ulnar articulation for two females: a 9-year-old female and a female of approximately 13 years of age. The trochlea (dashed arrow) and olecranon process (solid arrow) are highlighted in both the scans. The development of the trochlea and ulnar articulation (trochlear notch, olecranon process) is clear. Given that bone shape and size affect joint kinematics, the position and orientation of axes of rotation in children may differ from those identified in adults. The joint angle will also influence the relative contribution of joint structure to moment arm measurements. Thomis *et al.* (1997) observed that the genetic contribution to isometric elbow flexor torque was important at 170° but minimal at 50° (180° represented extension of the elbow). Bone structure may be expected to contribute more to joint torque when the elbow is more extended compared to when it is flexed. The angle dependency of the effect of bone characteristics on moment arm lengths and joint torque is also evident through studies seeking to relate anthropometric measurements to moment arm lengths. When examining knee extensor moment arms in children, knee depth and knee circumference may be expected to correlate with moment arm length measured in full extension but not in flexion (O'Brien *et al.* 2009a). In the latter position the moment arm is not in the same direction as knee depth. These studies highlight that the joint centre/axis of rotation cannot be assumed to be the same in children and adults, and it will also differ according to joint angle. Further, the relationship between surface anthropometry and moment arm length in children seems to be tendon-specific. In a recent study, Waugh *et al.* (2011) demonstrated that Achilles tendon moment arm (in contrast to quadriceps tendon moment arm) cannot be accurately predicted by surface anthropometry. Assumptions of planar motion, and identical and fixed centres/axes of rotation must be validated for the joint, range of motion (Lieber and Bodine-Fowler 1993) and participants being examined. This will enable accurate estimates of the relative contribution of moment arm lengths (using the geometric method) to joint torque to be examined. As described in Equation 4.1, if muscle moment arm lengths increase with age, an increase in joint torque would be measured for a given muscle force. Inter- and intra-individual variability in bone

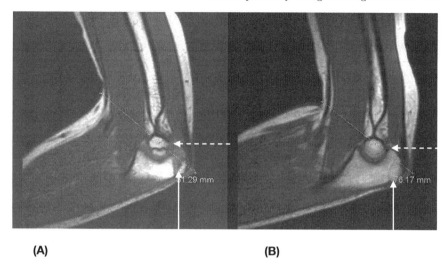

(A) **(B)**

Figure 4.2 Developmental changes in the trochlea of the humerus (dashed arrow), and olecranon process of the ulna (solid line) in a 10-year-old (A) and 13-year-old girl (B).

structure, development and joint kinematics suggest variability in muscle moment arm lengths which need to be accurately determined to understand age and sex differences in torque or muscle force-producing capability in children.

Different approaches have been used to model the muscle-tendon lines of action for moment arm calculations (Bonnefoy *et al.* 2007; Carmen and Milburn 2005; Jensen and Davy 1975; Pierrynowski 1995). If the muscle-tendon line of action is drawn through the centre of the muscle-tendon, muscle size and bone length will influence the moment arm (O'Brien *et al.* 2009a). Larger moment arms would be expected in individuals with greater muscle mass if centroidal measures are used. The reason for this relationship is that greater muscle mass would be associated with greater CSAs (Wood *et al.* 1989), which in turn would result in increased moment arms with the muscle-tendon line of action being further away from the centre of rotation (An *et al.* 1981). Therefore, changes in muscle size and bone development with growth and development may cause muscle moment arm lengths to vary. If this hypothesis holds, then we may propose that part of the sex-associated difference in force production during puberty may be attributed to the greater relative increase in muscle size in boys compared to girls and subsequent impact on the moment arm size.

Irrespective of the method used to estimate muscle moment arm lengths, the description of both measurement methods highlights the potential for change during growth and maturation. Developmental changes in moment arm lengths may therefore contribute to age- and sex-associated differences in joint torques. Mechanical advantage may also change if muscle moment arm(s) do not change in proportion to bone lever growth. Research examining muscle moment arms and mechanical advantage in children is very limited.

Age-related changes in muscle moment arms/mechanical advantage

Relatively few studies have examined moment arm lengths in children in relation to strength development (Garcia-Morales *et al.* 2003; Ikai and Fukanaga 1968; Morse *et al.* 2008; O'Brien *et al.* 2009a, 2009b, 2010; Parker 1989; Wood *et al.* 2006). Those studies that are available differ in terms of the study participants (age, sex, maturation), the joint/muscles/bones assessed, methods used to examine strength and the method of determining muscle moment arms. Parker (1989) estimated mechanical advantage using measurements taken on ulna, radius and tibia bone specimens. The skeletons were from 10 adults and seven children. For three of the child bones, age and sex were unknown. For the remaining four, the ages were 4–5, 12, 15 and 18 years. No clear trends in mechanical advantage were apparent from the measurements on the ulna and tibia. However, except for the 4–5-year-old child, the mechanical advantage for the radius was higher than in adults by two standard deviations. These data would indicate a greater relative increase in bone lever length in relation to moment arm length with age, implying an increased muscle force requirement to move a distally applied load. Ikai and Fukunaga (1968) measured isometric elbow flexion strength at a joint angle of 90° using a cable tensiometer. X-ray scans were used to estimate mechanical advantage in a small sample of five boys (aged approximately 13 years) and 10 men (20–30 years old). The biceps line of action was assumed to be perpendicular to the forearm. No significant differences in mechanical advantage were observed between boys and men. It is important to note that both Ikai and Fukunaga (1968) and Parker (1989) either made assumptions or did not consider the influence of muscle-tendon line of action or differences in the joint axis of rotation on the measurement of moment arms/mechanical advantage.

More recently, the limitations of earlier studies have been addressed. Morse *et al.* (2008) calculated the Achilles tendon moment arm in children using the tendon excursion method. The Achilles tendon moment arm lengths in 11 boys (mean age: 10.9± 0.3 years, early maturational stage – stage two secondary sex characteristics) were significantly smaller (25 per cent) compared to men. In another study, O'Brien *et al.* (2009b) calculated patellar tendon moment arm lengths using the geometric method (taking into account the muscle-tendon line of action with respect to the tibiofemoral contact point). They observed significant age-related differences in adults and prepubertal children. The patellar tendon moment arm lengths were significantly greater in men compared to both the boys and girls, and greater in the women compared to boys. Together, these studies suggest that, *in vivo*, measures of moment arm lengths increase with increasing age. For a given level of muscle force, this would imply increases in the joint torque and suggest increases in 'strength' with age.

Sex-related changes in muscle moment arms/mechanical advantage

Although there are limited data in children, the current available studies suggest that there are no significant sex differences in the muscle moment arm lengths of prepubertal children in the brachialis (Wood *et al.* 2006) or patella tendon (O'Brien *et al.* 2009b, 2010). In the latter studies, there were also no significant differences in moment arm lengths between adult men and women. This contrasts to the findings of Wretenberg *et al.* (1996) who observed significant sex differences in most knee muscle moment arms (hamstring tendons, sartorius, medial and lateral gastrocnemius and the patellar tendon) in the sagittal and frontal planes, with males demonstrating longer moment arm lengths than females. Both Wretenberg *et al.* (1996) and O'Brien *et al.* (2009b) measured moment arms at rest; however, the studies differed in the analyses undertaken (3D versus 2D, respectively, and calculation of the joint contact point). There is a lack of *in vivo* moment arm data on females and therefore further studies are required to examine sex differences in moment arms/mechanical advantage. Also, according to our knowledge, there are no data that encompass the period of adolescence when sex differences in force/torque production start to become significant. Whether sex differences in muscle moment arm lengths contribute to the divergence in strength during growth and maturation requires further longitudinal investigation encompassing puberty.

Anthropometric correlates of moment arm lengths

The relative lack of comprehensive longitudinal studies examining the age- and sex-associated changes in the muscle moment arm may in part be attributed to the cost of imaging technology along with access to facilities, equipment and appropriate expertise. These factors all represent challenges to obtaining accurate *in vivo* moment arm measurements throughout the joint range of motion utilized in strength assessment. In addition, when working with children who may be restless and anxious, radiographic data collection may not always be successful (Jenkins *et al.* 2003). It is therefore not surprising that anthropometric measurements to predict individual differences in moment arm lengths have been used with paediatric populations.

It must be noted that anthropometric determinants of moment arms appear to be muscle-tendon and joint angle specific. Nevertheless, when the patella tendon moment arm length measured in 3D was normalized to the width of the femoral condyle in adults, Krevolin *et al.* (2004) observed little difference across individuals. These findings would suggest that the moment arm lengths could be predicted by measurements of knee joint size. However, when Tsaopoulos *et al.* (2007) examined the ability to predict patella tendon moment arm length using measurements of height, body mass, body mass index, medial-lateral and anterior-posterior knee joint dimensions, knee circumference, femur length, tibia length and leg length, only knee circumference was found to be significant. Since this measurement could only account for ~14 per cent of the variance in the moment arm length, it was concluded that external anthropometric measurements could not be used to predict patella tendon moment arm lengths. These data are supported by a recent study which explores predicting Achilles tendon moment arms using a large range of surface anthropometric parameters in prepubertal children (Waugh *et al.* 2011). Waugh *et al.* (2011) reported that the range of surface anthropometric measures could only account for 49 per cent of the variance of moment arm length and that the error associated with the measurement was 14.5 per cent. Thus, while it is implicitly assumed that moment arms scale with size and can be normalized by segment lengths or limb circumferences, these scaling relationships between muscle moment arms and anthropometric dimensions appear to be poor and require further investigation with a range of ages (Murray *et al.* 2002). Further studies examining the relation of anthropometric variables to moment arm lengths are needed, especially in children.

Interpretation of age and sex differences in 'strength'

Interpretation of age- and sex-associated differences in strength is difficult when the underlying mechanical variables such as moment arms, mechanical advantage, muscle activation and PCSA have not been directly measured. Previous studies that have measured and compared net voluntary joint torques and distal forces (as opposed to tendon forces) are useful, indicating an individual's net torque/force potential encompassing the skeletal lever system and muscle development. These torques and forces are typically scaled using measurements of body size to enable 'strength' to be compared between boys and girls differing in size due to age and maturity. The scaling variables may be 'global' size measures (including stature, body mass, fat-free mass) or 'local' size measures (including limb segment length, ACSA, limb circumferences or skinfolds). Using these scaling techniques, normative data can be established as a reference tool for clinicians to aid understanding of pathology and injury risk. However, difficulty arises when these studies seek to identify the mechanisms underlying any age and sex differences in scaled/unscaled strength measures.

The majority of strength data have been scaled using the ratio standard (division of the 'strength' variable by the body size variable). In the longitudinal Saskatchewan child growth and development study, differences in isometric composite, upper and lower body strength between children of the same chronological age were found, but differences based on maturational age disappeared when adjusted for differences in body mass (Carron and Bailey 1974). Most commonly, differences in strength with age persist when normalized in relation to body mass and/or stature (Froberg and Lammert 1996; Housh *et al.* 1996; Nevill *et al.* 1998; Parker *et al.* 1990; Ramos *et al.* 1998; Sunnegardh *et al.* 1998). Sex differences in isometric and isokinetic strength are also typically observed when torque is normalized for body mass and/or stature (Castro *et al.* 1995; De Ste Croix *et al.* 1999; Docherty and Gaul 1991; Gilliam *et al.* 1979; Nevill *et al.* 1998; Ramos *et al.* 1998; Round *et al.* 1999; Sunnegardh *et al.* 1998; Wood *et al.* 2004).

Although growth of the long bones and weight-bearing activities are mechanical factors known to stimulate muscle development, global size measures may not parallel localized segmental growth (Enoka 1994; Malina 1986). This is particularly apparent during growth and maturation, when differential growth of the upper and lower extremity occurs (see Chapter 1). Measurement of body mass should not be viewed as a proxy for muscle development, as it includes both fat and fat-free mass and therefore does not allow differentiation according to body composition. Even if fat-free mass is used to normalize strength data, the muscle mass per unit fat-free mass will vary (Housh *et al.* 1996). Therefore, although strength data relative to global measures of body size have been used as references to define normative strength development, local size measures would be more appropriate.

When 'local' body size measures (limb segment length, ACSA, limb circumferences and skinfolds) have been used to scale strength data, significant age effects persist. Kanehisa *et al.* (1995b) normalized isometric force of the ankle dorsiflexors and plantarflexors to ACSA multiplied by lower leg length to approximate muscle volume and possibly PCSA. The 121 boys and 121 girls were separated into four age groups (7–9, 10–12, 13–15 and 16–18 years). The ratio of plantarflexion force to the product of ACSA and leg length was significantly higher in the 16–18-year-old boys compared to the other age groups. Similar findings were also observed in other muscle groups by the same research group (Kanehisa *et al.* 1994b). Kanehisa *et al.* (1994b) examined isokinetic knee extensor torque in 60 boys and girls (6–9 years) and 71 adult men and women over a range of velocities. Torques were divided by the dynamometer lever length to obtain force, and this was subsequently normalized to the product of ACSA and thigh length. A significant age effect was found for the two highest isokinetic velocities. When elbow and knee flexion and extension forces (torque divided by the dynamometer lever arm length) were normalized to ACSA in 7–18-year-old boys, the ratio of force to ACSA was significantly higher in the older children (Kanehisa *et al.* 1995a). Grip force normalized to the product of ACSA and forearm length also increased with age in 366 participants (6–23 years) (Neu *et al.* 2002). Longitudinal isometric and concentric elbow flexion and extension data were analysed by Wood *et al.* (2004). Interestingly, this study showed that the age effect was both action- and muscle-specific. Recent work by Falk *et al.* (2009) demonstrated no significant difference between prepubertal girls and women in isometric elbow flexion torque normalized to ACSA.

The above studies also analysed sex differences in strength in children with conflicting findings. Significant sex differences in isokinetic knee extensor forces normalized to the product of ACSA and thigh length have been found in adult men and women (Kanehisa *et al.* 1994a). However, no significant sex differences have been observed in isometric plantarflexor or dorsiflexor force in boys and girls aged 7–18 years (Kanehisa *et al.* 1995b); in grip force normalized to the product of ACSA and forearm length in 6–23-year-olds (Neu *et al.* 2002); in prepubertal children when isometric elbow flexion torque was scaled using ACSA (Wood *et al.* 2006); or in isometric/concentric elbow flexion and extension torque when accounting for differences in ACSA and a linear dimension (stature or arm length) in adolescent children over the course of a three-year longitudinal study (Wood *et al.* 2004).

The use of local measurements to scale strength data may be more appropriate than global measures since they account for the different growth rates of individual segments; however, they suffer from many of the same limitations of global size measures (assumptions of proportionality to PCSA and lever lengths). In addition, only direct measurement of muscle PCSA would be expected to identify differential growth rates of synergist muscle groups (e.g. elbow flexors/ extensors) during growth and maturation.

Unless otherwise stated, the above studies, which used global or local body size measures to normalize or scale strength data, measured the outcome of maximal voluntary actions. Therefore, differences in agonist muscle activation and/or antagonist coactivation may contribute to any sex and age group effects alongside the other study limitations. It is interesting to note that as more variables are considered (albeit even if they are not directly measured), age and sex differences in muscle tension development have not been observed. Davies (1985) ensured complete activation of the triceps surae in children (three groups of boys and girls: ~9.6 years, ~11.8 years and ~14.7 years) and adults (males and females) by using electrically evoked maximal twitch and tetanic contractions. Tendon tension was estimated using mechanical advantage data from previous studies and ACSA was adjusted to reflect PCSA. Irrespective of age and sex, the mean specific tension (33 N·cm^{-2}) was the same for the children and adults, suggesting that there is no age effect in estimated tendon tension.

To enable the mechanisms of differences in net joint torque to be examined in children, accurate estimates of muscle-tendon or fascicle forces are required. These consider variation in muscle activation, the lever system and muscle size. As more assumptions are made to consider the adjustment of torque/external forces for the differences in body size that accompany growth and maturation, the potential for identifying clear patterns and mechanisms of strength development in children declines.

Muscle-tendon and fascicle force development in children

Blimkie (1989) stated that 'The relative importance of gross muscle size in differentiating strength among children of the same age, across ages and between sexes remains to be determined' (p. 131). The studies reviewed in the previous section emphasize the difficulties in interpreting the mechanisms underlying age-related differences in strength (maximum torque) or distally applied force with age between boys and girls. Therefore, it is currently not known whether the differences reflect changes in all or any combination of muscle size, specific tension, motor unit recruitment and leverage factors. By evaluating the neuromusculoskeletal factors necessary to obtain accurate estimates of muscle-tendon forces in children (agonist and antagonist muscle activation, moment arms length(s)/ mechanical advantage and PCSA), recent studies (Morse *et al.* 2008; O'Brien *et al.* 2009b) have started to address the reason for differences in joint torques and tendon forces in children. These studies provide the most valid interpretation of strength development and may result in changes to our understanding of strength development in children by addressing many of the limitations of previous studies.

Morse *et al.* (2008) observed 41 per cent smaller Achilles tendon and lateral gastrocnemius fascicle forces in pre- and early pubertal boys compared to men. When the fascicle force was normalized to PCSA, the specific tension of the lateral gastrocnemius was 21 per cent higher in boys than in men (13.1 ± 2.0 N·cm^{-2} – men; 15.9 ± 2.7 N·cm^{-2} – boys). These results contrast to the findings of O'Brien *et al.* (2009b) who calculated quadriceps-specific tension in prepubertal boys and girls, and adult males and females. Although patella tendon and quadriceps forces were significantly greater in both the men and women when compared to those obtained for the boys and girls, there were no significant differences in quadriceps-specific tension (~55 N·cm^{-2}). These conflicting findings result in different interpretations relating to the adult–child strength (muscle-tendon/fascicle force) differences. The latter study suggests that increases in strength with age and maturity can be fully explained by increases in the level of voluntary activation, moment arm length and PCSA, while the study by Morse *et al.* (2008) would suggest alternative mechanisms underlying the observed higher force per unit area. Such mechanisms may include increased type II muscle fibre recruitment (Neu *et al.* 2002) and changes in tendon characteristics (Lambertz *et al.* 2003; Morse *et al.* 2008). These data also let us speculate about age-related differences in tendon specific tensions which may vary depending upon the muscle groups that are exerting force through that tendon (see Chapter 6 for more detail). Morse *et al.* (2008) also highlighted several limitations of their study, which may explain the unexpected findings including: measurement of moment arm lengths at rest rather than during maximal voluntary activation, the assumption of complete agonist motor unit activation during voluntary strength measurement and estimation of the percentage of total triceps surae PCSA occupied by the lateral gastrocnemius.

O'Brien *et al.* (2009b) currently provide the only muscle-tendon/fascicle force data on boys and girls, although we are aware of unpublished data from the study of Waugh *et al.* (2011). At least for the quadriceps, their results suggested that in agreement with measurements on adult men and women there were no significant sex differences in specific tension in prepubertal children. Further studies are required to confirm/refute these seminal studies by O'Brien *et al.* (2009b) and Morse *et al.* (2008), and to further examine the development of muscle tendon forces in boys and girls differing in age and maturity. It would also be beneficial to study upper extremity muscle groups where sex differences in torque are much greater than for the lower extremity (De Ste Croix *et al.* 2004).

Clinical and practical applications

Understanding of injury mechanisms in children

Compared to adults, bone structure is different in children since the layer of articular cartilage is thicker, and cartilage is present in the epiphyseal and apophyseal growth plates (Mafulli 2001). The cartilage present in immature bones is more susceptible to injury (Adirim and Cheng 2003; Micheli and Klein 1991), and growth of the long bone itself may constitute a risk factor for injury if there is asynchronous development of the associated muscle, tendon and ligament structures (Gerrard 1993). There is indirect evidence for musculoskeletal growth lags (see Chapter 1). Traction apophyseal injuries in children suggest increased tissue preload (passive tissue forces measured at rest) during growth (Hawkins and Metheny 2001). In addition to the proposition of a lag between the growth in length of the long bones and the growth in muscle-tendon length, muscle hypertrophy may also lag behind growth in muscle length (Kanehisa *et al.* 1995b; Xu *et al.* 2009). Hawkins and Metheny (2001) used biomechanical principles to consider the impact of growth on musculoskeletal loading. Utilizing ratios of limb length, mass and moment of inertia at age 14 years compared to age 6 years, 4.7 times the quadriceps muscular torque would be required to maintain a horizontal lower leg position when seated at 14 years compared to 6 years. Therefore, quadriceps muscle hypertrophy and increases in moment arm length would need to occur to support the growth of the lower leg. Any lag in these changes will induce a degree of overload on the muscle-tendon and insertional zones. Due to the rapid and differential growth of bones during the adolescent growth spurt (see Chapter 1) and the fact that muscles may span more than one joint, the potential for injury is clear. This also has implications for co-contraction and the muscles' ability to stabilize the joint during growth and maturation (see Chapter 11). Estimation of muscle specific tension using muscle fascicle forces and PCSA measurements in children alongside limb segment and muscle-tendon measures is needed to corroborate anecdotal evidence of musculoskeletal growth lags and overload.

Development of research-led training and coaching guidelines

In addition to injury mechanisms, the understanding of musculoskeletal growth rates also has important implications for the training of children (particularly during rapid growth phases). If growth in bone length induces a degree of overload on the muscle, tendons and insertion zones providing the stimulus for development, any increase in training load at this time would increase injury susceptibility. There may also be performance stasis/deterioration, while the individual adjusts to the perceptual, spatial, physiological and biomechanical changes induced by growth (see Chapter 1).

The earlier description of biomechanical principles also highlights the importance of children utilizing appropriate equipment and perhaps not changing equipment that may lead to increased musculoskeletal loading during rapid growth phases. Figure 4.3 illustrates that the resultant elbow flexion torque would need to increase to maintain isometric elbow flexion when holding racquets progressively increasing in length and mass. The greater elbow flexion torque would be achieved through greater elbow flexor force development and/or moment arm length, which occur with growth and maturation. However, a change in racquet size during and within a year following a rapid growth phase (when a degree of overload is already occurring due to growth) will induce further muscle overload, which potentially results in an increased risk of injury. This

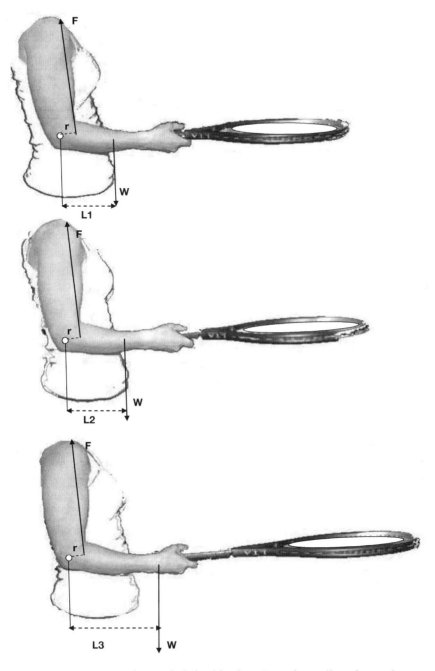

Figure 4.3 Racquet size and musculoskeletal loading. F: resultant elbow flexion force; r: resultant elbow flexion moment arm; W: mass of the forearm, hand and racquet multiplied by gravitational acceleration; L: distance from elbow joint axis of rotation to centre of mass of the forearm, hand and racquet.

example demonstrates the importance of understanding the mechanical as well as physiological changes that occur with growth and maturation to fully comprehend the impact of changes in training and/or equipment on injury/performance in children.

Conclusion and future directions

The assessment of all of the variables necessary to allow force estimation constitutes a challenge to researchers in terms of equipment, time, expertise and the difficulties which may uniquely apply to working with children. These studies present exciting opportunities for future research, including:

- Investigation of muscle-tendon/fascicle force development with age and maturation in boys and girls encompassing the pubertal growth spurt.
- The examination of muscle force development in relation to: growth in length and weight of long bones, muscle length and PCSA, tendon characteristics and apophyseal properties. This will enable identification of any growth lags and aid understanding of overuse injury mechanisms in children. It may also facilitate research-informed coaching/training guidelines (particularly during puberty when the rate of growth is rapid).
- The analysis of reciprocal muscle group function with implications for joint stability and control. Any differences in the development of musculature predominantly assuming antigravity or gravity-assisted roles can also be explored.
- The accumulation of accurate musculoskeletal data in children to facilitate the development of models for motion analysis and surgical planning (see Chapter 6).

Key points

- Traditional methods of assessing and reporting 'strength' data in children have resulted in different interpretations of age- and sex-related development of strength.
- Development of imaging techniques has allowed muscle-tendon and fascicle forces to be derived in children *in vivo*.
- Calculation of muscle specific tension using PCSA will allow the true force potential of muscles to be compared between boys and girls of differing age and maturity.
- Age- and sex-associated force development must be viewed concurrently with other known variables.
- Longitudinal data spanning the pubertal period and including the investigation of neuromuscular parameters are required.

References

Aagaard, P., Simonsen, E.B., Andersen, J.L., Magnusson, S.P., Bojsen-Møller, F. and Dyhre-Poulsen, P. (2000) 'Antagonist muscle coactivation during isokinetic knee extension', *Scandinavian Journal of Medicine and Science in Sports*, 10: 58–67.

Adirim, T.A. and Cheng, T.L. (2003) 'Overview of injuries in the young athlete', *Sports Medicine*, 33: 75–81.

An, K.N., Takahashi, K., Harrigan, T.P. and Chao, E.Y. (1984) 'Determination of muscle orientations and moment arms', *Journal of Biomechanical Engineering*, 106: 280–2.

An, K.N., Hui, F.C., Morrey, B.F., Linscheid, R.L. and Chao, E.Y. (1981) 'Muscles across the elbow joint: a biomechanical analysis', *Journal of Biomechanics*, 14: 659–69.

Beunen, G. and Malina, R.M. (1988) 'Growth and physical performance relative to the timing of the adolescent spurt', in K.B. Pandolf (ed.) *Exercise and Sport Sciences Reviews*, 16, London: Collier Macmillan, pp. 503–40.

Blimkie, C.J. (1989) 'Age and sex associated variation in strength during childhood: anthropometric, morphologic, neurologic, biomechanical, endocrinological, genetic and physical activity correlates', in C.V. Gisolfi and D.R. Lamb (eds) *Perspectives in Exercise Science and Sports Medicine, (Vol. 2), Youth, Exercise and Sport*, Indianapolis, IN: Benchmark Press, pp. 99–163.

Blimkie, C.J.R. and Macauley, D. (2000) 'Muscle strength', in N. Armstrong and W. Van Mechelen (eds) *Paediatric Exercise Science and Medicine*, 1st edn, Oxford: Oxford University Press, pp. 23–36.

Bonnefoy, A., Doriot, N., Senk, M., Dohin, B., Pradon, D. and Chéze, L. (2007) 'A non-invasive protocol to determine the personalized moment arms of knee and ankle muscles', *Journal of Biomechanics*, 40: 1776–85.

Carmen, A.B. and Milburn, P.D. (2005) 'Dynamic coordinate data for describing muscle-tendon paths: a mathematical approach', *Journal of Biomechanics*, 38: 943–51.

Carron, A.V. and Bailey, D.A. (1974) 'Strength development in boys from 10 through 16 years', *Monographs of the Society for Research in Child Development*, 39: 1–37.

Castro, M.J., McCann, D.J., Shaffrath, J.D. and Adams, W.C. (1995) 'Peak torque per unit cross-sectional area differs between strength-trained and untrained young adults', *Medicine and Science in Sports and Exercise*, 27: 397–403.

Davies, C.T.M. (1985) 'Strength and mechanical properties of muscle in children and young adults', *Scandinavian Journal of Sports Science*, 7: 11–15.

De Ste Croix, M.B.A. (2007) 'Advances in paediatric strength assessment: changing our perspective on strength development', *Journal of Sports Science and Medicine*, 6: 292–305.

De Ste Croix, M.B.A. (2009) 'Muscle strength', in N. Armstrong and W. Van Mechelen (eds) *Paediatric Exercise Science and Medicine*, Oxford: Oxford University Press, pp. 199–212.

De Ste Croix, M.B.A., Armstrong, N. and Welsman, J.R. (1999) 'Concentric isokinetic leg strength in pre-teen, teenage and adult males and females', *Biology of Sport*, 16: 75–86.

De Ste Croix, M.B.A., Deighan, M.A. and Armstrong, N. (2003) 'Assessment and interpretation of isokinetic strength during growth and maturation', *Sports Medicine*, 33: 727–43.

De Ste Croix, M.B.A., Deighan, M.A. and Armstrong, N. (2004) 'Time to peak torque for knee and elbow extensors and flexors in children, teenagers and adults', *Isokinetic and Exercise Science*, 12(2): 143–8.

Docherty, D. and Gaul, C.A. (1991) 'Relationship of body size, physique, and composition to physical performance in young boys and girls', *International Journal of Sports Medicine*, 12: 525–32.

Enoka, R.M. (ed.) (1994) *Neuromechanical Basis of Kinesiology*, Champaign, IL: Human Kinetics.

Falk, B., Brunton, L., Dotan, R., Usselman, C., Klentrou, P. and Gabriel, D. (2009) 'Muscle strength and contractile kinetics of isometric elbow flexion in girls and women', *Pediatric Exercise Science*, 21: 354–64.

Fath, F., Blazevich, A.J., Waugh, C.M., Miller, S.C. and Korff, T. (2010) 'Direct comparison of in vivo Achilles tendon moment arms obtained from ultrasound and MR scans', *Journal of Applied Physiology*, 109: 1644–52.

Froberg, K. and Lammert, O. (1996) 'Development of muscle strength during childhood', in O. Bar-Or (ed.) *The Encyclopedia of Sports Medicine VI, The Child and Adolescent Athlete*, London: Blackwell Science.

Fukunaga, T., Roy, R.R., Shellock, F.G., Hodgson, J.A. and Edgerton, V.R. (1996a) 'Specific tension of human plantar flexors and dorsiflexors', *Journal of Applied Physiology*, 80: 158–65.

Fukunaga, T., Ito, M., Ichinose, Y., Kuno, S., Kawakami, Y. and Fukashiro, S. (1996b) 'Tendinous movement of a human muscle during voluntary contractions determined by real-time ultrasonography', *Journal of Applied Physiology*, 81: 1430–3.

Garcia-Morales, P., Buschang, P.H., Throckmorton, G.S. and English, J.D. (2003) 'Maximum bite force, muscle efficiency and mechanical advantage in children with vertical growth patterns', *European Journal of Orthodontics*, 25: 265–72.

Gerrard, D.F. (1993) 'Overuse injury and growing bones: the young athlete at risk', *British Journal of Sports Medicine*, 27: 14–18.

Gilliam, T.B., Villanacci, J.F., Freedson, P.S. and Saday, S.P. (1979) 'Isokinetic torque in boys and girls ages 7–13: effect of age, height and weight', *Research Quarterly*, 50: 599–609.

Hawkins, D. and Metheny, J. (2001) 'Overuse injuries in youth sports: biomechanical considerations', *Medicine and Science in Sports and Exercise*, 33: 1701–7.

Herzog, W. (2000) 'Muscle properties and coordination during voluntary movement', *Journal of Sports Sciences*, 18: 141–52.

Housh, T.J., Johnson, G.O., Housh, D.J., Stout, J.R., Weir, J.P., Weir, L.L. and Eckerson, J.M. (1996) 'Isokinetic peak torque in young wrestlers', *Pediatric Exercise Science*, 8: 143–55.

Ikai, M. and Fukunaga, T. (1968) 'Calculation of muscle strength per unit cross-sectional area of human muscle by means of ultrasonic measurement', *Internationale Zeitschrift fur angewandte Physiologie, Einschliesslich Arbeitsphysiologie*, 26: 26–32.

Ikegawa, S., Tsunoda, N., Yata, H., Matsuo, A., Fukunaga, T. and Asami, T. (1985) 'The effect of joint angle on cross-sectional area and muscle strength of human elbow flexors', in D.A. Winter, R.W. Norman, R.P. Wells and K.C. Hayes (eds) *Biomechanics IX-A International Series on Biomechanics*, Champaign, IL: Human Kinetics.

Ito, M., Akimo, H. and Fukunaga, T. (2000) 'In vivo moment arm determination using B-mode ultrasonography', *Journal of Biomechanics*, 33: 215–18.

Jenkins, S.E.M., Harrington, M.E., Zavatsky, A.B., O'Connor, J.J. and Theologis, T.N. (2003) 'Femoral muscle attachment locations in children and adults, and their prediction from clinical measurement', *Gait & Posture*, 18: 13–22.

Jensen, R.H. and Davy, D.T. (1975) 'An investigation of muscle lines of action about the hip: a centroid line approach vs the straight line approach', *Journal of Biomechanics*, 8: 103–10.

Jones, D.A. and Round, J.M. (2000) 'Strength and muscle growth', in N. Armstrong and W. Van Mechelen (eds) *Paediatric Exercise Science and Medicine*, Oxford: Oxford University Press, pp. 133–42.

Kanehisa, H., Ikegawa, S. and Fukunaga, T. (1994a) 'Comparison of muscle cross-sectional area and strength between untrained women and men', *European Journal of Applied Physiology and Occupational Physiology*, 68: 148–54.

Kanehisa, H., Ikegawa, S., Tsunoda, N. and Fukunaga, T. (1994b) 'Strength and cross-sectional area of knee extensor muscles in children', *European Journal of Applied Physiology*, 68: 402–5.

Kanehisa, H., Ikegawa, S., Tsunoda, N. and Fukunaga, T. (1995a) 'Strength and cross-sectional areas of reciprocal muscle groups in the upper arm and thigh during adolescence', *International Journal of Sports Medicine*, 16: 54–60.

Kanehisa, H., Yata, H., Ikegawa, S. and Fukunaga, T. (1995b) 'A cross-sectional study of the size and strength of the lower leg muscles during growth', *European Journal of Applied Physiology*, 72: 150–6.

Kellis, E. (2003) 'Antagonist moment of force during maximal knee extension in pubertal boys: effects of quadriceps fatigue', *European Journal of Applied Physiology*, 89: 271–80.

Kellis, E. and Baltzopoulos, V. (1997) 'The effects of antagonist moment on the resultant knee joint moment during isokinetic testing of the knee extensors', *European Journal of Applied Physiology and Occupational Physiology*, 76: 253–9.

Krevolin, J.L., Pandy, M.G. and Pearce, J.C. (2004) 'Moment arm of the patellar tendon in the human knee', *Journal of Biomechanics*, 37: 785–8.

Kubo, K., Kawakami, Y. and Fukunaga, T. (1999) 'Influence of elastic properties of tendon structures on jump performance in humans', *Journal of Applied Physiology*, 87: 2090–6.

Lambertz, D., Mora, I., Grosset, J-F. and Pérot, C. (2003) 'Evaluation of musculotendinous stiffness in prepubertal children and adults, taking into account muscle activity', *Journal of Applied Physiology*, 95: 64–72.

Lieber, R.L. and Bodine-Fowler, S.C. (1993) 'Skeletal muscle mechanics: implications for rehabilitation', *Physical Therapy*, 73: 844–56.

Mafulli, N. (2001). 'The younger athlete', in P. Brukner and K. Khan (eds) *Clinical Sports Medicine*, 2nd edn, London: McGraw-Hill, pp. 651–73.

Maganaris, C.N. (2000) 'In vivo measurement-based estimations of the moment arm in the human tibialis anterior muscle-tendon unit', *Journal of Biomechanics*, 33: 375–9.

Maganaris, C.N. and Paul, J.P. (1999) 'In vivo human tendon mechanical properties', *Journal of Physiology*, 521: 307–13.

Maganaris, C.N., Baltzopoulos, V. and Sargeant, A.J. (2000) 'In vivo measurement-based estimations of the human Achilles tendon moment arm', *European Journal of Applied Physiology*, 83: 363–9.

Maganaris, C.N., Baltzopoulos, V., Ball, D. and Sargeant, A.J. (2001) 'In vivo specific tension of human skeletal muscle', *Journal of Applied Physiology*, 90(3): 865–72.

Malina, R.M. (1986) 'Growth of muscle tissue and muscle mass', in F. Falkner and J.M. Tanner (eds) *Human Growth, a Comprehensive Treatise: Post-Natal Growth, Neurobiology*, New York: Plenum Press, pp. 77–99.

Micheli, L.J. and Klein, J.D (1991) 'Sports injuries in children and adolescents', *British Journal of Sports Medicine*, 25: 6–9.

Morse, C.I., Tolfrey, K., Thom, J.M., Vassilopoulos, V., Maganaris, C.N. and Narici, M.V. (2008) 'Gastrocnemius muscle specific force in boys and men', *Journal of Applied Physiology*, 104: 469–74.

Murray, W.M., Buchanan, T.S. and Delp, S.L. (2000) 'The isometric functional capacity of muscles that cross the elbow', *Journal of Biomechanics*, 33: 943–52.

Murray, W.M., Buchanan, T.S. and Delp, S.L. (2002) 'Scaling of peak moment arms of elbow muscles with upper extremity bone dimensions', *Journal of Biomechanics*, 35: 19–26.

Murray, W.M., Delp, S.L. and Buchanan, T.S. (1995) 'Variation of muscle moment arms with elbow and forearm position', *Journal of Biomechanics*, 28: 513–25.

Neu, C.M., Rauch, F., Rittweger, J., Manz, F. and Schoenau, E. (2002) 'Influence of puberty on muscle development at the forearm', *American Journal of Physiology Endocrinology and Metabolism*, 283: 103–7.

Nevill, A.M., Holder, R.L., Baxter-Jones. A., Round, J.M. and Jones, D.A (1998) 'Modeling developmental changes in strength and aerobic power in children', *Journal of Applied Physiology*, 84: 963–70.

O'Brien, T.D., Reeves, N.D., Baltzopoulos, V., Jones, D.A. and Maganaris, C.N. (2009a) 'Moment arms of the knee extensor mechanism in children and adults', *Journal of Anatomy*, 215: 198–205.

O'Brien, T.D., Reeves, N.D., Baltzopoulos, V., Jones, D.A. and Maganaris, C.N. (2009b) 'In vivo measurements of muscle specific tension in adults and children', *Experimental Physiology*, 95: 202–10.

O'Brien, T.D., Reeves, N.D., Baltzopoulos, V., Jones, D.A. and Maganaris, C.N. (2010) 'Mechanical properties of the patellar tendon in adults and children', *Journal of Biomechanics*, 43: 1190–5.

Pandy, M.G. (1999) 'Moment arm of a muscle force', in J.O. Holloszy (ed.) *Exercise and Sport Sciences Reviews, 27*, Philadelphia, PA: Lippincott Williams & Wilkins.

Parker, D.F (1989) 'Factors controlling the development of strength of human skeletal muscle', unpublished thesis, University of London.

Parker, D.F., Round, J.M., Sacco, P. and Jones, D.A. (1990) 'A cross-sectional survey of upper and lower limb strength in boys and girls during childhood and adolescence', *Annals of Human Biology*, 17: 199–211.

Pierrynowski, M.R. (1995) 'Analytic representation of muscle line of action and geometry', in P. Allard, I.A.F. Stokes and J.-P. Blanchi (eds) *Three-Dimensional Analysis of Human Movement*, Champaign, IL: Human Kinetics, pp. 215–56.

Ramos, E., Frontera, W.R., Llopart, A. and Feliciano, D (1998) 'Muscle strength and hormonal levels in adolescents: gender related differences', *International Journal of Sports Medicine*, 19: 526–31.

Ravary, B., Pourcelot, P., Bortolussi, C., Konieczka, S. and Crevier-Denoix, N. (2004) 'Strain and force transducers used in human and veterinary tendon and ligament biomechanical studies', *Clinical Biomechanics*, 19: 433–47.

Round, J.M., Jones, D.A., Honour, J.W. and Nevill, A.M. (1999) 'Hormonal factors in the development of differences in strength between boys and girls during adolescence: a longitudinal study', *Annals of Human Biology*, 26: 49–62.

Spoor, C.W. and Van Leeuwen, J.L. (1992) 'Knee muscle moment arms from MRI and from tendon travel,' *Journal of Biomechanics*, 25: 201–6.

Spoor C.W., Van Leeuwen, J.L., Meskers, C.G.M., Titulaer, A.F. and Huson, A. (1990) 'Estimation of instantaneous moment arms of lower-leg muscles', *Journal of Biomechanics*, 23: 1247–59.

Sunnegardh, J., Bratteby, L.-E., Nordesjö, L.-O. and Nordgren, B. (1998) 'Isometric and isokinetic muscle strength, anthropometry and physical activity in 8 and 13 year old

Swedish children', *European Journal of Applied Physiology and Occupational Physiology*, 58: 291–7.

Thomis, M.A., Van Leemputte, M., Maes, H.H., Blimkie, C.J.R., Claessens, A.L., Marchal, G., Willems, E., Vlietinck, R.F. and Beunen, G.P. (1997) 'Multivariate genetic analysis of maximal isometric muscle force at different elbow angles', *Journal of Applied Physiology*, 82: 959–67.

Tsaopoulos, D.E., Maganaris, C.N. and Baltzopoulos, V. (2007) 'Can the patellar tendon moment arm be predicted from anthropometric measurements?', *Journal of Biomechanics*, 40: 645–51.

Tsirakos, D., Baltzopoulos, V. and Bartlett, R. (1997) 'Inverse optimization: functional and physiological considerations related to the force-sharing problem', *Critical Reviews in Biomedical Engineering*, 25: 371–407.

Waugh, C.M., Blazevich, A.J., Fath, F. and Korff, T. (2011) 'Can Achilles tendon moment arm be predicted from a single anthropometric measure in prepubescent children?', *Journal of Biomechanics*, 44: 1839–44.

Wood, J.E., Meek, S.G. and Jacobsen, S.C. (1989) 'Quantitation of human shoulder anatomy for prosthetic arm control-1. Surface modelling', *Journal of Biomechanics*, 22: 273–92.

Wood, L.E., Dixon, S., Grant, C. and Armstrong, N. (2004) 'Elbow flexion and extension strength relative to body or muscle size in children', *Medicine and Science in Sports and Exercise*, 36: 1977–84.

Wood, L.E., Dixon, S., Grant, C. and Armstrong, N. (2006) 'Elbow flexor strength, muscle size and moment arms in pre-pubertal boys and girls', *Paediatric Exercise Science*, 18: 457–69.

Wretenberg, P., Németh, G., Lamontagne, M. and Lundin, B. (1996) 'Passive knee muscle moment arms measured in vivo with MRI', *Clinical Biomechanics*, 11: 439–46.

Xu, L., Nicholson, P., Wang, Q., Alén, M. and Cheng, S. (2009) 'Bone and muscle development during puberty in girls: a seven-year longitudinal study', *Journal of Bone and Mineral Research*, 24: 1693–8.

5   Development of musculoskeletal stiffness

Anthony Blazevich, Charlie Waugh and Thomas Korff

Introduction

We move when our muscles generate forces that are transferred through tendons to the skeleton. Far from being a simple system, in which the muscles function like motors and the tendons as cables, both muscle and tendon can stretch and recoil as forces vary, so each movement requires a unique muscle activation sequence to account for the 'spring-like' properties of our musculoskeletal system. Understanding how differences in this spring-like property influence movement in children is vital so that we can determine which movement patterns are optimal for children and how these patterns should vary with normal growth and development. In this chapter we first discuss concepts relating to the mechanical properties of elastic tissues. Within this context, we describe the different relevant biomechanical measurement techniques. In the second part of the chapter we synthesize the literature relating to age-related changes in the elastic properties of the musculoskeletal system during childhood, and we derive practical implications from this body of knowledge.

Basic theoretical concepts

The physics of elastic tissues

Tissues such as our muscles and tendons deform when they are loaded. The amount of deformation depends on the tissue's stiffness and the load (force) imposed on it. This relationship is described by Hooke's Law:

$$F = kx \qquad (5.1)$$

where F is the applied force, k is the tissue stiffness and x is the deformation (i.e. stretch) of the tissue (actually, this equation is more correctly written $F = -kx$ to show that the recoil provides a negative, restoring, force). Essentially, this equation shows that tissues subjected to greater loading will stretch more. We can also rearrange the equation to show that the stiffness (k) of a tissue depends on how much it deforms (x) under a load (F):

$$k = \frac{F}{x} \tag{5.2}$$

Equation 5.2 illustrates that stiffer tissues require more force to be deformed. Tissues that are stretched store (elastic) potential energy, and much of this energy is regained during recoil to assist with movement production. The energy stored in an elastic tissue (E) is dependent upon the stiffness of the tissue and its deformation, such that:

$$E = \tfrac{1}{2}kx^2 \tag{5.3}$$

where k is the tissue stiffness and x is its deformation (stretch/elongation). From this equation we can see that stretching a stiffer tissue results in a greater energy storage ($\tfrac{1}{2}k$) but that stretching a tissue further has a more substantial impact (x^2). For this reason, tissues that are less stiff (i.e. more compliant: $1/k$) can store more energy for a given amount of force (see Figure 5.1). Therefore, differences in the stiffness of muscles and tendons between muscle groups or individuals impact significantly on the amount of energy stored and released during movement.

Nonetheless, stiffer tissues recoil faster when they work against a load. This is because a force is required to accelerate a mass, according to Newton's second law

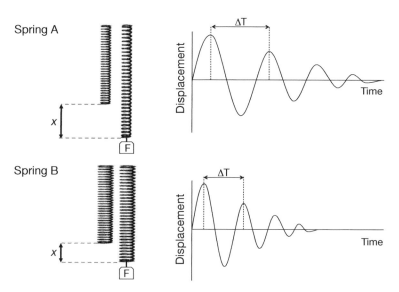

Figure 5.1 Speed of oscillation in a more compliant spring compared with a stiffer spring. When a mass (which applies a force, F) is hung from an elastic material (e.g. a spring), the system will oscillate if perturbed. The speed, or frequency, of this oscillation is slower for a more compliant spring (A) compared to a stiffer spring (B). Therefore, the period of oscillation (ΔT) is shorter for the stiffer spring. Notice that the number of oscillation cycles completed before the energy dissipates and the system stops oscillating is the same.

($F = ma$, where m is the mass and a is the acceleration). Further, the force required to elongate a tissue over a given distance increases with tissue stiffness, according to Hooke's Law (Equation 5.1). Given this interaction between force, stiffness and tissue elongation, if we were to hang a mass from a tissue and apply a brief force to start the system oscillating, the frequency of this oscillation (f) will be equal to:

$$f = \frac{1}{2}\pi \left(\sqrt{\frac{k}{m}} \right) \tag{5.4}$$

where k is the tissue stiffness and m is the mass (load) attached to the tissue. Notice that k is on top of the $\sqrt{\frac{k}{m}}$ part of the equation, so increasing k will increase f. The result of this relationship is that a system comprising stiffer tissues has a faster natural frequency, and therefore our ability to move rapidly is theoretically better if our limbs possess stiffer muscles and tendons. This is shown diagrammatically in Figure 5.1.

The tendency for a tissue to spring back into shape is called restitution. Essentially, the greater a tissue's restitution, the more 'elastic' it is. However, elasticity varies substantially between tissues as some of the energy is lost as heat. Such energy loss is called hysteresis, so a tissue that exhibits a large hysteresis loses a lot of energy and is 'less elastic'. For biological tissues such as tendons, hysteresis values have been shown to range from about 5 to 25 per cent (Kubo *et al.* 2001d; Lichtwark and Wilson 2005; Maganaris and Paul 2000), whereas for active muscles, hysteresis values can be significantly larger (~40 per cent) (Best *et al.* 1994). The smaller the energy loss from the tissues the more efficient the system. Thus, tendons tend to be more efficient than muscles at storing and recovering elastic potential energy. Within the context of human movement, it is important that elastic tissues such as muscles and tendons have low hysteresis so that our movements are performed with a high efficiency.

In fact, the natural frequency of a system is also an important factor when we consider the energy lost from an elastic system. This is because tissues such as muscles and tendons store and recover more energy (i.e. they will do more work) if external or muscular forces are applied at the tissue's natural frequency. This phenomenon is true for any oscillating system. For example, it requires less energy to push a child on a swing if the force is applied to the swing briefly at the top of its trajectory. If we applied forces either more or less frequently the swing would not move as far (i.e. less work output) for a given force application (i.e. energy input). So, if we want to use the stretch and recoil of elastic tissues to improve movement efficiency (work output per energy input), and the optimum rate at which forces should be applied changes with tissue stiffness, then our optimum movement frequency will vary with tissue stiffness.

In summary, our ability to store energy in elastic tissues might be greater when the tissues are more compliant because they elongate further under a given external or muscular force (load). However, the speed at which we can move using the recoil energy and with which we can achieve the greatest movement economy is greater if the tissues are stiffer. So the optimum tissue stiffness depends on the task to be performed.

Specific stiffness of muscles, tendons and whole limbs

Muscle stiffness

In the muscles, considerable elasticity resides in the aponeuroses (continuations of the tendon that form sheets onto which muscle fibres insert), actin and myosin filaments, the cross-bridges and the titin protein (which binds myosin to the Z-line of the sarcomere). These structures form part of the muscle's series elastic component (SEC), and they stretch when an active muscle is placed under load regardless of the length of the muscle relative to its resting length. The connective tissues that surround the muscle fibres, fascicles and whole muscle are also elastic. These structures form the parallel elastic component (PEC), and they stretch only when the muscle is lengthened considerably (past its slack length); the titin protein and the aponeuroses also contribute to the PEC. They only store elastic energy when the muscle is at a long length. In most human movements, active muscles tend to maintain lengths that are too short for energy storage in the PEC to be physiologically significant. Both SEC and PEC contribute to a muscle's stiffness, even when the muscle is not activated, but of course substantial increases in muscle stiffness also occur when the muscle is activated because the muscle resists lengthening better when more cross-bridges are in a strongly bound state (i.e. there is myosin–actin interaction). Thus, it is important to delineate between muscle stiffness being assessed in passive versus active conditions.

Tendon stiffness

The tendons themselves form a significant part of the SEC but are external to the muscle, so it is common to discuss muscle and tendon stiffness separately. This is important because tendon stretch and recoil occurs at the expense of muscle stretch-recoil in many human movements. Thus, changes in tendon length are not the same as those of muscle (Finni *et al.* 2003b; Ishikawa and Komi 2004; Lichtwark and Wilson 2006). Because of the greater energy storage capacity, higher recoil speed and lower hysteresis of tendons compared to muscles, it is important to understand the specific impact of tendon stiffness on movement performance, and therefore its measurement (described later) is now reasonably common.

Joint and limb stiffness

In addition to muscle and tendon stiffness, there are other measures of stiffness that are important. The stiffness of a whole limb (e.g. the leg as it makes contact with the ground during walking or running) is affected by both muscle and tendon stiffness as well as the elastic properties of joint structures and stiffness arising from muscular contraction (i.e. force generation). Joint stiffness is determined by the relationship between angular displacement and angular force (or moment), similar to the linear relationship described in Equation 5.1. Additionally, joint stiffness is influenced by the stiffness of muscles, tendons, ligaments and other connective tissue structures that surround a joint. Activation of agonist muscles plays a substantial role in setting limb stiffness (Agarwal and Gottlieb 1977) for two main reasons. First, increases in muscle activation increase muscle-specific stiffness (its resistance to stretch) because there are more cross-bridges strongly bound, thus generating greater resistive force. Second, tendons increase in stiffness as they are stretched because the tendon force-elongation relationship is non-linear, so increases in muscle force increase tendon stiffness simply because the tendons are stretched.

Of course, increases in agonist muscle activation without corresponding increases in antagonist activation could cause a considerable change in joint configuration. This is often unnecessary, or even undesirable. Therefore, agonist–antagonist coactivation often occurs as limb stiffness is increased. This coactivation has been shown to be important for limb stiffening prior to jump landings from a height (Arampatzis *et al.* 2001; Horita *et al.* 2002). Importantly, the increase in both muscle and tendon stiffness resulting from increases in muscle activation are substantial enough that the passive joint stiffness attributable to ligaments and other joint structures is often of little significance. Thus, understanding the impact of muscle, tendon and whole limb stiffness on movement performance has become an important scientific objective. For further details relating to joint stiffness, we refer the reader to Chapter 11.

Effect of tissue stiffness on movement performance

Many functional movements require the muscle-tendon unit (MTU) first to lengthen and then shorten without delay (e.g. walking, hopping, running and jumping). This pattern is referred to as a stretch-shorten cycle (SSC). While muscular power production is an important determinant of movement performance, the elastic tendon can modify performance because it is stretched and then subsequently recoils to assist with power production during the concentric part of the movement. In other words, the tendon acts as a power amplifier, so differences in tendon stiffness likely impact on higher-speed, concentric movement performance. The stiffness of both the muscle and tendon are important in SSC movements. For the tendon, a lesser stiffness is associated with a greater elongation, and thus more energy storage, for a given muscle force (Equation 5.3). This relationship can be important in many movements, including locomotion, as the greater the energy storage and recovery the less work that must be done by the muscles. In walking, for example, the plantar flexors are stretched during the early stance phase and much of this elongation occurs in the Achilles tendon (Lichtwark and Wilson 2006). The stored elastic energy is then recovered to provide propulsion during the later stance phase, allowing the muscles to perform less work and thus use less energy (Fukunaga *et al.* 2001; Lichtwark and Wilson 2006). Therefore, having a relatively compliant tendon (as long as it is stiff enough to transfer forces efficiently (Magnusson *et al.* 2003)) can be useful for improving efficiency during slower SSC movements.

Conversely, a relatively stiff tendon might be useful during faster SSC movements (Bojsen-Moller *et al.* 2005), as the natural frequency of a stiffer tissue is greater (Equation 5.4). For example, the ability to perform activities such as sprint running might be reduced in individuals with more compliant tendons. In fact, it has been shown that movement efficiency during moderate-speed running tends to be better in individuals who have stiffer Achilles tendons (Arampatzis *et al.* 2006). Similarly, maximal jumping performance appears to be better in individuals with greater vastus lateralis knee tendon stiffness (Bojsen-Moller *et al.* 2005). These findings demonstrate that tendon stiffness is an important determinant of SSC movement performance.

Muscle-specific stiffness is also an important consideration in SSC movements because many SSC movements are performed with the muscles doing work over relatively short ranges of motion while the tendons stretch and shorten over considerably larger ranges of motion. This distribution of muscle and tendon lengthening has been demonstrated experimentally in movements such as drop jumps, where there is relatively little elongation of the agonist muscles during ground contact but a substantial lengthening of the tendons (Finni *et al.* 2003a; Ishikawa and Komi 2004). In fact, this distribution of muscle and tendon lengthening has been shown to diverge even more as the loading intensity (i.e. the height of drop jump) is increased (Ishikawa and Komi 2004; Ishikawa *et al.* 2006). In these cases it is beneficial to have a stiffer muscle, because we would prefer more energy to be stored in the energy efficient (i.e. lower hysteresis) tendon. Having a stiffer muscle aids this because more energy is always stored in

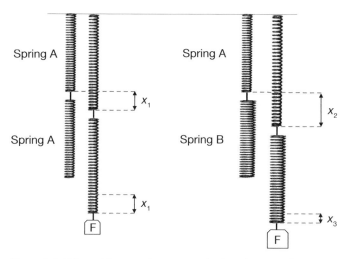

Figure 5.2 Effect of force application on the lengthening of two springs arranged in series. When two springs are placed in series (e.g. a muscle and a tendon), the more compliant spring will elongate more when a force is applied. Therefore, elongation of each spring (equal to x_1) is the same in the example on the left, but the elongation of the more compliant spring (x_2) is greater than the elongation of the stiffer spring (x_3) in the example on the right. Note: the greater force application in the example on the right is for illustrative purposes only.

the more compliant of two springs that are placed in series (Figure 5.2). Therefore, any discrepancies between individuals with respect to muscle stiffness might also impact on their abilities to perform SSC movements.

In addition to movement efficiency, the stiffness of muscles and tendons also influences the delay between the neural excitation of a muscle and the onset of the rise in force (electromechanical delay (EMD)) as well as the rate at which muscular forces are applied to the skeleton (rate of force development (RFD)). As muscle force rises, some energy is stored in the series elastic elements, which include the tendon as well as the muscle's SEC. A greater compliance of the SEC dramatically reduces the rate at which force is transferred (Wilkie 1950), increases EMD and reduces RFD. In fact, a moderate relationship has been found between tendon stiffness and both EMD (negative relation (Muraoka *et al.* 2004)) and RFD (positive relation (Bojsen-Moller *et al.* 2005)) in adults. Changes in the rate of force transfer ultimately influence motor performance, so any changes in the MTU's stiffness, and in particular tendon stiffness, can impact on movement coordination and performance.

Another mechanism by which movement coordination and balance can be influenced by muscle-tendon stiffness is through its role in altering afferent feedback. This is because feedback from muscle- and tendon-based receptors is affected by muscle-tendon stiffness. This feedback includes afferent information from muscle spindles about muscle lengths and their rate of change, which are important for the detection of limb position changes (Gandevia 1996; Goodwin *et al.* 1972; Proske *et al.* 2000). An example can be used to illustrate this. Take a single-joint system where a limb is held outstretched using only a muscle connected to a tendon that crosses the joint. To hold the limb in position, the muscle is required to be active and is therefore relatively stiff (remember, muscle stiffness increases as muscle activation increases). A small perturbation of the distal end of the limb would cause the limb to oscillate slightly, because the MTU is elastic. However, much of the muscle-tendon length change occurs in the tendon, assuming it is more compliant than the muscle. The muscle spindles, which are in parallel with the muscle fibres, might therefore not be stretched significantly, and they would thus provide relatively little feedback on the change in muscle length. If the tendon were stiffer, the muscle length change, and therefore the muscle spindle length change, would be greater. In this case the feedback from the muscle spindles would be more substantive. In this somewhat idealized example it can be seen that the feedback from important sensory organs, such as the muscle spindles, is influenced by both muscle and tendon stiffness. An understanding of this mechanism is important as it could affect movement accuracy during growth and development. For example, extremely low levels of tendon stiffness (or other series elastic structures) could result in insufficient afferent feedback from muscle spindles, which could result in difficulties in performing accurate movements and slow the rates of response to perturbations, such as those that occur when we fall. Given the above, we might speculate that any differences between individuals in tendon (or SEC) stiffness could influence their movement coordination and balance.

Measuring muscle-tendon stiffness in humans

There are several techniques commonly used to measure the stiffness of a specific tissue (e.g. tendon), joint or limb. A brief overview of the most common techniques is provided here.

Muscle-tendon stiffness estimation using the 'quick release' method

The quick release method characterizes the stiffness of the MTU by allowing the sudden and fast release of a contracted muscle (or, more specifically, the MTU) from an isometric position. The recoil (i.e. movement) speed in the very early period after release (before spinal reflex activity) can influence the movement and is dependent upon muscle-tendon stiffness, according to Equation 5.4. Therefore, measurement of the recoil speed allows the estimation of MTU stiffness. The fast release is achieved by releasing the movable parts of an ergometer in which the limb is placed at a greater velocity than the maximal shortening velocity of the muscle in question (Lambertz *et al.* 2001; Pérot *et al.* 1999) in order to produce a joint rotation. Muscle-tendon stiffness is then calculated from the limb's moment of inertia (I), its angular acceleration (α) and its angular displacement ($\Delta\theta$) during the period in which the SEC is assumed to recoil (Goubel and Pertuzon 1973), according to Equation 5.5.

$$stiffness = I \cdot \frac{\Delta\alpha}{\Delta\theta} \qquad (5.5)$$

It has been shown that the use of the quick release in its original form can be problematic, as it does not take the effect of antagonistic coactivation into account. Researchers have therefore extended the quick release method to calculate a stiffness index, which takes antagonistic coactivation into consideration (Cornu and Goubel 2001; Lambertz *et al.* 2003). Such a stiffness index has been proven to be useful to determine MTU stiffness in children, as it takes into account the relatively large amounts of coactivation found in children, which could influence the stiffness calculation by reducing limb acceleration upon release (Cornu and Goubel 2001). Developmental differences in MTU stiffness obtained from the quick release method are discussed later in this chapter.

Muscle-tendon stiffness estimation using sinusoidal perturbation

The elastic properties of a muscle-tendon-joint system can be characterized by observing the oscillatory (i.e. bounce) response of a limb system to perturbation of a specific frequency or when weighed by a specific load. Perturbing the system at a specific frequency (sinusoidal perturbation technique) allows stiffness to be calculated based on the concept of mechanical impedance (i.e. how much a body resists motion when a force is applied). This impedance varies with the frequency at which a force is applied to the system, and so the natural limb frequency, or resonant frequency, of the system is the frequency associated with the lowest impedance. As the stiffness of a mechanical system is related to its natural frequency (Equation 5.4), we can estimate it from the system's mechanical behaviour at its resonant frequency. More specifically, muscle-tendon stiffness is obtained from the relationship between the frequency of limb oscillation and the phase lag between joint position and moment generated (Cornu and Goubel 2001), which is least when the limb is perturbed at its natural frequency.

An alternative is to perturb a limb system and let it oscillate freely at its resonant frequency under a specific load (free oscillation technique). In this case, stiffness is calculated directly from the oscillation frequency, which is greater if the system is stiffer (Equation 5.4). The faster the measured natural frequency of the system, the stiffer the system must be.

Ultrasonographic measurement of tendon-specific stiffness

According to Equation 5.2, tendon stiffness can be calculated as the ratio of change in tendon force to tendon elongation. Using ultrasonography, tendon elongation can be estimated by clearly visualizing and monitoring displacement of the muscle-tendon junction (MTJ) of interest during a voluntary, typically maximal isometric, muscle action or during passive stretch of a MTU. Tendon force is calculated as the ratio of the active muscle moment recorded during the contraction or passive stretch (typically obtained via isokinetic dynamometry) to the moment arm of the muscle or tendon of interest (Baltzopoulos 1995; Maganaris 2000; Visser *et al.* 1990). The stiffness of the tendon is then calculated by determining the gradient of the relationship between tendon elongation (MTJ displacement) and tendon force (Fukashiro *et al.* 1995; Magnusson *et al.* 2003). This process is shown in Figure 5.3.

Muscle-tendon stiffness in children and adults

It is clear that muscle-tendon stiffness is an important determinant of movement performance and efficiency. In a developmental context, the question thus arises as to whether muscle-tendon stiffness changes as a function of age in children and whether such changes are associated with age-related differences in movement performance and efficiency.

Muscle-tendon stiffness during childhood

There is some debate as to whether muscle-tendon stiffness increases as a function of age, or whether it varies with some other factor such as body mass or muscle force production potential. For example, Lambertz *et al.* (2003) used the quick release method to demonstrate that plantar flexor muscle-tendon stiffness increases with age in typically developing 7–10-year-old children. This finding is in agreement with results from Cornu and Goubel (2001), who used the quick release method to show that muscle-tendon stiffness of the elbow flexors increased with age in children aged between 9 and 15 years. Both studies also

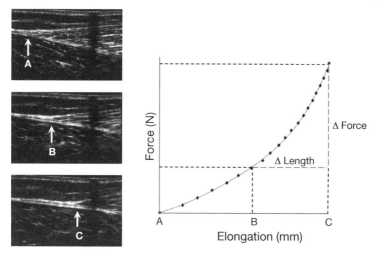

Figure 5.3 Ultrasound image of the MTJ tracking and graph of tendon elongation vs. tendon force. During an isometric muscular contraction, ultrasound imaging can be used to track the elongation of the tendon. In this example, elongation of the Achilles tendon from A to C can be seen when an ultrasound probe is placed over the muscle-tendon junction of the gastrocnemius medialis (in the calf muscle). By calculating the muscle force (active joint moment divided by muscle-tendon moment arm, measured in newtons) and plotting it against tendon elongation (millimetres), the stiffness of the tendon can be calculated as the gradient of the force-elongation relationship. Thus, the tendon stiffness is equal to ΔForce per ΔElongation. Note: the letters A, B and C on the abscissa correspond with points A, B and C on the ultrasound images.

found muscle-tendon stiffness in adults to be significantly greater than in children. However, as muscle-tendon stiffness is influenced by the levels of muscle force, the authors concluded that the observed age-related changes in muscle-tendon stiffness should be partially attributed to age-related differences in muscular force production capabilities. In their study, Cornu and Goubel (2001) also report muscle-tendon stiffness at the elbow (using the sinusoidal oscillation technique) to be substantially lower in 9–15-year-old children than adults. As muscle-tendon stiffness obtained from the sinusoidal oscillation technique is not influenced by active muscle forces when the resting limb is forcibly oscillated, the authors concluded that both active and passive stiffness components increase as children grow older. This speculation was substantiated by Grosset *et al.* (2005), who quantified muscle-tendon stiffness of the soleus-ankle joint complex (using sinusoidal perturbations) in prepubescent children at various intensities of muscle contraction, which also allowed an estimation of stiffness at zero muscular contraction intensity by extrapolation of the data (Grosset *et al.* 2007). Stiffness was found to increase significantly with age under both passive and active conditions. Furthermore, stiffness was significantly lower in children than in adults under all conditions. Lebiedowska and Fisk (1999) assessed the stiffness of the knee joint in 87 boys and girls ranging in age from 5 to 18 years using the free oscillation technique. They found that stiffness increased to the fifth power of body stature. The stiffness of the eldest children was comparable to previously published values for adults (Mansour and Audu 1986). The authors concluded that stiffness increases as a function of growth and maturation rather than age *per se*.

In summary, age-related changes in muscle-tendon stiffness are likely to be influenced by changes in body mass and muscle strength as well as level of muscle contraction, but the relative contributions of these factors is currently unknown. Our recent experiments relating to tendon stiffness (see section below) substantiate this speculation. However, more research is needed to fully understand the relative contributions of mass, strength and other mechanical factors on muscle-tendon stiffness. Regardless, experimental evidence suggests that both active and passive stiffness increase as children grow older. In addition, the development of stiffness appears to be joint specific. Further research is required to determine the extent to which these various parameters/factors contribute to the age- and sex-related increase in muscle-tendon stiffness.

Tendon-specific stiffness during childhood

The literature is consistent in demonstrating that tendon stiffness increases with age during childhood. Kubo *et al.* (2001c) found that vastus lateralis tendon stiffness obtained from ultrasound measures was greater in 15-year-old compared to 11-year-old boys. Similarly, O'Brien *et al.* (2010) found that patellar tendon stiffness was greater in adults than in 8–10-year-old children. In our laboratory, we have recently found that Achilles tendon stiffness increases significantly in children between 5 and 12 years of age. While the literature is consistent in demonstrating the age-related increases in tendon stiffness in young children, the question about the mechanisms associated with such changes arises. We have recently found that in 5–12-year-old children, body mass is more strongly correlated with Achilles tendon stiffness than age (Waugh *et al.* 2011a). Thus, a substantial part of the age-related increase in stiffness might result from an increase in body mass (i.e. mechanical adaptations in response to the requirement to tolerate higher loads). However, our findings further suggest that tendon stiffness increases in 5–12-year-old children even if changes in body size are accounted for. This finding is indicative of a growth-related change in the tendon's intrinsic mechanical properties. In summary, the literature is consistent in demonstrating that tendon stiffness increases with age during childhood. However, similar to the findings relating to age-related changes in muscle-tendon stiffness, the exact mechanisms underlying this increase are not presently clear. Determining these more clearly and linking changes in tendon stiffness to changes in motor development should be a goal of future research. As there are limited data in older children, especially during adolescence, future research should also focus on examining sex differences in tendon stiffness during the pubertal years.

Effects of age on tendon hysteresis during childhood

Minimizing hysteresis (i.e. energy dissipation) in tendons could be an important mechanism by which children might improve their movement efficiency. A few studies have assessed tendon hysteresis in adults using ultrasound to measure tendon elongation and recoil during loading–unloading, either during isometric force generation and relaxation (Kubo *et al.* 2001b; Maganaris and Paul 2000) or during more complex movements (Lichtwark and Wilson 2005). It has also been shown that tendon hysteresis might be changeable with training practices such as stretching (Kubo *et al.* 2002a, 2002b) and plyometrics (Foure *et al.* 2010). However, to our knowledge, no studies have examined hysteresis in children. Thus, it is not clear whether the efficiency of elastic potential energy recovery from tendons in children is similar to that of adults and if it can adapt to physical training. Knowing that children often perform motor tasks less efficiently than adults (Frost *et al.* 1997; Schepens *et al.* 2001), understanding the impact of changes in tendon hysteresis in the context of movement efficiency is important for furthering our understanding of the mechanisms underlying age- and sex-related differences in movement efficiency and motor performance.

Effects of exercise training on muscle-tendon, tendon-specific and muscle-specific stiffness in children

In adults, it has been shown that certain forms of strength or flexibility training can lead to increases or decreases in muscle-tendon stiffness, respectively (Lindstedt *et al.* 2001; Wilson *et al.* 1992). These findings have important implications for clinicians or exercise specialists who seek to formulate interventions that modify the muscle-tendon properties in their patients or clients. In a developmental context, knowledge of training adaptations in children's muscle-tendon properties are vital in order to be able to optimize movement efficiency, accelerate motor skill acquisition (in both typically developing children or those with developmental delay), or improve the movement capacity of children with disabilities that impact on muscle-tendon stiffness such as cerebral palsy or spasticity.

In adults, it is well established that stiffness increases in response to high intensity loading, such as heavy resistance training (Arampatzis *et al.* 2007; Kubo *et al.* 2001a, 2001b, 2002b), and decreases in response to unloading, such as bed rest (Kubo *et al.* 2000). To our knowledge, no published research has documented the effect of physical training on muscle-tendon properties in children. In our laboratory, we recently conducted a study to investigate the effects of a ten-week resistance training study on Achilles tendon stiffness in prepubertal children (age = 8.9 ± 0.2 years). We found that tendon stiffness increased significantly (~35 per cent) in the experimental group while there was no change in a non-training control group (Waugh *et al.* 2011b). These findings are important as they demonstrate that tendon properties can be altered by physical training in prepubertal children, which has potential implications for motor development in both sporting and clinical contexts. However, to understand the interaction between training, muscle-tendon mechanics and motor development more fully, it is also important to understand developmental changes in tendon hysteresis and muscle-specific stiffness. The interactions between growth, exercise training and muscle stiffness are likely to have a profound effect on performance in SSC activities such as walking, running and jumping. Understanding these interactions might also have direct implications for injury prevention in children (see Chapter 11). In summary, there is a clear need for research to examine the effects of physical training on muscle, tendon and muscle-tendon stiffness and hysteresis in response to exercise interventions in children, and to determine the influence of such changes on movement performance.

Relationship between stiffness and motor development in children and adults

Using the limb's spring-like properties

One way to determine whether a person is behaving like a simple spring-mass system is to correlate the ground reaction force and the centre of mass displacement during the ground contact phase of the task of interest. The more linear this relationship the more the system is behaving according to the spring-mass model (Farley *et al.* 1991). Farley *et al.* (1991) used this model to show that adults behaved like a simple spring-mass system when performing two-legged hopping at their preferred hopping frequency but did not behave like a simple spring-mass system at slower hopping frequencies, as the correlations between ground reaction forces and centre of mass displacements were significantly lower.

To determine whether children behave like a simple spring-mass system when performing hopping tasks, Korff *et al.* (2009) performed a mechanical analysis of hopping in 11–18-year-old children. The authors found very strong correlations between the vertical ground reaction force and the vertical centre of mass displacement during the ground contact phase of hopping (average correlation coefficient, $r = 0.97$), indicating that, during hopping, children as young as 11 years of age behave like a simple spring-mass system. More recently, we have observed that this correlation is also strong in 9-year-old children but lower than in their older peers ($r = 0.91$) (Bennett *et al.* 2011). Together, these results suggest that children behave more like a simple spring-mass system as they develop during childhood and adolescence, which implies that they acquire the ability to take advantage of the elastic properties of their mechanical systems.

Modulating limb stiffness to optimize movement performance

As discussed earlier in this chapter, limb stiffness is dictated by both active (muscular contraction and co-contraction) and passive (elasticity of tissues) components. It is therefore not a simple reflection of, nor necessarily correlated with, tendon (Rabita *et al.* 2008) or joint stiffness (Kubo *et al.* 2007). For example, Rabita *et al.* (2008) showed that leg stiffness during two-legged hopping in healthy adults did not correlate with whole muscle-tendon stiffness when examined using either the oscillation or quick release method. The authors thereby demonstrated that healthy adults must actively modify their leg stiffness to optimize the use of elastic energy during hopping.

From a developmental perspective, the question arises as to whether children modulate their leg stiffness in a similar fashion to adults. Korff *et al.* (2009) found that hopping stiffness in children between the ages of 11 and 18 years did not differ once stiffness was normalized by body weight. We have also recently observed normalized leg stiffness to be greater in 9-year-old children than in adults (Bennett *et al.* 2011). These data are supported by Schepens *et al.* (2004) who found that mass-specific leg stiffness during running decreased during development from 2 to 12 years; thereafter, it did not change substantially. These results allow us to speculate that developing children learn to adjust, or maintain, their leg stiffness during hopping or running in a way to optimize the storage and release of elastic energy, with the ultimate goal of moving more efficiently. This speculation is also consistent with findings from Schepens *et al.* (2001) who demonstrated that children run more efficiently as they grow older. For further insights into the role of stiffness within the context of the development of locomotion, we refer the reader to Chapter 8.

Results from studies on walking by Holt *et al.* (1991, 1995) give us additional insights into the relationship between leg stiffness and movement efficiency in children. These authors used a force-driven inverted pendulum (a slightly more sophisticated version of the simple spring-mass model as described above for hopping and running) to model the energetics of gait and showed that adults' preferred stride frequencies coincide with their most efficient stride frequency (Holt *et al.* 1995). These results imply that adults take advantage of their body's elastic properties to maximize efficiency during gait. Holt *et al.* (1991) and Jeng *et al.* (1997) subsequently demonstrated that the force-driven inverted pendulum could accurately predict the preferred step frequency during walking in 3–12-year-old children, and Jeng *et al.* (1996) demonstrated that typically developing children between 7 and 12 years of age preferred to use a stride frequency that optimized walking efficiency. Together, these results suggest that children as young as 7 years of age use their body's elastic properties appropriately to maximize walking efficiency.

While the above results clearly show that children take advantage of the storage and release of elastic energy in the later stages of locomotor skill acquisition, it is important to understand that this ability develops with age and practice. Holt *et al.* (1995), for example, showed that very early walkers do not behave like a simple inverted pendulum, indicating that new walkers do not fully exploit their elastic resources. One explanation for this is that new walkers have an increased demand for stability, which results in a suboptimal walking pattern with regard to both mechanical energy exchange and metabolic efficiency (Hallemans *et al.* 2005).

Holt *et al.* (2006) further explored whether infants could take advantage of spring dynamics immediately after learning to walk. The authors tested seven toddlers on seven occasions over the six-month period after first learning to walk and analysed the magnitude, timing and direction of the centre of mass acceleration during stance and swing phases to make indirect inferences about the efficiency of elastic energy use. They found that toddlers demonstrated appropriately timed and directed centre of mass accelerations only one month after the onset of walking. This strongly suggests that toddlers at this early age already take some advantage of pendular spring mechanics, and can utilize stored elastic energy.

In conclusion, children possess the ability to take advantage of the body's passive elastic properties to optimize efficiency in the early stages of locomotor skill acquisition. This ability increases as children grow older, although the developmental schedule depends on the task to be performed (e.g. walking vs. running). The ability to take advantage of the storage and release of elastic energy during walking in very young children increases as a function of age and experience due to a decreased demand for stability. Whether this improvement can be augmented with specific task practice or physical training is yet to be determined.

Influence of muscle-tendon stiffness on high-speed force production (power) in children

It is well known that increased muscle-tendon stiffness is associated with improved power production during multi-joint tasks in adults (Arampatzis *et al.* 2006; Bojsen-Moller *et al.* 2005). In a developmental context, then, it is interesting to consider whether the reduced stiffness of children compared to adults affects their ability to produce muscular force or power.

In terms of multi-joint tasks, little is known about how age-related changes in whole muscle-tendon or leg stiffness affect the development of maximum power production. Wang *et al.* (2004) found that both leg stiffness and jump performance increased with age and speculated that increases in stiffness could have been a causative factor. Results from Korff *et al.* (2009) substantiated this speculation by showing that the relationship between leg stiffness and maximum power production during vertical jumping was stronger in 16–18-year-old adolescents than in 11–13-year-old pre-adolescents, suggesting that children acquire the ability to use their body's elastic resources more effectively during adolescence to enhance maximum power production.

A potential explanation for the association between stiffness and maximum power production is that stiffer tendons allow a faster transfer of forces from the muscles to the skeleton. Significant differences in EMD and RFD have been reported between young children and adults in the elbow flexors and plantar flexors, respectively (Asai and Aoki 1996; Grosset *et al.* 2005). Previous research in adults has shown that both EMD (Muraoka *et al.* 2004) and RFD (Bojsen-Moller *et al.* 2005) are influenced by tendon stiffness. However, neither Asai and Aoki (1996) nor Grosset *et al.* (2005) specifically measured tendon, or limb, stiffness and they therefore could not explore whether there is a specific link between stiffness and EMD or RFD. In children, we have recently examined the relationship between Achilles tendon stiffness and EMD in prepubescent boys and girls (5–12 years) (Korff *et al.* 2011). We found the two variables to be well correlated (r = 0.74), suggesting that tendon stiffness may influence EMD in children. Our results let us speculate that compliant tendons of younger children might slow the rate of force transmission and result in a greater time for muscular forces to be transferred to the skeleton. This in turn may decrease the knee joint stability and increase the relative risk of injury. However, our data do not provide incontrovertible evidence for this speculation because other age-dependent factors may have influenced the relationship. Therefore, further research is required to examine the causal link between these two variables. Thus, while our findings are a useful first step towards understanding how stiffness impacts on force production during maximum power multi-joint tasks, this area warrants further research to understand more fully age- and sex-related increases in stiffness and maximum power production during multi-joint tasks.

Influence of muscle-tendon stiffness on balance in children

Balance during standing or complex movement performance is difficult for humans because multiple segments with a high centre of mass must be supported over a small base of support. Balance control in children improves with practice over the first ~7 years, to become similar to adults (Roncesvalles *et al.* 2001). It is well recognized that there is a substantial perceptual-motor contribution to static and dynamic balance acquisition (Hatzitaki *et al.* 2002), and that balance training can promote changes in the neurological control of balance (Taube *et al.* 2008). Nonetheless, it has been shown that, particularly in research investigating the control of balance during standing, muscle-tendon (and joint) stiffness plays a significant role in stability control (Fitzpatrick *et al.* 1992; Loram and Lakie 2002; Loram *et al.* 2007). For example, Loram *et al.* (2007) demonstrated that passive muscle-tendon stiffness at the ankle is important for minimizing the requirement for muscle activity, and that passive ankle torque alone is sufficient to provide at least unstable balance (i.e. muscle activity is required only for larger balance corrections). Given the importance of muscle-tendon stiffness, and considering that the relative centre of mass in children is higher and their base of support is smaller, muscle-tendon stiffness could be considered an important parameter contributing to balance performance.

To our knowledge, no research has examined the relationship between muscle-tendon properties and balance performance in static or dynamic tasks in children. It is also not clear whether the learning of balance performance is influenced by changes in stiffness, or rather whether optimum balance control is achieved solely through neuromuscular mechanisms. In order to better determine the factors influencing balance control and understand the mechanisms underpinning balance deficiencies in some children, there is a need for research aiming to link stiffness modifications to balance control.

Influence of muscle-tendon stiffness on movement economy (endurance performance) in children

It is well known that movement economy in pursuits such as running is reduced in children when compared to adults and that this difference increases with increasing movement speed (Davies 1980; Rowland and Green 1988). Several physiological and anthropometric factors are associated with prolonged movement in children, including higher levels of oxygen consumption and ventilation, higher heart rates, inferior breathing efficiency (Davies 1980; Montoye *et al.* 1985; Rowland and Green 1988; Rowland *et al.* 1987) and a greater surface area:body mass ratio possibly contribute to a greater need for oxygen transport to maintain body temperature (Bar-Or 1983; Rowland and Green 1988; Taylor *et al.* 1970).

However, some data in adults are suggestive that the mechanical properties of the MTUs, and in particular the tendons themselves, might impact on movement economy. For example, Arampatzis *et al.* (2006) found that runners with a better running economy had stiffer Achilles tendons and more compliant (when measured at low force levels) vastus lateralis tendons than runners with lower economy. However, although some researchers have measured limb stiffness during movements such as running (Schepens *et al.* 1998) and hopping (Korff *et al.* 2009), to our knowledge there are presently no data examining the relationship between muscle-tendon properties and movement economy in children. Given that muscle-tendon stiffness and hysteresis are important determinants of movement speed and efficiency, further research is required to examine the importance of it in children and whether adaptations are directly associated with changes in movement economy.

Clinical and practical applications

The literature is consistent in demonstrating that both passive and active muscle-tendon stiffness increases with age through to adulthood. The empirical evidence suggests that such age-related increases in MTU stiffness are a result of both growth and maturation. However, it is not completely known whether optimum movement frequencies should change with growth in children and it is not known whether changes in muscle-tendon stiffness underpin changes in movement efficiency. The stiffer tissues of adults, however, are speculatively considered to be a contributor to movement performance and increased movement efficiency when compared to children, and may assist them in tasks requiring fast force application. Some recent evidence exists suggesting that resistance training can increase tendon stiffness in children, although it is not yet known whether this influences movement performance (especially at high speed) or balance during complex tasks. Such information would be useful for coaches and teachers who seek to improve children's motor performance. It is also not known whether any forms of training influence muscle-tendon hysteresis in children, even though there is some evidence that stretching (Kubo *et al.* 2001d, 2002b) or plyometric training (Foure *et al.* 2010) might be useful in adults. Such an effect would have implications for teachers, youth coaches and paediatric clinicians who seek to improve movement efficiency, especially in clinical populations (e.g. cerebral palsy) where alterations in muscle-tendon stiffness dramatically affect movement performance. In summary, despite the substantial impact of muscle-tendon stiffness on movement performance demonstrated in adult populations, little is known about how we might develop it in children, and whether this might improve movement outcomes in both normally developing and slowly developing or clinical populations.

Key points

- Muscle-tendon stiffness is an important variable related to the storage and release of elastic energy, which enhances mechanical power production and efficiency of human movement.
- Hysteresis is the loss of stored elastic energy as heat, and minimizing it is important for improving movement efficiency.
- The ability for an individual to move with a movement frequency that matches the natural frequency of their elastic limbs is important in order to minimize the energetic cost of movement.
- There is an improvement in the ability to use the spring-like properties of elastic tissues to improve movement performance (and efficiency/economy) during walking, running and jumping tasks from early childhood through to adolescence.
- Children have both a lower muscle-tendon and tendon-specific stiffness than adults. However, these differences (1) appear at least partly due to differences in body mass and muscle force capability that directly influence muscle-tendon stiffness and (2) disappear with age into adolescence.
- No research has compared tissue (particularly tendon) hysteresis values between adults and children. Thus, it is not known whether differences exist and whether they might contribute to movement performance differences.
- Few data exist documenting the effects of physical training on stiffness in children. Some recent evidence suggests that resistance training in children increases tendon stiffness. It is not yet known how training-related changes in muscle-tendon stiffness influence movement performance in children.
- The scientific evidence regarding the interactions between stiffness, maximum power production, movement efficiency and strength or flexibility training has implications for teachers, youth coaches and paediatric clinicians, and needs to be the focus of an intense research effort.

Acknowledgement

The preparation of this chapter was supported by the Engineering and Physical Sciences Research Council under grant EP/E013007/1.

References

Agarwal, G.C. and Gottlieb, G.L. (1977) 'Oscillation of the human ankle joint in response to applied sinusoidal torque on the foot', *Journal of Physiology*, 268: 151–76.

Arampatzis, A., Bruggemann, G.P. and Klapsing, G.M. (2001) 'Leg stiffness and mechanical energetic processes during jumping on a sprung surface', *Medicine and Science in Sports and Exercise*, 33: 923–31.

Arampatzis, A., Karamanidis, K. and Albracht, K. (2007) 'Adaptational responses of the human Achilles tendon by modulation of the applied cyclic strain magnitude', *Journal of Experimental Biology*, 210: 2743–53.

Arampatzis, A., De Monte, G., Karamanidis, K., Morey-Klapsing, G., Stafilidis, S. and Bruggemann, G.P. (2006) 'Influence of the muscle-tendon unit's mechanical and morphological properties on running economy', *Journal of Experimental Biology*, 209: 3345–57.

Asai, H. and Aoki, J. (1996) 'Force development of dynamic and static contractions in children and adults', *International Journal of Sports Medicine*, 17: 170–4.

Baltzopoulos, V. (1995) 'A videofluoroscopy method for optical distortion correction and measurement of knee-joint kinematics', *Clinical Biomechanics*, 10: 85–92.

Bar-Or, O. (1983) *Pediatric Sports Medicine for the Practitioner*, New York: Springer-Verlag.

Bennett, F., Waugh, C. and Korff T. (2011) 'Differences in hopping mechanics between children and adults', paper presented at the BASES student conference, University of Chester, April.

Best, T.M., McElhaney, J., Garrett, W.E., Jr and Myers, B.S. (1994) 'Characterization of the passive responses of live skeletal muscle using the quasi-linear theory of viscoelasticity', *Journal of Biomechanics*, 27: 413–19.

Bojsen-Moller, J., Magnusson, S.P., Rasmussen, L.R., Kjaer, M. and Aagaard, P. (2005) 'Muscle performance during maximal isometric and dynamic contractions is influenced by the stiffness of the tendinous structures', *Journal of Applied Physiology*, 99: 986–94.

Cornu, C. and Goubel, F. (2001) 'Musculo-tendinous and joint elastic characteristics during elbow flexion in children', *Clinical Biomechanics*, 16: 758–64.

Davies, C.T. (1980) 'Metabolic cost of exercise and physical performance in children with some observations on external loading', *European Journal of Applied Physiology and Occupational Physiology*, 45: 95–102.

Farley, C.T., Blickhan, R., Saito, J. and Taylor, C.R. (1991) 'Hopping frequency in humans: a test of how springs set stride frequency in bouncing gaits', *Journal of Applied Physiology*, 71: 2127–32.

Finni, T., Ikegawa, S., Lepola, V. and Komi, P.V. (2003a) 'Comparison of force-velocity relationships of vastus lateralis muscle in isokinetic and in stretch-shortening cycle exercises', *Acta Physiologica Scandinavica*, 177: 483–91.

Finni, T., Hodgson, J.A., Lai, A.M., Edgerton, V.R. and Sinha, S. (2003b) 'Nonuniform strain of human soleus aponeurosis-tendon complex during submaximal voluntary contractions in vivo', *Journal of Applied Physiology*, 95: 829–37.

Fitzpatrick, R.C., Taylor, J.L. and McCloskey, D.I. (1992) 'Ankle stiffness of standing humans in response to imperceptible perturbation: reflex and task-dependent components', *Journal of Physiology*, 454: 533–47.

Foure, A., Nordez, A. and Cornu, C. (2010) 'Plyometric training effects on Achilles tendon stiffness and dissipative properties', *Journal of Applied Physiology*, 109: 849–54.

Frost, G., Dowling, J., Dyson, K. and Bar-Or, O. (1997) 'Cocontraction in three age groups of children during treadmill locomotion', *Journal of Electromyography and Kinesiology*, 7: 179–86.

Fukashiro, S., Itoh, M., Ichinose, Y., Kawakami, Y. and Fukunaga, T. (1995) 'Ultrasonography gives directly but noninvasively elastic characteristic of human tendon in vivo', *European Journal of Applied Physiology and Occupational Physiology*, 71: 555–7.

Fukunaga, T., Kubo, K., Kawakami, Y., Fukashiro, S., Kanehisa, H. and Maganaris, C.N. (2001) 'In vivo behaviour of human muscle tendon during walking', *Proceedings: Biological Sciences, The Royal Society*, 268: 229–33.

Gandevia, S.C. (1996) 'Kinaesthesia: roles for afferent signals and motor commands', in J.B. Rowell and J.T. Shepard (eds) *Handbook of Physiology*, New York: Oxford University Press, pp. 128–72.

Goodwin, G.M., McCloskey, D.I. and Matthews, P.B. (1972) 'The contribution of muscle afferents to kinaesthesia shown by vibration induced illusions of movement and by the effects of paralysing joint afferents', *Brain*, 95: 705–48.

Goubel, F. and Pertuzon, E. (1973) 'Évaluation de l'élasticité du muscle *in situ* par une méthode de quick-release [Evaluation of the elasticity of muscle in situ by the quick-release method]', *Archives Internationales de Physiologie et de Biochimie*, 81: 697–707.

Grosset, J.F., Mora, I., Lambertz, D. and Perot, C. (2005) 'Age-related changes in twitch properties of plantar flexor muscles in prepubertal children', *Pediatric Research*, 58: 966–70.

Grosset, J.F., Mora, I., Lambertz, D. and Perot, C. (2007) 'Changes in stretch reflexes and muscle stiffness with age in prepubescent children', *Journal of Applied Physiology*, 102: 2352–60.

Hallemans, A., De Clercq, D., Otten, B. and Aerts, P. (2005) '3D joint dynamics of walking in toddlers: a cross-sectional study spanning the first rapid development phase of walking', *Gait & Posture*, 22: 107–18.

Hatzitaki, V., Zisi, V., Kollias, I. and Kioumourtzoglou, E. (2002) 'Perceptual-motor contributions to static and dynamic balance control in children', *Journal of Motor Behavior*, 34: 161–70.

Holt, K.G., Jeng, S.F. and Fetters, L. (1991) 'Walking cadence of 9 year olds is predictable as the resonant frequency of a force-driven harmonic oscillator', *Pediatric Exercise Science*, 3: 121–8.

Holt, K.J., Jeng, S.F., Ratcliffe, R. and Hamill, J. (1995) 'Energetic cost and stability during human walking at the preferred stride velocity', *Journal of Motor Behavior*, 27: 164–78.

Holt, K.G., Saltzamn, E., Ho, C.L., Kubo, M. and Ulrich, B.D. (2006) 'Discovery of the pendulum and spring dynamics in the early stages of walking', *Journal of Motor Behavior*, 38: 206–18.

Horita, T., Komi, P.V., Nicol, C. and Kyrolainen, H. (2002) 'Interaction between pre-landing activities and stiffness regulation of the knee joint musculoskeletal system in the drop jump: implications to performance', *European Journal of Applied Physiology and Occupational Physiology*, 88: 76–84.

Ishikawa, M. and Komi, P.V. (2004) 'Effects of different dropping intensities on fascicle and tendinous tissue behavior during stretch-shortening cycle exercise', *Journal of Applied Physiology*, 96: 848–52.

Ishikawa, M., Komi, P.V., Finni, T. and Kuitunen, S. (2006) 'Contribution of the tendinous tissue to force enhancement during stretch-shortening cycle exercise depends on the prestretch and concentric phase intensities', *Journal of Electromyography and Kinesiology*, 16: 423–31.

Jeng, S.F., Holt, K.G., Fetters, L. and Certo, C. (1996) 'Self-optimization of walking in nondisabled children and children with spastic hemiplegic cerebral palsy', *Journal of Motor Behavior*, 28: 15–27.

Jeng, S.F., Liao, H.F., Lai, J.S. and Hou, J.W. (1997) 'Optimization of walking in children', *Medicine and Science in Sports and Exercise*, 29: 370–6.

Korff, T., Horne, S.L., Cullen, S.J. and Blazevich, A.J. (2009) 'Development of lower limb stiffness and its contribution to maximum vertical jumping power during adolescence', *Journal of Experimental Biology*, 212: 3737–42.

Korff, T., Waugh, C.M., Fath, F. and Blazevich A.J. (2011) 'Determinants of muscular force production in pre-pubertal children: The roles of tendon stiffness and muscle activation', paper presented at European College of Sport Sciences, Liverpool, July.

Kubo, K., Kanehisa, H. and Fukunaga, T. (2001a) 'Effects of different duration isometric contractions on tendon elasticity in human quadriceps muscles', *Journal of Physiology*, 536: 649–55.

Kubo, K., Kanehisa, H. and Fukunaga, T. (2002a) 'Effect of stretching training on the viscoelastic properties of human tendon structures in vivo', *Journal of Applied Physiology*, 92: 595–601.

Kubo, K., Kanehisa, H. and Fukunaga, T. (2002b) 'Effects of resistance and stretching training programmes on the viscoelastic properties of human tendon structures in vivo', *Journal of Physiology*, 538: 219–26.

Kubo, K., Kanehisa, H., Ito, M. and Fukunaga, T. (2001b) 'Effects of isometric training on the elasticity of human tendon structures in vivo', *Journal of Applied Physiology*, 91: 26–32.

Kubo, K., Kanehisa, H., Kawakami, Y. and Fukanaga, T. (2001c) 'Growth changes in the elastic properties of human tendon structures', *International Journal of Sports Medicine*, 22: 138–43.

Kubo, K., Kanehisa, H., Kawakami, Y. and Fukunaga, T. (2001d) 'Influence of static stretching on viscoelastic properties of human tendon structures in vivo', *Journal of Applied Physiology*, 90: 520–7.

Kubo, K., Morimoto, M., Komuro, T., Tsunoda, N., Kanehisa, H. and Fukunaga, T. (2007) 'Influences of tendon stiffness, joint stiffness, and electromyographic activity on jump performances using single joint', *European Journal of Applied Physiology and Occupational Physiology*, 99: 235–43.

Kubo, K., Akima, H., Kouzaki, M., Ito, M., Kawakami, Y., Kanehisa, H. and Fukunaga, T. (2000) 'Changes in the elastic properties of tendon structures following 20 days bedrest in humans', *European Journal of Applied Physiology and Occupational Physiology*, 83: 463–8.

Lambertz, D., Mora, I., Grosset, J.F. and Perot, C. (2003) 'Evaluation of musculotendinous stiffness in prepubertal children and adults, taking into account muscle activity', *Journal of Applied Physiology*, 95: 64–72.

Lambertz, D., Perot, C., Kaspranski, R. and Goubel, F. (2001) 'Effects of long-term spaceflight on mechanical properties of muscles in humans', *Journal of Applied Physiology*, 90: 179–88.

Lebiedowska, M.K. and Fisk, J.R. (1999) 'Passive dynamics of the knee joint in healthy children and children affected by spastic paresis', *Clinical Biomechanics*, 14: 653–60.

Lichtwark, G.A. and Wilson, A.M. (2005) 'In vivo mechanical properties of the human

Achilles tendon during one-legged hopping', *Journal of Experimental Biology*, 208: 4715–25.

Lichtwark, G.A. and Wilson, A.M. (2006) 'Interactions between the human gastrocnemius muscle and the Achilles tendon during incline, level and decline locomotion', *Journal of Experimental Biology*, 209: 4379–88.

Lindstedt, S.L., LaStayo, P.C. and Reich, T.E. (2001) 'When active muscles lengthen: properties and consequences of eccentric contractions', *News in Physiological Sciences*, 16: 256–61.

Loram, I.D. and Lakie, M. (2002) 'Direct measurement of human ankle stiffness during quiet standing: the intrinsic mechanical stiffness is insufficient for stability', *Journal of Physiology*, 545: 1041–53.

Loram, I.D., Maganaris, C.N. and Lakie, M. (2007) 'The passive, human calf muscles in relation to standing: the non-linear decrease from short range to long range stiffness', *Journal of Physiology*, 584: 661–75.

Maganaris, C.N. (2000) 'In vivo measurement-based estimations of the moment arm in the human tibialis anterior muscle-tendon unit', *Journal of Biomechanics*, 33: 375–9.

Maganaris, C.N. and Paul, J.P. (2000) 'Hysteresis measurements in intact human tendon', *Journal of Biomechanics*, 33: 1723–7.

Magnusson, S.P., Hansen, P. and Kjaer, M. (2003) 'Tendon properties in relation to muscular activity and physical training', *Scandinavian Journal of Medicine and Science in Sports*, 13: 211–23.

Mansour, J.M. and Audu, M.L. (1986) 'The passive elastic moment at the knee and its influence on human gait', *Journal of Biomechanics*, 19: 369–73.

Montoye, H.J., Ayen, T., Nagle, F. and Howley, E.T. (1985) 'The oxygen requirement for horizontal and grade walking on a motor-driven treadmill', *Medicine and Science in Sports and Exercise*, 17: 640–5.

Muraoka, T., Muramatsu, T., Fukunaga, T. and Kanehisa, H. (2004) 'Influence of tendon slack on electromechanical delay in the human medial gastrocnemius in vivo', *Journal of Applied Physiology*, 96: 540–4.

O'Brien, T.D., Reeves, N.D., Baltzopoulos, V., Jones, D.A. and Maganaris, C.N. (2010) 'Mechanical properties of the patellar tendon in adults and children', *Journal of Biomechanics*, 43: 1190–5.

Pérot, C., Bosle, J.P., Delanaud, S. and Goubel, F. (1999) 'Un ergomètre multi-modalités dédié à l'étude des propriétés mécaniques musculo-articulaires chez l'enfant préadolescent', *Revue européenne de biotechnologie médicale*, 21: 212–17.

Proske, U., Wise, A.K. and Gregory, J.E. (2000) 'The role of muscle receptors in the detection of movements', *Progress in Neurobiology*, 60: 85–96.

Rabita, G., Couturier, A. and Lambertz, D. (2008) 'Influence of training background on the relationships between plantarflexor intrinsic stiffness and overall musculoskeletal stiffness during hopping', *European Journal of Applied Physiology and Occupational Physiology*, 103: 163–71.

Roncesvalles, M.N., Woollacott, M.H. and Jensen, J.L. (2001) 'Development of lower extremity kinetics for balance control in infants and young children', *Journal of Motor Behavior*, 33: 180–92.

Rowland, T.W. and Green, G.M. (1988) 'Physiological responses to treadmill exercise in females: adult-child differences', *Medicine and Science in Sports and Exercise*, 20: 474–8.

Rowland, T.W., Auchinachie, J.A., Keenan, T.J. and Green, G.M. (1987) 'Physiologic responses to treadmill running in adult and prepubertal males', *International Journal of Sports Medicine*, 8: 292–7.

Schepens, B., Willems, P.A. and Cavagna, G.A. (1998) 'The mechanics of running in children', *Journal of Physiology*, 509: 927–40.

Schepens, B., Bastien, G.J., Heglund, N.C. and Willems, P.A. (2004) 'Mechanical work and muscular efficiency in walking children', *Journal of Experimental Biology*, 207: 587–96.

Schepens, B., Willems, P.A., Cavagna, G.A. and Heglund, N.C. (2001) 'Mechanical power and efficiency in running children', *Pflugers Archives (European Journal of Physiology)*, 442: 107–16.

Taube, W., Gruber, M. and Gollhofer, A. (2008) 'Spinal and supraspinal adaptations associated with balance training and their functional relevance', *Acta Physiologica*, 193: 101–16.

Taylor, C.R., Schmidt-Nielsen, K. and Raab, J.L. (1970) 'Scaling of energetic cost of running to body size in mammals', *American Journal of Physiology*, 219: 1104–7.

Visser, J.J., Hoogkamer, J.E., Bobbert, M.F. and Huijing, P.A. (1990) 'Length and moment arm of human leg muscles as a function of knee and hip-joint angles', *European Journal of Applied Physiology and Occupational Physiology*, 61: 453–60.

Wang, L.I., Lin, D.-C. and Huang, C. (2004) 'Age effect on jumping techniques and lower limb stiffness during vertical jump', *Research in Sports Medicine*, 12: 209–19.

Waugh, C.M., Blazevich, A.J., Fath, F. and Korff, T. (2011a) 'Age-related changes in mechanical properties of the Achilles tendon during childhood', submitted for publication.

Waugh, C.M., Korff, T., Fath, F. and Blazevich, A.J. (2011b) 'Resistance training increases tendon stiffness and influences rapid force production in prepubertal children', paper presented at European College of Sport Sciences, Liverpool, July.

Wilkie, D.R. (1950) 'The relation between force and velocity in human muscle', *Journal of Physiology*, 110: 249–80.

Wilson, G.J., Elliott, B.C. and Wood, G.A. (1992) 'Stretch shorten cycle performance enhancement through flexibility training', *Medicine and Science in Sports and Exercise*, 24: 116–23.

6 Paediatric biomechanical modelling techniques

Thomas Korff and Florian Fath

Introduction

When conducting research to understand the mechanisms underlying age-related changes in motor behaviour, the analysis of muscular forces and torques (rotational forces) can give us unique insights into the nervous system's strategies to achieve the task of interest (Winter and Eng 1995). As the direct measurement of internal muscle forces is invasive and unethical, researchers use musculoskeletal models to estimate them indirectly. In general, two particular modelling techniques (inverse and forward dynamics) are common tools, which help researchers to estimate muscle forces and torques indirectly. The goal of both inverse and forward dynamics is to estimate muscle forces or torques from non-invasive biomechanical measures such as joint kinematics (typically obtained from motion analysis) and ground reaction forces (typically obtained from force platforms or similar devices). Both techniques require a range of input parameters relating to the properties of the musculoskeletal system. These parameters can be broadly divided into those relating to segmental inertia and those relating to muscle-tendon mechanics. In order to obtain valid outputs from biomechanical models, it is important to use accurate input parameters. In this chapter, we discuss the basic concepts of both inverse and forward dynamics, and we give an overview of the input parameters required for each modelling technique. We will then discuss developmental aspects of these input parameters and their implications relating to developmental research.

Basic theoretical concepts

Equations of motion

Before we discuss the specifics of biomechanical modelling, it is important to understand the concept of equations of motion, which are relevant for both inverse and forward dynamics. Equations of motion provide us with specific mathematical relationships between forces and the resulting movement. The simplest equation of motion is the well-known Newton's second law (force equals mass times acceleration):

$$F = m \cdot a \tag{6.1}$$

The angular equivalent of this equation is:

$$T = I \cdot \alpha \tag{6.2}$$

where:

> T is a torque (rotational force)
> I is the system's moment of inertia
> α is its angular acceleration.

Equations 6.1 and 6.2 provide a specific relationship between the movement (in this case the accelerations a or α) and the force or torque required to produce this movement. Expanding on Equation 6.2, it is important to acknowledge that within the context of human movement the torque T is the result of muscular and non-muscular influences. Non-muscular torques include gravitational and other external contributions (e.g. resistive forces applied by a strength-testing machine or a stationary bicycle). Thus, we can then expand and rearrange Equation 6.2 to:

$$I \cdot \alpha = T_{MUS} + T_{GRA} + T_{EXT} \tag{6.3}$$

where:

> T_{MUS} represents muscular torques
> T_{GRA} represents gravitational torques
> T_{EXT} represents external torques.

When analysing complex mechanical systems, so-called motion-dependent influences also contribute to the overall equation. These motion-dependent influences arise from the movement itself and include centripetal and Coriolis forces. Thus, we can expand Equation 6.3 to:

$$I \cdot \alpha = T_{MUS} + T_{GRA} + T_{EXT} + T_{MDT} \tag{6.4}$$

where:

> T_{MDT} represents motion-dependent torques.

Acknowledging that muscular torques are the product of muscle moment arms and muscular forces, we can reformulate Equation 6.4 to:

$$I \cdot \alpha = \Sigma (F_{MUS} \cdot MA) + T_{GRA} + T_{EXT} + T_{MDT} \tag{6.5}$$

where:

MA represents a muscle moment arm

F_{MUS} represents a muscle force

$\Sigma(F_{MUS} \cdot MA)$ represents the sum of all muscle forces multiplied by the respective muscle moment arms.

There are many ways to derive and express equations of motion, and for complex mechanical systems, the equations of motion are much more complex than illustrated above. A more detailed description would be beyond the scope of this chapter. However, the general concepts outlined here will help us understand the basics of inverse and forward dynamics.

Inverse dynamics

Inverse dynamics is a technique that estimates muscular torques from kinematic and kinetic measures. Kinematic measures include positions, velocities and accelerations of joints and segmental centres of mass. Positional data are typically obtained from motion capture systems or potentiometers and can be mathematically differentiated with respect to time taken to derive velocities and accelerations. Kinetic data used for inverse dynamics are commonly obtained from reaction forces that act on the human body. Measurement devices that allow us to measure reaction forces (e.g. force platforms or force pedals) are typically equipped with force transducers based on piezo-electric or strain gauge technology. Such force transducers measure an electrical charge in response to applied loads, which allows us to make inferences about the reaction forces applied to the human body.

The process of inverse dynamics can be best illustrated using the simplified Equation 6.3. Rearranging this equation yields:

$$T_{MUS} = T_{GRA} + T_{EXT} - I \cdot \alpha \tag{6.6}$$

Using inverse dynamics, we apply Equation 6.6 to each body segment, starting with the most distal one (typically the segment, at which the measured reaction force is applied). T_{GRA}, T_{EXT} and $I \cdot \alpha$ can all be obtained from experimentally derived kinematic and kinetic measures:

T_{GRA} is a function of joint angles

T_{EXT} is a function of reaction force measures

$I \cdot \alpha$ is a function of angular acceleration, which can be obtained from double-differentiating joint positions.

In order to obtain the specific values for T_{GRA}, T_{EXT} and $I \cdot \alpha$ we also need to know the inertial parameters of the segment of interest. These inertial characteristics include segmental masses, centre of mass locations and moments of inertia. They are typically derived from tables provided in the literature, which is discussed later in this chapter. Once we know the values for T_{GRA}, T_{EXT} and $I \cdot \alpha$, we can substitute them into Equation 6.6 to obtain the muscular torque about the proximal joint of the segment of interest. This process can then be used iteratively to calculate the muscular torques about the more proximal joints.

In developmental research, inverse dynamics techniques have been used to study the development of balance control (Roncesvalles *et al.* 2001), walking (Ganley and Powers 2005, 2006), cycling (Brown and Jensen 2003, 2006), reaching (Schneider *et al.* 1989) and kicking (Jensen *et al.* 1995; Schneider *et al.* 1990). For more details, we refer the reader to Chapters 7–9.

Forward dynamics

When using forward dynamics, we model the mechanical processes in the order in which they occur (in contrast to inverse dynamics where we start with the movement and calculate the torques that must have caused it). A detailed description of forward dynamics techniques is beyond the scope of this chapter. The information provided here is limited to what is relevant in a developmental context. For a more detailed description of forward dynamics modelling, we refer the reader to the excellent reviews by Buchanan *et al.* (2004) and Zajac (1989).

When using forward dynamics, we start with muscle forces or torques and use the equations of motion to calculate the system's acceleration. For a given point in time, the acceleration can then be numerically integrated, which yields velocities and positions at the next time step. Applying this concept iteratively yields a full mechanical description of the movement of interest. Information gained from such forward dynamics analyses are often more insightful than that obtained from inverse dynamics. For example, it allows us to quantify individual muscle forces more readily and their individual contributions to aspects of the movement.

When performing forward dynamics analyses, it is important that the simulated movement is realistic. Therefore, researchers typically validate their models by determining the simulation that approximates the experimentally obtained data most closely. This so-called 'tracking solution' is found by using optimization techniques to find the input (i.e. muscle forces) that minimizes the difference between the simulated and experimentally measured kinematics. One challenge that arises is the determination of the initial value for the input (i.e. muscular force or torque). There are several ways of finding this so-called 'initial guess', the crudest being to randomly estimate it. However, more commonly, researchers are partial to additional information about their experimental set-up to allow for a more informed guess. A more sophisticated way of estimating muscle forces is the inclusion of neural and physiological processes into the forward model, which requires the integration of additional differential equations. Such models are driven by neural inputs, which represent the excitation of the muscle by the nervous system. The mathematical implementation of such a model requires specific muscle-tendon parameters, which are discussed later in this chapter.

Bearing in mind that complex movements require appropriately timed muscle activations of numerous muscles, the implementation of forward dynamics models is complex and can require significant computational resources. A variety of (commercial and open source) software packages that allow for the implementation of forward dynamics models exist. While the use of such software is relatively intuitive, it is important for the user to be aware of the assumptions relating to the models and the source of their input parameters. In a developmental context, this means that attention should be given to the appropriate scaling of the model's input parameters. The assumptions about various model input parameters and the issues relating to them being suitable for paediatric forward dynamics models are discussed in the next section.

The importance of musculoskeletal input parameters and implications for motor development

Segmental inertial parameters

For both inverse and forward dynamics valid and reliable determination of anthropometric parameters (inertial segmental characteristics) is vital. These parameters include segmental lengths, masses, centre of mass locations and moments of inertia. In most cases, segment lengths can be measured by the experimenter using appropriate anatomical landmarks (Norton and Olds 1996). Obtaining segmental masses, centre of mass locations and moments of inertia is less straightforward. These parameters are commonly estimated using regression equations presented in the literature. With the help of such equations, the relevant parameters can be estimated from more measurable variables such as body mass, stature and segmental length (e.g. Winter 1990). For adults, numerous studies have been performed to derive such equations (de Leva 1996; Dempster 1955; Hatze 1980; Zatsiorsky and Seluyanov 1983). Such studies have used a variety of techniques to measure the relevant parameters directly, including X-ray (e.g. Ganley and Powers 2004), magnetic resonance imaging (MRI) (e.g. Mungiole and Martin 1990) and cadavers (e.g. Dempster 1955).

When using musculoskeletal models in a developmental context, it is important to take developmental changes in anthropometric characteristics into consideration. Developmental changes in segment lengths are easily accounted for, as they can be measured directly (Norton and Olds 1996). When determining segmental masses, centre of mass locations and moments of inertia, it is important to use appropriate regression equations which estimate inertial segmental parameters from easily measurable variables, such as body mass, segment length, segment circumference or age. Such regression equations quantifying changes in inertial segmental characteristics from infancy to adulthood are available in the literature (Jensen 1989; Schneider and Zernicke 1992).

Schneider and Zernicke (1992) demonstrated that inertial limb characteristics change significantly between 1 and 18 months of age. These authors report regression equations that quantify segmental masses and moments of inertia change as a function of body mass, segment length, segmental circumference and age. Sun and Jensen (1994) conducted a similar study and reported regression equations quantifying segmental inertial characteristics as a function of age in infants between 2 and 15 months of age. Although the two studies are not directly comparable, their results are consistent in demonstrating that age-related increases in segmental inertial characteristics during infancy are non-linear and segment-specific.

During childhood, inertial segmental characteristics also change in a non-linear fashion. In a series of studies, Jensen quantified how these parameters change between 4 years of age and adulthood (Jensen 1981a, 1986, 1988, 1989; Jensen and Nassas 1988). The most recent of these studies provided a comprehensive list of regression equations which allow the reader to estimate segmental masses, centre of mass locations and moments of inertia from segmental lengths, body mass and age (Jensen 1989).

Most noteworthy, Jensen (1989) reported that the segmental mass of the thigh increases from 8 to 12 per cent of total body mass between the ages of 4 and 20 years. Furthermore, the author reported that the relative centre of mass of the shank moves to a more proximal position during this time. Jensen's (1989) work signifies an invaluable contribution to paediatric biomechanical modelling. However, one limitation of Jensen's work is that all his participants were male. Thus, the question arises as to whether Jensen's equations are appropriate for females. Ganley and Powers (2004) quantified developmental changes in inertial characteristics between 7 and 13 years of age for both male and female participants. These authors' results are comparable to those of Jensen (1989). Therefore, it seems appropriate to use Jensen's equations for both male and female children. However, future research should specifically quantify age-related changes in inertial segmental parameters in girls during infancy and childhood. Furthermore, it must be acknowledged that the inclusion of chronological age as a predictor of segmental growth bears its own limitations as chronological age is only one among several predictors of skeletal maturity (see Chapter 1).

Given that the development of segmental inertial characteristics during infancy and childhood is multifactorial and segment-specific, the question arises as to whether age-related changes in inertial segmental characteristics influence muscular force production and the acquisition of new motor tasks. Several studies have been conducted to gain insights into these interactions. Jensen *et al.* (1997) conducted a longitudinal study to quantify how changes in inertial segmental characteristics during infancy would change gravitational torques about various joints within the trunk. Unsurprisingly, they found that gravitational torques increase between 9 and 60 weeks of age. More interestingly however, they also found a decrease in the rate of change of the gravitational torque. This decrease was dependent on the joint about which the gravitational torque was calculated. Bearing in mind that gravitational torques need to be overcome by muscular torques to lift the corresponding segments or body parts, the authors concluded that the observed joint-specific decrease in the slope of gravitational torque gains is an essential contributor to the acquisition of lifting and controlling head and trunk segments during infancy.

Age-related changes in anthropometry also affect muscular force and power production throughout childhood during more complex fundamental motor tasks such as walking, jumping or cycling. Chester and Jensen (2005) investigated changes in segmental inertial properties of head and trunk segments between 28 and 68 weeks of age within the context of walking and postural control. They found that the inertias of these segments increase at different rates. The authors concluded that these segment-specific non-linear changes in segmental inertia control the acquisition of walking and postural control. Jensen (1989) performed a kinetic analysis of a standing long jump and found that substituting adult's inertial segmental characteristics to child-appropriate ones resulted in significantly greater muscular torques at the neck and the hip to perform the same movement. Jensen (1981b) confirmed these findings experimentally, by demonstrating that age-matched boys of different body types (endomorph, mesomorph and ectomorph) produce different kinetic patterns to produce standing long jumps. For further information relating to the importance of body composition within the context of motor development, we refer the reader to Chapter 1.

During submaximal cycling, the effects of developmental changes in anthropometry are also significant. Brown and Jensen (2003) compared force production during cycling between children and adults. They decomposed the pedal reaction force measured during the cycling action into muscular and non-muscular components and found greater muscular and lower non-muscular forces in children when compared to adults. Bearing in mind that non-muscular forces are directly influenced by the anthropometry of the performer, the authors speculated that these differences were a result of differences in anthropometric characteristics between children and adults. In a follow-up study, these authors added mass to the children's legs, which resulted in a more adult-like distribution between muscular and non-muscular forces (Brown and Jensen 2006).

Korff and Jensen (2008) expanded on these results by quantifying the effect of differences in segmental inertial characteristics on muscular power production. To isolate this effect they created a torque-driven forward dynamics simulation and changed segmental mass proportions, relative centre of mass locations and moments of inertia of the model to represent those of children of different ages. Further, they constrained the simulation to produce identical kinematics. The changes in anthropometric characteristics significantly affected muscular power production. This finding confirms that observed developmental differences in muscular force or power production can be functional muscular adjustments to non-linear segmental growth.

Muscle-tendon parameters

As discussed earlier in this chapter, the implementation of forward dynamics simulation requires input parameters to muscle-tendon mechanics. These parameters include tendon slack length, tendon stiffness (strain), optimal fibre length, pennation angle at optimal fibre length, muscle moment arm as well as maximal isometric force. When creating paediatric musculoskeletal models, it is important to take age-related changes of these parameters into consideration. In the following sections, we discuss how some of these parameters are experimentally determined and how they change during childhood. Unfortunately, the literature relating to the development of muscle-tendon mechanics is in its infancy, and much work is needed to fully describe age-related changes in muscle-tendon parameters. Here, we give an overview of what is known, and point out areas requiring further research.

Muscle moment arm

Muscle-tendon forces act at a distance from a joint's centre of rotation (COR) in order to create a rotational movement. The perpendicular distance between the muscle-tendon's line of force action and the COR of the corresponding joint is termed the 'moment arm'. A torque about a joint is the product of the muscle-tendon force and the moment arm. The moment arm of a muscle-tendon unit can be estimated using a variety of imaging techniques such as MRI (Maganaris *et al.* 1998), X-ray videofluoroscopy (Tsaopoulos *et al.* 2009) and ultrasonography (Fath *et al.* 2010).

In a developmental context, researchers have quantified moment arm lengths in children for the Achilles tendon (Morse *et al.* 2008; Waugh *et al.* 2011), the brachialis muscle (Wood *et al.* 2006) and the patellar tendon (O'Brien *et al.* 2009). Unsurprisingly, results from all these studies are consistent in demonstrating that moment arm lengths are smaller in children than in adults. More interestingly however, age-related changes in moment arm length seem to be muscle-specific once the effects of growth are taken into account. For example, Wood *et al.* (2006) found that brachialis moment arms in 10-year-old children were not significantly different from those of adults, once normalized by forearm length. This finding suggests that forearm length may be an appropriate measure to account for growth-related differences in brachialis moment arm. O'Brien *et al.* (2009) quantified the relationship between superficial anthropometric characteristics and patellar tendon moment arm. These authors found strong correlations between tibial length and patellar tendon moment arm length in 9-year-old children (~80 per cent of accounted variance). However, the addition of adult data to the regression analysis yielded much weaker correlations between these measures. These results demonstrate that prepubertal segmental growth accounts for a large proportion of age-related increases in patellar tendon moment arms. However, they also lead to the conclusion that patellar tendon moment arms in children cannot be predicted by scaling adult values to superficial anthropometrics, which has obvious implications for the development of paediatric musculoskeletal models. Waugh *et al.* (2011) quantified the relationship between Achilles tendon moment arm length and various anthropometric characteristics in 5–12-year-old children. These authors found only a moderate relationship between surface anthropometry (i.e. foot length) and Achilles tendon moment arm (~40 per cent of accounted variance) and concluded that Achilles tendon moment arm length is not accurately predictable from surface anthropometry in children. It should therefore be determined for individual participants where possible.

The degree to which moment arms can be related to surface anthropometry appears to be dependent on the muscle of interest. While the interactions between muscle-tendon moment arm, age and surface anthropometry give us some idea about how to scale moment arms appropriately when creating paediatric musculoskeletal models, more research is needed to describe growth-related changes in muscle-tendon moment arms more comprehensively. The role of muscle moment arm in a more general context is discussed in Chapter 4.

Pennation angle

Within a muscle-tendon unit, muscle fibres and their contractile elements are responsible for muscular force production. In many muscles, muscle fibres are not linearly aligned with the tendon but insert at an angle (Zajac 1989). This angle between the muscle fibres and the tendon is called the 'pennation angle'. From a structural point of view, the advantage of muscle fibres branching off the tendinous structures at an angle (compared to being arranged in series) is that more muscle fibres can be arranged in parallel, resulting in greater force production capabilities of the muscle. However, a consequence of muscles being pennate is that not all the force produced by the muscle fibres is transferred to the tendon. This relationship can be described by the equation below:

$$F_T = F_M \cdot cos(\theta) \tag{6.7}$$

where:

F_T is the tendon force
F_M is the muscle force
θ is the muscle's pennation angle.

At first glance, the relationship between muscle force, tendon force and pennation angle appears to be a disadvantage as the greater the pennation angle the smaller the proportion of muscle force being transferred to the tendon. However, this seeming disadvantage is outweighed by the aforementioned advantage of an increased number of muscle fibres being arranged in parallel (Woittiez *et al.* 1984).

The pennation angle of superficial muscles can be determined *in vivo* using ultrasound imaging (Kawakami *et al.* 2006). Several authors have quantified age-related changes in pennation angle of various muscles during childhood (Binzoni *et al.* 2001; Kawakami *et al.* 2006; Morse *et al.* 2008; O'Brien *et al.* 2010b). In a benchmark study, Binzoni *et al.* (2001) documented changes in pennation angle of the gastrocnemius medialis in 162 individuals ranging from 0 to 70 years of age. These authors demonstrated that the pennation angle of this muscle increases significantly during childhood, especially during the early years of life. Binzoni *et al.* (2001) further demonstrated that pennation angle is linearly related to muscle thickness, and argued that the growth-related increase in pennation angle is an important adaptation to a muscle's increased requirement to produce and tolerate high forces. In a subsequent study, Kawakami *et al.* (2006) quantified the relationship between pennation angle and muscle thickness in the triceps brachii, vastus lateralis and gastrocnemius medialis (n = 711; age range: 3–92 years). For all muscles, the authors reported significant positive relationships between pennation angle and muscle thickness, which supports Binzoni *et al.*'s (2001) argument. They further demonstrated that the relationship between muscle thickness and pennation angle was stronger for the triceps brachii than for the vastus lateralis and gastrocnemius medialis. Together, these findings suggest that the development of pennation angle is muscle specific.

In addition to this muscle-specific development of pennation angle, developmental increases in pennation angle appear to be non-linear. Morse *et al.* (2008) did not detect any significant differences in pennation angle of the gastrocnemius lateralis between 12-year-old boys and male adults. These results are in agreement with findings from O'Brien *et al.* (2010b), who also found only small differences in pennation angle between 9-year-old children and adults. Although not apparently obvious, results from Morse *et al.* (2008) and O'Brien *et al.* (2010b) are consistent with the data presented by Binzoni *et al.* (2001), who found age-related increases in pennation angle to become less pronounced later during childhood. Together, these findings suggest that the most significant developmental changes in pennation angle occur before 10 years of age. Nevertheless, more research is needed to document age- and sex-related differences in muscle pennation angles more comprehensively.

Optimal fibre length

Another input parameter for musculoskeletal models is the muscle's optimal fibre length, which is defined as the length at which the muscle's force output is maximal (i.e. the peak of the force–length relationship). At its core, the optimal fibre length is related to the optimal sarcomere length (i.e. the sarcomere's length at which there is maximal overlap of the contractile elements). However, the optimal fibre length is often inferred via the length of a muscle fascicle (bundle of muscle fibres), which can be measured using ultrasonography (Maganaris *et al.* 1998). Several authors have quantified muscle fascicle length in children, and it is not surprising that children's fascicles are shorter than those of adults (Kannas *et al.* 2010; Morse *et al.* 2008; O'Brien *et al.* 2010a). More interestingly, however, once normalized to muscle length or tendon length, such age-related differences in fascicle length disappear (Morse *et al.* 2008; O'Brien *et al.* 2010b).

The fact that muscle growth is mainly achieved by the addition of sarcomeres in series, while the (absolute) length of sarcomeres stays relatively constant (Goldspink 1964), allows sarcomeres to stay at their optimal length during muscle growth. From these findings, we can deduce that optimal fibre length (relative to total muscle or tendon length) stays relatively constant between 10 years of age and adulthood. Implicit in this conclusion is that optimal fibre length can be scaled in proportion to total muscle or tendon length when creating paediatric musculoskeletal models. One limitation to this recommendation is that it may be possible that age-related changes in optimal fibre length occur earlier during childhood (i.e. before 10 years of age). It is possible that the aforementioned age-related changes in pennation angle during infancy and early childhood (Binzoni *et al.* 2001) are accompanied by changes to the ratio of fascicle to muscle-tendon length, which could affect the optimal fibre length. Future research should specifically address this question.

Tendon slack length

When a muscle contracts and shortens, a muscular force is transferred via the tendon to the bone. In order for the tendon to be able to transfer muscular forces, it needs to be stretched beyond a certain threshold (i.e. the tendon slack length). Tendon slack length can be estimated by means of MR imaging or ultrasonography (Manal and Buchanan 2004). Tendon slack length is an important input parameter, and the outputs of musculoskeletal models are very sensitive to slight variations in tendon slack length. Unfortunately, we know very little about developmental changes in tendon slack length. The fact that the ratio of muscle and tendon length for the knee extensor muscles does not differ between 9-year-old children and adults (O'Brien *et al.* 2010a) suggests that tendon slack length scales proportionally to the length of the whole muscle-tendon unit during childhood. However, caution should be taken when choosing tendon slack length as an input parameter for a paediatric musculoskeletal model due to the aforementioned sensitivity of the model outputs to slight variations in tendon slack length. When creating paediatric musculoskeletal models, it is advisable to determine tendon slack length individually or perform appropriate sensitivity analyses to quantify the potential effects of an erroneous estimation of tendon slack length.

Tendon strain and tendon stiffness

Tendon strain can be defined as the change in tendon length relative to tendon slack length:

$$\varepsilon^T = (L^T - L^T_S) / L^T_S \tag{6.8}$$

where:

ε^T is tendon strain
L^T is tendon length
L^T_S is tendon slack length.

Tendon strain changes as a function of force, and the force-strain relationship is another important input parameter for musculoskeletal models (Zajac 1989). Related to tendon strain is tendon stiffness, which can be defined as the ratio of tendon force and tendon elongation (see Chapter 5):

$$k^T = F^T/(L^T - L^T{}_S)$$ (Equation 6.9)

where:

k^T is tendon stiffness
L^T is tendon length
F^T is tendon force
$L^T{}_S$ is tendon slack length.

Combining Equations 6.8 and 6.9 yields that tendon strain is inversely proportional to tendon stiffness. As we have seen in Chapter 5, tendons become stiffer during childhood (Kubo *et al.* 2001; Lambertz *et al.* 2003; O'Brien *et al.* 2010c), which implies that tendon strain is greater in children than in adults. Indeed, O'Brien *et al.* (2010c) found a significantly greater patellar tendon elongation for a given force output in 9-year-old children when compared to adults, which is indicative of greater tendon strain. These findings demonstrate that force-strain relationships should be adjusted appropriately when creating paediatric musculoskeletal models.

Maximum isometric force

Maximum isometric force is a crucial parameter when developing musculoskeletal models. It is well documented that age-related changes in muscular strength are non-linear and muscle-specific (Blimkie 1989), which should be taken into consideration when developing paediatric musculoskeletal models. For a detailed discussion of age-related changes in muscular strength, we refer the reader to Chapter 4.

Clinical and practical implications

To create paediatric biomechanical models, knowledge of many input parameters relating to the musculoskeletal system is required. The accuracy of these parameters determines the accuracy of the model's output. Further, it needs to be acknowledged that several input variables (as opposed to a single one) bear a greater (cumulative) potential for error. Thus, the development of forward musculoskeletal paediatric models is a particular challenge, as it requires a large number of input parameters. Another important consideration within this context is the reliability of the input parameters, as it will influence the reproducibility of the model's results. Reliability is of particular importance in paediatric populations due to the aforementioned growth-related changes in body composition (see Chapter 1) and muscle-tendon mechanics (see Chapter 5).

Nevertheless, information relating to the development of the musculoskeletal system during childhood exists, and it should be used to its full potential when creating paediatric musculoskeletal models. For paediatric inverse dynamics models, anthropometric parameters need to be adjusted appropriately. Such adjustments are relatively straightforward, as developmental changes in segmental parameters are well documented in the literature. Creating a paediatric musculoskeletal model is more challenging as it requires a much larger number of input parameters relating to muscle-tendon mechanics, some of which are not readily available in the literature. As we have seen in the sections above, we have limited knowledge of the development of muscle moment arms, pennation angle as well as muscle and tendon lengths, especially in girls. When creating paediatric musculoskeletal models this information should be used to maximize the validity of the model. Nevertheless, the limitations of the aforementioned studies should be considered, and appropriate sensitivity analyses should be performed to quantify the effect of potential errors in the estimation of muscle-tendon parameters on the model output. It is also clear that much more research is needed to document age-related changes in muscle and tendon mechanics comprehensively. If a particular parameter is not available to the developmental biomechanists, they may wish to use appropriate scaling techniques to derive the parameter from adult values. However, caution must be taken when scaling adult values to those of children, and the appropriate sensitivity analyses should be performed.

Finally, specific caution should be taken when using biomechanical models in clinical paediatric populations. Often, age-related changes in muscle-tendon parameters are reported as a function of chronological age, which is not always the best representation of biological maturity (see Chapter 1). Furthermore, the properties of the musculoskeletal system in paediatric clinical populations can differ immensely from those of typically developing children (see Chapters 10, 12 and 13). Researchers should be mindful of these factors when creating paediatric clinical biomechanical models.

Key points

- Estimates of muscular forces and torques can be obtained by means of inverse dynamics and forward dynamics modelling.
- When using paediatric kinetic models, it is important to choose child-appropriate inertial parameters, as inaccurate estimates of inertial characteristics can result in significant errors in the estimation of muscular forces or torques.
- Non-linear and segment-specific increases in segmental inertia are important contributors to the acquisition of new motor tasks during infancy and childhood.
- Observed age-related changes in muscular force production do not necessarily reflect an immature neuro-motor system, but can be indicative of functional muscular adjustments necessary to account for anthropometry-driven differences in non-muscular forces.
- The study of muscle-tendon mechanics in children is in its infancy. Based on the results of the relatively few relevant studies, it appears that the development of muscle-tendon parameters such as moment arms, pennation angles, tendon stiffness and maximum muscular force production is non-linear and

muscle-specific. Such non-linear changes should be taken into consideration when creating paediatric musculoskeletal models.

- When using child-specific input parameters for biomechanical models (derived from the literature or by scaling adult values), appropriate sensitivity analyses must be performed to quantify the effect of potential errors in the estimation of such parameters on the model's output.
- More research is needed to provide developmental biomechanists with the appropriate parameters to develop accurate and reliable paediatric musculo-skeletal models.

Acknowledgement

The preparation of this chapter was supported by the Engineering and Physical Sciences Research Council under grant EP/E013007/1.

References

Binzoni, T., Bianchi, S., Hanquinet, S., Kaelin, A., Sayegh, Y., Dumont, M. and Jequier, S. (2001) 'Human gastrocnemius medialis pennation angle as a function of age: from newborn to the elderly', *Journal of Physiological Anthropology and Applied Human Science*, 20: 293–8.

Blimkie, C.J. (1989) 'Age and sex associated variation in strength during childhood: anthropometric, morphologic, neurologic, biomechanical, endocrinological, genetic and physical activity correlates', in C.V. Gisolfi and D.R. Lamb (eds) *Perspectives in Exercise Science and Sports Medicine, (Vol. 2), Youth, Exercise and Sport*, Indianapolis, IN: Benchmark Press, pp. 99–163.

Brown, N.A. and Jensen, J.L. (2003) 'The development of contact force construction in the dynamic-contact task of cycling [corrected]', *Journal of Biomechanics*, 36: 1–8.

Brown, N.A. and Jensen, J.L. (2006) 'The role of segmental mass and moment of inertia in dynamic-contact task construction', *Journal of Motor Behavior*, 38: 313–28.

Buchanan, T.S., Lloyd, D.G., Manal, K. and Besier, T.F. (2004) 'Neuromusculoskeletal modeling: estimation of muscle forces and joint moments and movements from measurements of neural command', *Journal of Applied Biomechanics*, 20: 367–95.

Chester, V.L. and Jensen, R.K. (2005) 'Changes in infant segment inertias during the first three months of independent walking', *Dynamic Medicine*, 4: 1–9.

de Leva, P. (1996) 'Adjustments to Zatsiorsky-Seluyanov's segment inertia parameters', *Journal of Biomechanics*, 29: 1223–30.

Dempster, W.T. (1955) 'Space requirements of the seated operator', in *WADC-TR-55-159*, Aerospace Medical Research Laboratories, Ohio.

Fath, F., Blazevich, A.J., Waugh, C.M., Miller, S.C. and Korff, T. (2010) 'Direct comparison of in vivo Achilles tendon moment arms obtained from ultrasound and MR scans', *Journal of Applied Physiology*, 109: 1644–52.

Ganley, K.J. and Powers, C.M. (2004) 'Anthropometric parameters in children: a comparison of values obtained from dual energy X-ray absorptiometry and cadaver-based estimates', *Gait & Posture*, 19: 133–40.

Ganley, K.J. and Powers, C.M. (2005) 'Gait kinematics and kinetics of 7-year-old children: a comparison to adults using age-specific anthropometric data', *Gait & Posture*, 21: 141–5.

Ganley, K.J. and Powers, C.M. (2006) 'Intersegmental dynamics during the swing phase of gait: a comparison of knee kinetics between 7-year-old children and adults', *Gait & Posture*, 23: 499–504.

Goldspink, G. (1964) 'Increase in length of skeletal muscle during normal growth', *Nature*, 204: 1095–6.

Hatze, H. (1980) 'A mathematical model for the computational determination of parameter values of anthropomorphic segments', *Journal of Biomechanics*, 13: 833–43.

Jensen, J.L., Thelen, E., Ulrich, B.D., Schneider, K. and Zernicke, R.F. (1995) 'Adaptive dynamics of the leg movement patterns of human infants: III. Age-related differences in limb control', *Journal of Motor Behavior*, 27: 366–74.

Jensen, R.K. (1981a) 'The effect of a 12-month growth period on the body moments of inertia of children', *Medicine and Science in Sports and Exercise*, 13: 238–42.

Jensen R.K. (1981b) 'The effect of differences in body segment inertias on the segmental force vector patterns for children jumping', in A. Moreclu, K. Fidelus, K. Kedzior and A. Wit (eds) *Biomechanics VII-B*, Baltimore, MD: University Park Press, pp. 277–83.

Jensen, R.K. (1986) 'The growth of children's moment of inertia', *Medicine and Science in Sports and Exercise*, 18: 440–5.

Jensen, R.K. (1988) 'Developmental relationships between body inertia and joint torques', *Human Biology*, 60: 693–707.

Jensen, R.K. (1989) 'Changes in segment inertia proportions between 4 and 20 years', *Journal of Biomechanics*, 22: 529–36.

Jensen, R.K. and Nassas, G. (1988) 'Growth of segment principal moments of inertia between four and twenty years', *Medicine and Science in Sports and Exercise*, 20: 594–604.

Jensen, R.K., Sun, H., Treitz, T. and Parker, H.E. (1997) 'Gravity constraints in infant motor development', *Journal of Motor Behavior*, 29: 64–71.

Kannas, T., Kellis, E., Arampatzi, F. and de Villarreal, E.S. (2010) 'Medial gastrocnemius architectural properties during isometric contractions in boys and men', *Pediatric Exercise Science*, 22: 152–64.

Kawakami, Y., Abe, T., Kanehisa, H. and Fukunaga, T. (2006) 'Human skeletal muscle size and architecture: variability and interdependence', *American Journal of Human Biology*, 18: 845–8.

Korff, T. and Jensen, J.L. (2008) 'Effect of relative changes in anthropometry during childhood on muscular power production in pedaling: a biomechanical simulation', *Pediatric Exercise Science*, 20: 292–304.

Kubo, K., Kanehisa, H., Kawakami, Y. and Fukanaga, T. (2001) 'Growth changes in the elastic properties of human tendon structures', *International Journal of Sports Medicine*, 22: 138–43.

Lambertz, D., Mora, I., Grosset, J.F. and Perot, C. (2003) 'Evaluation of musculotendinous stiffness in prepubertal children and adults, taking into account muscle activity', *Journal of Applied Physiology*, 95: 64–72.

Maganaris, C.N., Baltzopoulos, V. and Sargeant, A.J. (1998) 'Changes in Achilles tendon moment arm from rest to maximum isometric plantarflexion: in vivo observations in man', *Journal of Physiology*, 510: 977–85.

Manal, K. and Buchanan, T.S. (2004) 'Subject specific estimates of tendon slack length: a numerical method', *Journal of Applied Biomechanics*, 20: 195–203.

Morse, C.I., Tolfrey, K., Thom, J.M., Vassilopoulos, V., Maganaris, C.N. and Narici, M.V. (2008) 'Gastrocnemius muscle specific force in boys and men', *Journal of Applied Physiology*, 104: 469–74.

Mungiole, M. and Martin, P.E. (1990) 'Estimating segment inertial properties: comparison of magnetic resonance imaging with existing methods', *Journal of Biomechanics*, 23: 1039–46.

Norton, K. and Olds, T. (eds) (1996) *Anthropometrica*, Sydney: UNSW Press.

O'Brien, T.D., Reeves, N.D., Baltzopoulos, V., Jones, D.A. and Maganaris, C.N. (2009) 'Moment arms of the knee extensor mechanism in children and adults', *Journal of Anatomy*, 215: 198–205.

O'Brien, T.D., Reeves, N.D., Baltzopoulos, V., Jones, D.A. and Maganaris, C.N. (2010a) 'In vivo measurements of muscle specific tension in adults and children', *Experimental Physiology*, 95: 202–10.

O'Brien, T.D., Reeves, N.D., Baltzopoulos, V., Jones, D.A. and Maganaris, C.N. (2010b) 'Muscle-tendon structure and dimensions in adults and children', *Journal of Anatomy*, 216: 631–42.

O'Brien, T.D., Reeves, N.D., Baltzopoulos, V., Jones, D.A. and Maganaris, C.N. (2010c) 'Mechanical properties of the patellar tendon in adults and children', *Journal of Biomechanics*, 43: 1190–5.

Roncesvalles, M.N., Woollacott, M.H. and Jensen, J.L. (2001) 'Development of lower extremity kinetics for balance control in infants and young children', *Journal of Motor Behavior*, 33: 180–92.

Schneider, K. and Zernicke, R.F. (1992) 'Mass, center of mass, and moment of inertia estimates for infant limb segments', *Journal of Biomechanics*, 25: 145–8.

Schneider, K., Zernicke, R.F., Schmidt, R.A. and Hart, T.J. (1989) 'Changes in limb dynamics during the practice of rapid arm movements', *Journal of Biomechanics*, 22: 805–17.

Schneider, K., Zernicke, R.F., Ulrich, B.D., Jensen, J.L. and Thelen, E. (1990) 'Understanding movement control in infants through the analysis of limb inter-segmental dynamics', *Journal of Motor Behavior*, 22: 493–520.

Sun, H. and Jensen, R. (1994) 'Body segment growth during infancy', *Journal of Biomechanics*, 27: 265–75.

Tsaopoulos, D.E., Baltzopoulos, V., Richards, P.J. and Maganaris, C.N. (2009) 'A comparison of different two-dimensional approaches for the determination of the patellar tendon moment arm length', *European Journal of Applied Physiology and Occupational Physiology*, 105: 809–14.

Waugh, C.M., Blazevich, A.J., Fath, F. and Korff, T. (2011) 'Can Achilles tendon moment arm be predicted from a single anthropometric measure in prepubescent children?', *Journal of Biomechanics*, 44: 1839–44.

Winter, D.A. (1990) *Biomechanics and Motor Control of Human Movement*, New York: John Wiley & Sons.

Winter, D.A. and Eng, P. (1995) 'Kinetics: our window into the goals and strategies of the central nervous system', *Behavioural Brain Research*, 67: 111–20.

Woittiez, R.D., Huijing, P.A., Boom, H.B. and Rozendal, R.H. (1984) 'A three-dimensional muscle model: a quantified relation between form and function of skeletal muscles', *Journal of Morphology*, 182: 95–113.

Wood, L., Dixon, S.J., Grant, C. and Armstrong, N. (2006) 'Elbow flexor strength, muscle size, and moment arms in prepubertal boys and girls', *Pediatric Exercise Science*, 18: 457–69.

Zajac, F.E. (1989) 'Muscle and tendon: properties, models, scaling, and application to biomechanics and motor control', *Critical Reviews in Biomedical Engineering*, 17: 359–411.

Zatsiorsky, V. and Seluyanov, V. (1983) 'The mass and inertia characteristics of the main segments of the human body', in H. Matsui and K. Kobayashi (eds) *Biomechanics VIII-B*, Champaign, IL: Human Kinetics, pp. 1152–9.

Part III

Biomechanical aspects of the development of postural control and selected fundamental motor skills

7 Biomechanical aspects of the development of postural control

Jody Jensen and Renate van Zandwijk

Introduction

Biomechanics refers to biological systems and how their physical properties interact with the physical environment. In this chapter we consider the effect of the changing biomechanics on skill acquisition in the developing child. Specifically, this analysis will focus on the development of postural control because of its prominence in early development and the critical role of postural control in the acquisition of other skills. Hurdles in developing postural control include the management of total body mass under both static and dynamic conditions as well as producing muscle forces appropriate to the external environment of gravity, contact and motion-dependent forces. Postural skill in particular requires control of the body's mass and inertia: resistance to and task-appropriate exploitation of gravity as well as the accommodation to the effects of changing external forces. Understanding the developmental course of postural control and the biomechanics that underlie such skills is important to the early identification of atypical developmental processes and early identification of children in need of intervention.

Getting upright

Sitting in a chair, reaching for the morning cup of coffee, seems routine and automatic. Standing in conversation with someone requires our attention to the topic, not the task. For most of us, configuring and maintaining our multi-segmented bodies in a desired posture, like the well-practised postures of sitting or standing, takes little conscious effort. These are skills we acquired long ago. But indeed, these skills were acquired. Our goal in this chapter is to review the developmental progression from novice to expert in the demonstration and control of upright postures. In this review, we focus on the biomechanics of postural control, the interface between biology and engineering. Sitting and standing are behaviours that require an alignment of body-building blocks in such a way as to minimize the collapsing effort of gravity. Sitting and standing are postures – static formations of segmental configurations – that underlie purposeful movement. Sitting and standing as tasks are dynamic as the postures are maintained under conditions of movement or perturbation. How do we go from the complete lack of purposeful movement as a newborn infant to the expertise of head-up control while being bounced in the arms of a parent, staying upright while riding a bicycle, or traversing the billowy obstacle course of a moonwalk? The infant demonstrates independent sitting by 6 months of age (50th percentile, 95th percentile at 8 months) and shows independent standing by 11 months of age (50th percentile, 95th at 14 months). But do not confuse competence with mastery. The 6- and 11-month references are the average ages (rounded to whole months), at which infants acquire these basic skills (Martorell *et al.* 2006). Getting from the onset of a skill to being expert at its performance is its own story.

The structure of this chapter is as follows. First we address the transition from lack of ability to the onset of a reliable (repeatable) performance. What factors and features of the biomechanical system must change to support the expression of a new behaviour? Second, we look at the transition from competence to expertise. Again, what changes, biological and/or mechanical, occur to support the wider diversity of movement competence? Increasing competence in sitting and standing skills means success at accomplishing these skills across more challenging environments – and that means the transition from static to dynamic conditions.

Basic theoretical concepts

Developmental biomechanics

The human body is a biological system that can be described by the terms of mechanics – masses, levers, forces and inertia. The body's composition of bones, joints and muscle can alternatively be conceived of as levers, axes and forces. The appeal of studying human movement from the perspective of biomechanics lies in the evolving understanding of how the mechanics of movement facilitate (or hinder) our actions, independent of neural control. The modifier 'developmental' refers to the passage of time and changing states of the biological system. Consider the developmental changes in the body in the first year of life. On average, body weight increases 280 per cent between birth and the first birthday, body length increases 50 per cent, and head circumference increases an average of 30 per cent (Kuczmarski *et al.* 2000). It is in this same first year that the infant transitions to toddler status by acquiring the species-typical upright bipedal locomotion. The mechanical features of the human body are changing dramatically during the period in which core movement skills are being gained. Any account of movement skill acquisition must also account for such significant change in the moving system.

Dependent measures

Postural skill development has been described by a multitude of dependent measures, commonly categorized as kinematics and kinetics (or dynamics). It is typical to first describe change by kinematic measures – the descriptors of movement quantified as position, velocity and acceleration. Position descriptors are used to define the alignment of body segments (Forssberg and Hirschfeld 1994), ranges of motion (Wong *et al.* 1998) or the position of the centre of gravity with respect to the perimeters of the base of support (Usui *et al.* 1995). Velocity measures have been used to characterize the rate of change in segmental angles or the speed of postural sway (Harbourne *et al.* 1993). Estimating head acceleration has been used to relate sensory system contributions (see Chapter 2) to postural control (e.g. the vestibular system is responsive to acceleration (Allum and Pfaltz 1985)). To understand the cause of movement, we look to the domain of kinetics (dynamics). Forces and torques create changes in the kinematics of bodies. Improving skill may be revealed in the task with the appropriate matching of forces and torques. For example, early efforts at reaching for a toy are characterized by muscle torques produced to stabilize posture and the movement of the arm on a stable base (Thelen and Spencer 1998). Skilful reaching is characterized by faster reaches and fewer trajectory corrections, both of which require the production of muscle forces necessary to construct the movement and manage the passive reactive forces associated with movement (Thelen *et al.* 1996). The limits of postural control are often tested by applying forces or torques to the body to quantify and evaluate the adaptive response necessary to avoid a fall and recover standing balance (Bothner and Jensen 2001; Brown *et al.* 2001; Hall *et al.* 1999).

Beyond these broad categories of dependent measures, there are some measures that are historically associated with assessments of postural control. The centre of pressure (COP) is one of these measures. COP refers to a location – the location at the interface between the body and the support surface at which the resultant ground reaction force is applied. This resultant ground reaction force is the equal and opposite force of a weighted average of all applied forces at the contact surface (Prieto *et al.* 1996; Winter 1990). It is a summary variable that reveals the stability/variability of a quiescent system (e.g. during quiet sitting or quiet stance (Cignetti *et al.* 2011; Deffeyes *et al.* 2009b; Kirshenbaum *et al.* 2001)) or information about the dynamic control of posture or gait (Winter and Eng 1995).

COP has been used as a marker of stability, postural vulnerability and movement control (Inman 1966; Riach and Hayes 1987). Under conditions of quiescence, the analysis can be undertaken as a series of static trials. In this case, the location of the COP can roughly be aligned with the location of the centre of mass (COM) (Winter *et al.* 1998) and inferences about the rigidity or compliance of standing postural control may be made. Under more dynamic circumstances, the horizontal distance between the COP and COM bears a direct relationship to the linear acceleration experienced by the COM (Jian *et al.* 1993). Mann *et al.* (1979) reported step initiation to be characterized by a backward movement of the COP with the resulting torque about the whole body centre of mass initiating a forward fall and progression. COP trajectories may be further partitioned into anterior– posterior (A/P) and medial–lateral (M/L) excursions to provide further detail about postural control (Winter *et al.* 1993, 1996). COP trajectories, then, may be interpreted in the context of maintaining postural control or balance recovery adjustments (Kirshenbaum *et al.* 2001).

COP is easily quantified and is sensitive to conditions of development, impairment and training (Hasan *et al.* 1996; Prieto *et al.* 1996; Winter and Eng 1995). This variable has been shown to reveal differences between healthy and impaired subjects (e.g. patients with autism, cerebral palsy or stroke (Deffeyes *et al.* 2009a; Fournier *et al.* 2010; Niam *et al.* 1999)). COP measurements can be used as a tool for early measures of motor development pathology (Deffeyes *et al.* 2009b). Further, COP sway measures may be used to study the influences of training on balance control (Bayouk *et al.* 2006; Harbourne *et al.* 2010).

Studies in the past have used different statistical derivatives of COP-based sway measures to analyse balance. These measures can be divided into linear and non-linear sway measurements. Linear techniques quantify the amount of movement and/or variability of the COP. These measurements document an overall pattern in balance control but they do not provide any temporal information (Kyvelidou *et al.* 2010). Commonly used linear sway measures in the study of the development of balance include sway path, area and velocity (Chen *et al.* 2007; Kyvelidou *et al.* 2010). The sway path quantifies the distance that the COP travels in a certain time period. The travelled path can be calculated in each direction separately or by the total excursion in both directions. Prieto *et al.* (1996) quantified the sway trajectory by calculating average sway (the accumulated distance between adjacent samples divided by the number of samples or the root mean square (RMS) associated with deviation from a static point). Sway area is an estimation of the area that is enclosed by the path of the COP. It is dependent on the distance from the mean sway path and the distance travelled by the COP (Prieto *et al.* 1996). The velocity is calculated by quantifying the position change between adjacent samples relative to the corresponding time difference. This velocity is then averaged over the entire trajectory (Prieto *et al.* 1996). High velocity of the COP is considered an indication of poor balance, as the boundaries of support may be reached more quickly. High velocity is acceptable if the COP is also characterized by high frequencies (van Dieën *et al.* 2010), since this combination of features marks rapid adjustment consistent with good balance.

In addition to these linear measurements, non-linear measures provide further insights into the properties of a dynamical system, which is represented by the time series of a biological signal (e.g. the COP excursion). They allow us to make inferences about the control strategies underlying the observed (descriptive) differences in (linear) kinematic or kinetic variables (Stergiou 2004). Non-linear techniques provide measures of variability in the COP movement across time (Kyvelidou *et al.* 2009a). Several non-linear techniques exist for the analysis of postural movement. These measures include the computation of the approximate entropy (ApEn), the Lyapunov exponent (LyE) or correlation dimension (CoD). The ApEn estimates the signal's complexity by quantifying its predictability. A more complex system is less predictable than a less complex system. The LyE estimates how rapidly nearby trajectories within biological signals diverge. When nearby trajectories diverge quickly, the system is more unstable. Thus, LyE measures the local stability of a dynamical system. The CoD is a way to identify the number of degrees of freedom within a dynamical system. It quantifies how many dimensions are needed to describe a signal, which in turn provides us with information about its complexity (Stergiou 2004). All these complex, non-linear techniques result in a single dependent variable that quantifies the regularity or predictability of a signal: the higher the value, the less predicable the signal (or the more complex the system); the smaller the value, the more predictable the signal (or the less complex the system).

Non-linear measures have been used by several authors to study postural control (Harbourne and Stergiou 2003, 2009; Newell *et al.* 1997). Such studies have provided us with more information about the mechanisms underlying observed age-related differences in the (linear) behaviour. For example, Harbourne and Stergiou (2003) studied the development of sitting by quantifying age-related changes in ApEn, LyE and CoD of COP excursions during sitting in infants. The authors found that ApEn and LyE significantly decreased as infants gained more sitting experience, which indicates that they became more stable. The CoD decreased initially and increased thereafter, which suggests that infants first froze their mechanical degrees of freedom (which is common when learning new tasks) and released them thereafter (which is beneficial in becoming adaptable in response to perturbations, which are described later in this chapter).

The developmental biomechanics – the principles

Progressive skill acquisition for the developing child follows three general principles that define emerging motor control in the context of biomechanics (Jensen and Bothner 1998):

1   Progression in skill development shows an increasing ability to control greater proportions of total body mass.
2   Progression in skill development shows an increasing ability to oppose and when appropriate, exploit the context of gravity.
3   Progressive skill development shows an increasing ability to match muscular with non-muscular forces in task appropriate ways.

Figure 7.1 shows an approximate timeline of emerging postural skills. The use of a stacked Venn diagram is functional, as the skills clustered in each smaller circle are similar in mechanical demand. Each larger circle signifies the control of larger proportions of body mass. Within each circle, the progression of skills (bottom to top) demonstrates a recurring theme: skill is gained under static, then dynamic conditions.

Development of sitting

Lifting the head

At birth, the infant is incapable of voluntary, purposeful movement. One of the earliest challenges is simply expressing sufficient strength to lift her head up from the mattress. The newborn's head is just shy of one-quarter the length of the body

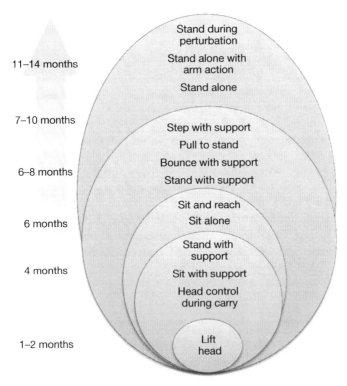

Figure 7.1 Schematic representation of the development of postural control.

(and as an adult will comprise one-seventh of body stature (Farris 1961)). Having lifted the head with purpose, the infant is challenged to maintain head control while being carried in a parent's arms. Imagine the bouncing and dipping of the parent's knees, the rotation experiences as mother or father turns to look in another direction. All of these movements are transmitted to the child and the last distal link in the kinetic chain – the head. Parents and caregivers are notably cautious with young infants, but soon they accelerate their movements to elicit smiles and noises of joy from the child.

Stacking the blocks

In early development, gaining skill is about aligning body parts and overcoming the collapsing effect of gravity. Thus, extending this period of developing head control leads to the emerging practice phase for independent sitting. To sit, the infant must stabilize the head, neck and trunk. The trunk is supported by 24 vertebrae, each articulating with another bone. Each articulation is a potential site of collapse. The infant's first experience of supported sitting is in an infant seat, carrier or sling, or with the parent's hands wrapped around its trunk. At 1 month of age, infants show little head control, maintaining an upright head posture only at approximately 3 months, and even at 3 months of age, extensive head wobble is observed (Hedberg *et al.* 2005).

The first attempts at creating a sitting posture are characterized by disorganized firings of the trunk and hip muscles, and a susceptibility to position-dependent effects of gravity (Hedberg *et al.* 2005; Kyvelidou *et al.* 2009b). At 1 month of age, infants show direction-specific muscle responses to perturbations of sitting better than 70 per cent of the time (Hedberg *et al.* 2004). By the time the infant is 7–8 months of age, these direction-specific responses are stable and present virtually 100 per cent of the time (Hedberg *et al.* 2005). Forssberg and Hirschfeld (1994) proposed that these approximate muscle synergies evident in the 1-month-old infant are pre-adapted organizational circuits upon which a second level of control is built for refined motor control. This second level of control emerges as a function of experience (Hadders-Algra 2005).

The expression of developing control can be seen as well in the changing kinematics of sitting. The height of the COM above the base of support and the configuration of the body segments stacked upon the base combine to provide varying degrees of stability for the sitting infant. The base of support during sitting is much larger than during standing with the COM close (vertically) to the base of support (Chen *et al.* 2007). The typical sitting posture is the 'ring-sit' position with legs extended, hips externally rotated, knees slightly flexed and feet together, creating the ring of legs over which the trunk leans. Thus, great stability is available in the forward direction. Consequently, little stability is available for backward leaning (Brogren *et al.* 1998).

Postural configuration influences muscle synergy activation. Harbourne *et al.* (1993) studied sitting attempts in infants 2–5 months of age. In this study, infants were held in an upright position, and then the support was released. With experience (as associated with age), disorganized muscle responses gave way to predictable synergies among the trunk extensor paraspinal muscles, the hip extensor hamstrings and the hip/trunk flexor quadriceps muscles. The more sophisticated muscle synergies emerge from the practice of a variety of postural configurations (Harbourne *et al.* 1993; Hedberg *et al.* 2005).

The kinetic and kinematic descriptors of postural control of sitting complement our understanding of the corresponding neural changes in the early months. Infants experience a range of postural support conditions from sitting with support, to independent sitting for a few seconds, to independent stable sitting. Interestingly, linear measurements of the COP do not show much change between the different conditions (Harbourne and Stergiou 2003). Measurements such as the length of the path, excursion in x and y directions and the RMS in the A/P direction show little change as the infant gains sitting control. The one clear kinematic change was in the M/L direction (Cignetti *et al.* 2011). This finding is not unexpected given the ring-sit posture of the infant with forward trunk lean. Further, perturbations of this posture are likely to occur in the M/L direction where the lean of the body must be accommodated by extensions of the arm in the opposite direction.

As described earlier, the non-linear measures relating to the COP reveal more information about the development of postural control. During the transition from novice to experienced sitter, infants have to gain control of the degrees of freedom of their multi-segmented body (Cignetti *et al.* 2011; Harbourne and Stergiou 2003). The control of the degrees of freedom is accompanied by increasing stability and the regulation of the strategy during mastering the sitting task (Harbourne and Stergiou 2003). By 5 months, the infant has managed to stack the blocks of head and sequential vertebrae on top of the sacrum and pelvis to maintain, momentarily, a vertical column aligned with gravity (Hedberg *et al.* 2005). Each joint is described by multiple degrees of freedom (e.g. the joints between the vertebrae each have six degrees of freedom) (Harbourne and Stergiou 2003). Muscle coordination improves such that the child requires only slight external support to maintain the upright posture, but the need for support is the observable effect of limited control of the degrees of freedom. During this stage, the balance of the infant is unstable and very complex as evidenced by high LyE and ApEn values, respectively (Cignetti *et al.* 2011; Harbourne and Stergiou 2003).

Independent sitting is achieved by 6 months with 95 per cent of children sitting alone by 8 months of age (Martorell *et al.* 2006). These developmental milestones, however, refer to quiet sitting, not sitting while doing. When the infant learns to sit for a few seconds, they are more (local) stable, as in preparation for other skills, such as reaching. The increased stability is due to the aforementioned freezing of the degrees of freedom. With increased practice, the sway patterns of the infant become less complex, indicating an improved strategy for sitting, and once infants are able to sit independently, they release the degrees of freedom, which increases the repertoire of responses to eventual perturbations (Harbourne and Stergiou 2003).

The development of postural control during sitting in infants depends on the sway direction (Cignetti *et al.* 2011). The stability increases to a greater extent in the A/P direction when compared to the M/L direction. The reason for this direction-dependence of the development of postural control lies in its functional relevance: stability in the A/P direction is important to reach and grasp objects. Therefore, the infant is subjected to challenges to postural control in the A/P direction more frequently (when compared with the M/L), which encourages them to explore the solutions to the task to a greater extent. As a result, postural control in the A/P direction is typically acquired first.

Once the infant sits stably and independently, they still do not show adult-like sway patterns. Throughout infancy and childhood, children continue to free their degrees of freedom and explore the different solutions to the task (Cignetti *et al.* 2011). Adults are better in controlling their degrees of freedom, which makes them more adaptable to changes in the environment (Chen *et al.* 2007). Furthermore, the development of sitting control is 'disrupted' by other motor milestones that require postural control. For example, Chen *et al.* (2007) observed deterioration in the postural control of independent sitting just before the onset of walking. The authors argue that independent walking requires a 're-calibration of an internal model for the sensorimotor control of posture' (p. 16), which temporarily impairs postural control of independent sitting.

Clinical and practical applications: sitting postural control

The emergence of independent sitting is a major motor milestone. Upon achievement, the infant's hands are freed for manipulation activities. As a result, the infant spends more time in the upright position, gaining strength in opposing gravity, and the possibilities for the direction of gaze and attention expand tremendously. The corollary is that failure to achieve independent upright sitting control is associated with a cascade of developmental delays. Early intervention is important for minimizing the consequences of developmental delay (see Chapters 12 and 13).

Developmental delay in the acquisition of sitting postural control is, at a mechanical level, an inability to manage the mass of the body in a gravitational environment. A review of the biomechanics of sitting postures revealed that linear kinematic descriptors and non-linear techniques have the potency to discriminate early between typical and delayed developmental groups (Deffeyes *et al.* 2009a, 2009b) with non-linear techniques providing a more fine-grained analysis. These techniques are reliable within and between sessions for measuring balance during sitting, especially for the measures of RMS and LyE (Kyvelidou *et al.* 2010). These measures are sufficiently well established as to be discriminating among typical and atypical development and useful in the design of individual interventions (Deffeyes *et al.* 2009b; Harbourne *et al.* 2010; Kyvelidou *et al.* 2010). For example, Fallang and Hadders-Algra (2005) evaluated infants born prematurely and compared their postural development to infants who reached full term. COP analyses at 4 and 6 months showed preterm infants to be restricted in COP movements. The excursion of the COP was significantly reduced, as was the maximum COP velocity. These assessments were performed while the infant was lying supine and reaching for a suspended toy. Functionally, the COP restrictions were also associated with fewer reaches to the toy. The importance of these data was revealed in a follow-up study of the preterm infants when they reached 6 years of age (Fallang *et al.* 2005). In this study COP restriction at 6 months of age was found to be associated with less favorable neuromotor function at school age. Consequently, the authors identified abnormal COP patterns in infants as a potential marker for mild brain dysfunction.

As important as the detection of developmental delay is the potential for intervention. Early intervention can lead to improvement of motor skill execution. Harbourne *et al.* (2010) showed that an eight-week perceptual-motor intervention in infants with motor delays can improve sitting postural control. After the intervention, both linear and non-linear measures of COP showed infants with an initial developmental delay to more closely approximate the behaviour of typically developing infants.

Intervention also provides the opportunity for creating a synthesis of neural system response in the context of the existing biomechanics. Hadders-Algra *et al.* (1996) were able to demonstrate strengthening of the muscle activation patterns for sitting as a function of balance training in typically developing infants. Repeated perturbations of sitting led to the increased probability of observing consistent muscle synergies across multiple body segments. An intervention with children diagnosed with cerebral palsy also showed some efficacy in improving control in the basic direction-specific muscle synergies (de Graaf-Peters *et al.* 2007). However, children who had not achieved independence in sitting by 18 months lacked the basic level of postural control and fared significantly worse at later ages. Such findings argue for the importance of early intervention.

Basic theoretical concepts: standing postural control

Sitting to standing

Sitting postural control is the stable base upon which functional movements are built. From the control of a stable column, to control of a column that has moving appendages, increasing skill means being able to match muscular and non-muscular forces in a context appropriate way (Jensen 2005; see also Chapter 6). In early skill development, the non-muscular force constraining motion is gravity. Infants not yet skilful in reaching for objects show immediate improvement when provided with postural support (Bertenthal and Von Hofsten 1998; Rochat 1992). An infant given postural support around the trunk can perform alternate steps on a moving treadmill (Thelen 1986). The step kinematics of toddlers just taking their first independent steps look much more mature when provided with light hand support (Clark *et al.* 1988; Ivanenko *et al.* 2005). What this postural support provides is assistance with managing larger proportions of total body mass and assistance with opposing gravity.

Imagine the task of an infant making the transition from sitting to standing. Some 68 per cent of the infant's mass is above the hip joint. The legs are segmented and the knee, when flexed, is prone to collapse. The typical infant can support 100 per cent of total body weight at 4 months of age (Roncesvalles and Jensen 1993) on stiffly extended legs. Within a few weeks, she can bounce by alternately flexing and extending at the hips and knees. By 6–7 months of age, she will gain vigor and produce 1.6–2.0 times body weight in leg thrusts. She will self-pull/push into a standing posture by 9–10 months of age and take voluntary steps while holding on to the coffee table or stroller. At 11–12 months of age, she will stand alone and become so stable as to be able to perform a knee bend to retrieve a dropped toy from the ground (Martorell *et al.* 2006). This developmental trajectory demonstrates progressive accomplishment in mastering the changing biomechanics of morphology and movement.

Once upright, the task remains one of stacking the blocks one atop the other and keeping alignment with gravity. First attempts at independent stance are tenuous. In a classic study, Thelen *et al.* (1982) demonstrated the power of the gravitational environment in shaping the production of movement. The presence of the infant stepping reflex shortly after birth is one of the hallmarks of the intact nervous system and the heavily reflexive-based movements displayed by the newborn. It is not long after birth, however, that the stepping reflex becomes harder and harder to elicit, ultimately disappearing from the infant's motor repertoire. Debate as to the reason for this apparent loss of skill ranged from explanations based in cognition and information processing (Zelazo 1983), to simple physics. Thelen *et al.* (1982) noted an inverse relationship between the number of steps performed by infants and their weight gain. This correlation, manifested in the reduction of steps performed during attempts to elicit the stepping reflex, might also have been accompanied by central nervous system reorganization. An explicit test of the role of physics was required. Thus, the authors undertook a test of the hypothesis that the disappearance of the stepping reflex was a function of a reduction in the strength-to-weight ratio that occurred during a window of significant weight gain by the infant. Using infants between 2 and 8 weeks of age, Thelen *et al.* (1982) manipulated the requirements of strength necessary for stepping by testing the stepping reflex during conditions of added weight and reduced weight. For the added weight condition, small ankle weights were attached to the infant's limbs. The reduced weight condition was performed while the infants were supported chest-deep in water. Buoyancy in the immersion test functionally reduced the weight of the infant limbs. The authors observed that stepping was significantly reduced in the added weight condition (not unexpectedly), but significantly increased in the water condition. This experiment made clear that the neural pathways for stepping had not been disrupted or dissolved, and no significant shift in cognition was required. Rather, the expression of infant stepping was being suppressed because of an environmental constraint.

We see in the expanding movement repertoire a number of behaviours in which the infant is gaining control over the muscles to oppose and exploit, the effects of gravity. Long before attempts at standing, infants engage in play that helps them work out their interaction with gravity. It begins with kicking. Thelen and colleagues argued that supine kicking was the practice arena for sustaining the limb coordination that was first revealed in the stepping reflex and which would re-emerge in independent walking (Thelen 1985; Thelen *et al.* 1982). The flexion and extension phases as well as the inter-kick intervals shared temporal features of the early steps of supported walking.

Roncesvalles and Jensen (1993) continued with this line of reasoning by showing that infant kicks (kicking while held in an upright posture; air stepping) also served as the unloaded condition in which to practise forceful leg extension. In this study, the leg movement patterns of 2-week-, 3-month-, 5-month-, 7-month- and 9-month-old infants were compared under kicking and bouncing conditions. The purpose of this comparison was to determine how infants met the demands of the two tasks. While the kinematics of kicking and bouncing are quite similar (nearly synchronous flexion and extension at the hip and knee), the muscular force requirements are quite different. Kicking requires a lift of the leg against the effect of gravity, but the extension phase may be passive – a 'giving in' to gravity. Thus, the flexion phase of the movement is effortful, but the extensor phase of the movement is, in many cases, relatively passive. It is not uncommon during the extensor phase to find a net flexor muscle torque at the hip counterbalancing the extensor effects of gravity (Jensen *et al.* 1994). To bounce, however, the active and passive phases of the movement are reversed. In the bounce it is the flexion, or knee bend, portion of the movement that is passive (controlled by eccentric, or lengthening muscle contraction). Its performance requires a release of extensor tone so that gravity may pull the system into a knee bend. The effortful phase of the bounce is the extensor phase as the infant must develop sufficient muscular force to overcome the downward acceleration due to gravity and reverse the direction of the body's motion. The development of kicking sets the stage for the acquisition of weight-bearing skills and the emergence of bouncing. There is no intrinsic requirement of kicking that demands a forceful or vigorous extension phase, yet such vigour in extension is required for the successful bounce. The addition of the bounce to the movement repertoire is a demonstration of learning to alternately use, then oppose gravity. What begins at this age is the exploration of the push and relax, the extension and flexion cycling ('bouncing') of the spring-like behaviour inherent in the musculoskeletal system of the lower extremities (see Chapters 5 and 8).

This bouncing behaviour provides a unique window on the acquisition of bipedal locomotor abilities as it represents some of the infant's earliest experiences with the regulation of their own weight in an upright posture (Roncesvalles and Jensen 1993). Two-week-old infants typically support approximately half of their body weight. The proportion of body weight managed by self-support increases to two-thirds of body weight at 3 months. Beyond 4 months of age, infants are fully capable of full body weight support, during quiet stance. A slight bend of the knees, a slight reduction in extensor tone, and the body collapses. By 5 months of age, however, the infant begins to play with flexion and extension. The child relaxes the extensor tone and then pushes up with a force equalling 110 per cent of body weight. By 7 months, the infant is able to absorb and generate forces equivalent to 130 per cent of body weight, and by 9 months is regularly pushing forces equal to or greater than 160 per cent. It is not unusual for a 9-month-old to produce extension forces greater than 2.5 times their body weight during vigorous bounces.

The neural strategy underlying these behavioural changes is the earlier recruitment of the extensor musculature (quadriceps), such that sufficient force is developed in a timely way to overcome the downward acceleration driven by gravity (Roncesvalles and Jensen 1993). For example, a 4-month-old who performed few successful bounces activated the quadriceps an average of 119 ms prior to the time of maximum ground reaction force. By 7 months of age, this same infant showed an average onset of 186 ms prior to reaching the depth of the knee bend (temporally associated with maximum vertical force).

The neural tuning and escalating force production is associated with the transition from standing with support to bouncing with support. This expansion of movement repertoire demonstrates the transition from managing larger proportions of whole body mass to learning how to work with and against gravity. These are the underlying skills that support cruising, free standing and standing while bending to pick up dropped objects.

Standing independently requires the added skills of managing the total body mass and managing the reactive forces associated with body movement (lifting the arms to reach, turning the head to look towards a sound or toy). The direction-specific muscle synergies (ventral and dorsal) built up to support independent sitting continue to support the trunk, neck and head in standing. Automatic postural responses (Horak and Nashner 1986; Nashner 1976) among the lower extremity muscles are invoked to control body sway. From distal to proximal, the triceps surae, hamstrings and paraspinal muscles respond in sequence to forward sway, and the tibialis anterior, quadriceps and abdominal muscles respond to backward sway. These postural synergies are present in rudimentary form (Hadders-Algra 2005) and highly variable (Sveistrup and Woollacott 1996) as independent standing is first expressed. Newly standing infants show greater COP excursions and sway closer to their limits of balance than children who have the experience of walking and running (van Zandwijk *et al.* 2010).

Greater stability in postural control is gained by tuning the nervous system's response to the non-muscular forces associated with movement and context. This tuning can be driven by training. Sveistrup and Woollacott (1996) gave newly standing infants multiple perturbations of balance (forward and backward translation of the support surface) across five days. This training increased the probability of a more complex synergistic response to the threat to balance.

Standing: from static to dynamic

When rigid objects are perturbed, they fall. When compliant objects are perturbed, they sway. Improvement in balance control can be described as a shift from a rigid to a compliant mechanical system – the aforementioned releasing of the degrees of freedom. This is the challenge then for emerging balance control – how to manage the perturbation energy or forces that might precipitate a fall.

A standard testing paradigm in balance-control research is the moving platform. The participant stands on the platform and, without warning, the platform moves forward or backward some distance and with some rate of speed. The greater the displacement and the higher the speed, the more challenging it is to retain balance without falling or taking a step to regain balance. These tests are synonymous with balance challenges of slips, trips or sudden stops of a bus in which you are standing. Using this paradigm, we have a robust description of the infant/toddler/child's transition from a rigid to compliant mechanical system. Roncesvalles and colleagues described the developmental trajectory of emerging balance control (Roncesvalles *et al.* 2001; Sundermier *et al.* 2001). In these studies, infants and toddlers were grouped by locomotor ability; new standers (who could not yet walk), new walkers (initiated independent walking within the past two weeks), intermediate walkers (up to one month of walking experience) advanced walkers (up to three months of walking experience) and other more advanced performers who could run and jump, hop and gallop and skip. These children were tested on the moving platform, which induced threats to balance. Emerging balance control was associated with increasing extensor torques, applied in more time-appropriate ways. With experience came more consistent muscle torque production and task-matched torque magnitudes.

Standing balance control is generally described by the behaviours associated with balance recovery. In response to small threats to balance, the behavioural response is likely to be limited to the sway of a rigid body about a single pivot point – the ankle. This action is referred to as the ankle strategy (Horak *et al.* 1990). In this case, the focal point of action is the ankle with rotation of the body about the frontal axis and minimal movement about the knees and hips. The response to larger threats to balance is characterized by the hip strategy (Horak and Nashner 1986), in which there is rapid forward rotation of the trunk on the hips (usually in rapid succession to an attempt at an ankle strategy). And, if the threat to balance cannot be managed by the hip strategy, a step is executed to regain balance. Of course, if one or more of these behavioural strategies cannot be performed, then the child falls.

In another study by Roncesvalles *et al.* (2000), children showed an array of responses that followed the expected age/experience trend. The children performed either a feet-in-place recovery action such as an ankle or hip strategy, took a step or fell. The most experienced children – those who could gallop and skip – absorbed the perturbation energy and used a feet-in-place response (ankle or hip). At the other end of the skill range, the infants who could stand, but not walk, typically fell. Unexpectedly, the new walkers also fell. The new walkers had clearly shown the ability to incorporate the voluntary step into their locomotor behaviour, but seemed unable to utilize this motor pattern in constructing a response to unexpected perturbations. McCollum and Leen (1989) hypothesized that infant balance control was poor, in part, because of the inherent mechanics of their small stature.

The McCollum and Leen hypothesis was built on two observations (McCollum and Leen 1989). The first observation was that muscle response onsets are relatively slow in the infant and young child. A successful recovery of balance requires timely activation of muscles to produce counterbalancing forces to oppose the gravitational force driving the fall. Onset times for lower extremity postural muscles in infants who can stand but not yet walk are on the order of just over 100 ms for the gastrocnemius, 147 ms for the hamstrings and 167 ms for the trunk extensors (Sveistrup and Woollacott 1996). For new walkers, the onsets are more rapid for the hamstrings and trunk extensors, but the onset latency for the gastrocnemius remains longer than 100 ms (104 ms) (Sundermier *et al.* 2001). The 100 ms onset latency is a key point for McCollum and Leen (1989). Their second observation was derived from a model of the young child. Considering the natural sway frequency, the damping and stiffness of a second-order system (an inverted pendulum – see Chapter 8), McCollum and Leen (1989) argued that the hip oscillation frequency of a perturbed system would be too fast to be managed by hip muscle responses longer than 100 ms. Thus, should the perturbation be large enough to require the more vigorous hip strategy, the infant or toddler would be unable to execute the recovery and would fall. This hypothesis appeared to have support in the findings of Roncesvalles *et al.* (2000) as the new walkers typically fell when perturbed.

The moving platform has been a dominant paradigm for testing standing balance control. The outcome measurements have been used to infer developmental progress in the neuromotor control of balance. To interpret developing neuro-motor control appropriately, however, one must also understand the mechanics of the testing paradigm. The moving platform contributes energy to the platform-person system. In doing so, balance is threatened. So what part of the balance recovery is active and what part is a passive construction of the induced movement? Bothner and Jensen (2001) showed that passive torques (from the perspective of the performer) induce destabilizing rotations of leg segments. Trunk motion is the result of passive (mechanical) motion-dependent torques arising from movement of adjacent body segments as well as active muscle responses. Most of the focus in moving platform paradigms is on the magnitude of the displacement and the acceleration creating the disturbance. Equally important are the characteristics of platform movement acceleration and cessation (Brown *et al.* 2001). As shown by Brown *et al.* (2001) and Bothner and Jensen (2001), the stopping of platform movement creates forces that contribute to the recovery of balance.

Returning to balance recovery strategies, understanding platform mechanics helps to identify active from passive recovery. In a test of the McCollum and Leen (1989) hypothesis, Roncesvalles *et al.* (2004) examined the kinetics of the hip strategy following perturbation. The balance recovery strategies were successful in averting a fall for all trials included in the analysis. The authors observed the hip strategy in 21 per cent of cases for walkers, and between 16 and 32 per cent of cases for hoppers, gallopers and skippers. Thus, even for the youngest and least skilled children, the hip strategy was present. The young walkers, however, were not as successful overall in averting a fall. What was revealed in the kinetic analysis was that hip strategy for new walkers was a fortuitous event of passive torques creating the hip response and thus a successful recovery. The range of perturbations that could be managed, however, was limited. It was the emergence of an active hip strategy (associated with active abdominal muscle activation) that characterized the increasingly successful balance strategies of older, more experienced, children.

The expansion of postural competence lasts throughout childhood. It is not until children have mastered the fundamental locomotor skills (i.e. running, hopping, galloping and skipping) that their postural recovery responses match adult responses in kinematic features and maturity of control (Roncesvalles *et al.* 2001). The mature response is characterized by a more rapid recovery of static balance and a reduction in the cumulative path of the COP. The changes that support the behavioural improvement include increasingly complete mus-cle response synergies and task-appropriate matching of the muscle torques (Sundermier *et al.* 2001). The generated muscle torques rise smoothly and more quickly to maximum and then decline, much like the force-tension curve of a mechanical spring (see Chapter 5). This is in contrast to the multimodal torque curve associated with the novice performer.

Just as in the development of sitting posture, the first priority is to gain static control over the mass of the system. As competence improves, the perturbations of volitional movement are managed. Increasingly, the passive–reactive forces associated with movement and the external forces of the environment require a successfully tuned neuromotor response to achieve the goal. This is the process of development.

Clinical and practical applications: standing postural control

Dependent measures associated with the centre of pressure (linear and non-linear measures) have been shown to be discriminating in identifying not only immaturity in balance control, but risk for developmental delay. As the infant gains upright postural control, the opportunities for motion expand and the ability to create task appropriate matching of muscular and non-muscular forces expand and define progress in skill development. What distinguishes postural competence as a stander from competence as a sitter is the increasing complexity of the required response.

Postural training has been shown to be effective. Sveistrup and Woollacott (1996) demonstrated the responsiveness of typically developing infants to postural perturbation training. Non-walking infants were supported on a translating platform. Five days of exposure led to more organized synergistic muscle responses. This paradigm was repeated by Shumway-Cook *et al.* (2003) and Woollacott *et al.* (2005) as an intervention for balance recovery in a sample of children with cerebral palsy (see Chapter 13). One hundred trials on each of five days led to improvement in recovery time, reduced area of excursion of the centre of pressure (Shumway-Cook *et al.* 2003), faster muscle activation times, stronger distal to proximal synergistic responses and improved modulation of muscle activity to match the perturbation (Woollacott *et al.* 2005). Importantly, these changes were maintained in a 30-day follow-up. Such findings have direct clinical applications with respect to atypically developing children. Rine *et al.* (2004), for example, demonstrated postural training to be effective with children with sensorineural hearing loss and vestibular impairment, highlighting the importance of postural training within clinical contexts.

In sum, the literature contains support for the efficacy of balance training for infants and children. There is some evidence for the maintenance of training effects after cessation of training, but additional work is needed in this area to strengthen support for the retention of training benefits. Importantly, there is a need for research to establish the connection between early intervention during infancy and childhood, and improved functional abilities later in life.

Key points

- Understanding biomechanics is important to understanding developing skill and the contextual environment in which development takes place.
- Kinematics (e.g. displacements and velocity, COP excursions and velocities), kinetics (e.g. muscle torques, gravitational forces) and muscle activity (e.g. direction-specific responses) provide useful variables for understanding the development of postural control.

- Non-linear analyses, in ways not available through linear measures, allow us to understand changing complexity in developing postural skills.
- Increasing postural control is akin to learning to stack body parts above an increasingly smaller base of support (e.g. from lying to sitting and from sitting to standing) and gaining skill in balancing these blocks under static and dynamic conditions (e.g. standing and walking).
- Even though infants are able to support their body weight by the age of 4 months, they are not able to control all the degrees of freedom that are associated with the support task. Early in the development of postural control, infants begin by freezing multiple degrees of freedom in order to gain control over the body. A release of degrees of freedom characterizes increasing postural control, as skill is gained and the body becomes more compliant.
- The biomechanics of postural control is helpful in early detection of development delay. Early detection is critical to early intervention to minimize the consequences of development delay.

References

Allum, J.H. and Pfaltz, C.R. (1985) 'Visual and vestibular contributions to pitch sway stabilization in the ankle muscles of normals and patients with bilateral peripheral vestibular deficits', *Experimental Brain Research*, 58: 82–94.

Bayouk, J.F., Boucher, J.P. and Leroux, A. (2006) 'Balance training following stroke: effects of task-oriented exercises with and without altered sensory input', *International Journal of Rehabilitation Research*, 29: 51–9.

Bertenthal, B. and Von Hofsten, C. (1998) 'Eye, head and trunk control: the foundation for manual development', *Neuroscience and Biobehavioral Reviews*, 22: 515–20.

Bothner, K.E. and Jensen, J.L. (2001) 'How do non-muscular torques contribute to the kinetics of postural recovery following a support surface translation?', *Journal of Biomechanics*, 34: 245–50.

Brogren, E., Hadders-Algra, M. and Forssberg, H. (1998) 'Postural control in sitting children with cerebral palsy', *Neuroscience and Biobehavioral Reviews*, 22: 591–6.

Brown, L.A., Jensen, J.L., Korff, T. and Woollacott, M.H. (2001) 'The translating platform paradigm: perturbation displacement waveform alters the postural response', *Gait & Posture*, 14: 256–63.

Chen, L.C., Metcalfe, J.S., Jeka, J.J. and Clark, J.E. (2007) 'Two steps forward and one back: learning to walk affects infants' sitting posture', *Infant Behavior and Development*, 30: 16–25.

Cignetti, F., Kyvelidou, A., Harbourne, R.T. and Stergiou, N. (2011) 'Anterior–posterior and medial–lateral control of sway in infants during sitting acquisition does not become adult-like', *Gait & Posture*, 33: 88–92.

Clark, J.E., Whitall, J. and Phillips, S.J. (1988) 'Human interlimb coordination: the first 6 months of independent walking', *Developmental Psychobiology*, 21: 445–56.

de Graaf-Peters, V.B., Blauw-Hospers, C.H., Dirks, T., Bakker, H., Bos, A.F. and Hadders-Algra, M. (2007) 'Development of postural control in typically developing children and children with cerebral palsy: possibilities for intervention?', *Neuroscience and Biobehavioral Reviews*, 31: 1191–200.

Deffeyes, J.E., Harbourne, R.T., Kyvelidou, A., Stuberg, W.A. and Stergiou, N. (2009a)

'Nonlinear analysis of sitting postural sway indicates developmental delay in infants', *Clinical Biomechanics*, 24: 564–70.

Deffeyes, J.E., Harbourne, R.T., DeJong, S.L., Kyvelidou, A., Stuberg, W.A. and Stergiou, N. (2009b) 'Use of information entropy measures of sitting postural sway to quantify developmental delay in infants', *Journal of Neuroengineering and Rehabilitation*, 6: 34.

Fallang, B. and Hadders-Algra, M. (2005) 'Postural behavior in children born preterm', *Neural Plasticity*, 12: 175–82.

Fallang, B., Øien, I., Hellem, E., Saugstad, O.D. and Hadders-Algra, M. (2005) 'Quality of reaching and postural control in young preterm infants is related to neuromotor outcome at 6 years', *Pediatric Research*, 58(2): 347–53.

Farris, E.J. (1961) *Art Students' Anatomy*, New York: Dover Publications.

Forssberg, H. and Hirschfeld, H. (1994) 'Postural adjustments in sitting humans following external perturbations: muscle activity and kinematics', *Experimental Brain Research*, 97: 515–27.

Fournier, K.A., Kimberg, C.I., Radonovich, K.J., Tillman, M.D., Chow, J.W., Lewis, M.H., Bodfish, J.W. and Hass, C.J. (2010) 'Decreased static and dynamic postural control in children with autism spectrum disorders', *Gait & Posture*, 32: 6–9.

Hadders-Algra, M. (2005) 'Development of postural control during the first 18 months of life', *Neural Plasticity*, 12: 99–108.

Hadders-Algra, M., Brogren, E. and Forssberg, H. (1996) 'Training affects the development of postural adjustments in sitting infants', *Journal of Physiology*, 493: 289–98.

Hall, C.D., Woollacott, M.H. and Jensen, J.L. (1999) 'Age-related changes in rate and magnitude of ankle torque development: implications for balance control', *Journals of Gerontology. Series A, Biological Sciences and Medical Sciences*, 54: M507–13.

Harbourne, R.T. and Stergiou, N. (2003) 'Nonlinear analysis of the development of sitting postural control', *Developmental Psychobiology*, 42: 368–77.

Harbourne, R.T. and Stergiou, N. (2009) 'Movement variability and the use of nonlinear tools: principles to guide physical therapist practice', *Physical Therapy*, 89: 267–82.

Harbourne, R.T., Giuliani, C. and Neela, J.M. (1993) 'A kinematic and electromyographic analysis of the development of sitting posture in infants', *Developmental Psychobiology*, 26: 51–64.

Harbourne, R.T., Willett, S., Kyvelidou, A., Deffeyes, J. and Stergiou, N. (2010) 'A comparison of interventions for children with cerebral palsy to improve sitting postural control: a clinical trial', *Physical Therapy*, 90: 1881–98.

Hasan, S.S., Robin, D.W., Szurkus, D.C., Ashmead, D.H., Peterson, S.W. and Shiavi, R.G. (1996) 'Simultaneous measurement of body center of pressure and center of gravity during upright stance. Part I: Methods', *Gait & Posture*, 4: 1–10.

Hedberg, A., Forssberg, H. and Hadders-Algra, M. (2004) 'Postural adjustments due to external perturbations during sitting in 1-month-old infants: evidence for the innate origin of direction specificity', *Experimental Brain Research*, 157: 10–17.

Hedberg, A., Carlberg, E.B., Forssberg, H. and Hadders-Algra, M. (2005) 'Development of postural adjustments in sitting position during the first half year of life', *Developmental Medicine and Child Neurology*, 47: 312–20.

Horak, F.B. and Nashner, L.M. (1986) 'Central programming of postural movements: adaptation to altered support-surface configurations', *Journal of Neurophysiology*, 55: 1369–81.

Horak, F.B., Nashner, L.M. and Diener, H.C. (1990) 'Postural strategies associated with somatosensory and vestibular loss', *Experimental Brain Research*, 82: 167–77.

Inman, V.T. (1966) 'Human locomotion', *Canadian Medical Association Journal*, 94: 1047–54.

Ivanenko, Y.P., Dominici, N., Cappellini, G. and Lacquaniti, F. (2005) 'Kinematics in newly walking toddlers does not depend upon postural stability', *Journal of Neurophysiology*, 94: 754–63.

Jensen, J.L. (2005) 'The puzzles of motor development: how the study of developmental biomechanics contributes to the puzzle solutions', *Infant and Child Development*, 14: 501–511.

Jensen, J.L. and Bothner, K.E. (1998) 'Revisiting infant motor development schedules: the biomechanics of change', in E. van Praagh (ed.) *Pediatric Anaerobic Performance*, Champaign, IL: Human Kinetics, pp. 23–43.

Jensen, J.L., Schneider, K., Ulrich, B.D., Zernicke, R.F. and Thelen, E. (1994) 'Adaptive dynamics of the leg movement patterns of human infants: I. The effects of posture on spontaneous kicking', *Journal of Motor Behavior*, 26: 303–12.

Jian, Y., Winter, D.A., Ishac, M.G. and Gilchrist, L. (1993) 'Trajectory of the body COG and COP during initiation and termination of gait', *Gait & Posture*, 1: 9–22.

Kirshenbaum, N., Riach, C.L. and Starkes, J.L. (2001) 'Non-linear development of postural control and strategy use in young children: a longitudinal study', *Experimental Brain Research*, 140: 420–31.

Kuczmarski, R.J., Ogden, C.L., Grummer-Strawn, L.M., Flegal, K.M., Guo, S.S., Wei, R., Mei, Z., Curtin, L.R., Roche, A.F. and Johnson, C.L. (2000) 'CDC growth charts: United States', *Advance Data*, 1–27.

Kyvelidou, A., Harbourne, R.T., Shostrom, V.K. and Stergiou, N. (2010) 'Reliability of center of pressure measures for assessing the development of sitting postural control in infants with or at risk of cerebral palsy', *Archives of Physical Medicine and Rehabilitation*, 91: 1593–601.

Kyvelidou, A., Harbourne, R.T., Stuberg, W.A., Sun, J.F. and Stergiou, N. (2009a) 'Reliability of center of pressure measures for assessing the development of sitting postural control', *Archives of Physical Medicine and Rehabilitation*, 90: 1176–84.

Kyvelidou, A., Stuberg, W.A., Harbourne, R.T., Deffeyes, J.E., Blanke, D. and Stergiou, N. (2009b) 'Development of upper body coordination during sitting in typically developing infants', *Pediatric Research*, 65: 553–8.

Mann, R.A., Hagy, J.L., White, V. and Liddell, D. (1979) 'The initiation of gait', *Journal of Bone and Joint Surgery. American Volume*, 61: 232–9.

Martorell, R., de Onis, M., Martines, J., Black, M., Onyango, A. and Dewey, K.G. (2006) 'WHO Motor Development Study: windows of achievement for six gross motor development milestones', *Acta Paediatrica. Supplement*, 450: 86–95.

McCollum, G. and Leen, T.K. (1989) 'Form and exploration of mechanical stability limits in erect stance', *Journal of Motor Behavior*, 21: 225–44.

Nashner, L.M. (1976) 'Adapting reflexes controlling the human posture', *Experimental Brain Research*, 26: 59–72.

Newell, K.M., Slobounov, S.M., Slobounova, E.S. and Molenaar, P.C. (1997) 'Stochastic processes in postural center-of-pressure profiles', *Experimental Brain Research*, 113: 158–64.

Niam, S., Cheung, W., Sullivan, P.E., Kent, S. and Gu, X. (1999) 'Balance and physical impairments after stroke', *Archives of Physical Medicine and Rehabilitation*, 80: 1227–33.

Prieto, T.E., Myklebust, J.B., Hoffmann, R.G., Lovett, E.G. and Myklebust, B.M. (1996) 'Measures of postural steadiness: differences between healthy young and elderly adults', *IEEE Transactions on Biomedical Engineering*, 43: 956–66.

Riach, C.L. and Hayes, K.C. (1987) 'Maturation of postural sway in young children', *Developmental Medicine and Child Neurology*, 29: 650–8.

Rine, R.M., Braswell, J., Fisher, D., Joyce, K., Kalar, K. and Shaffer, M. (2004) 'Improvement of motor development and postural control following intervention in children with sensorineural hearing loss and vestibular impairment', *International Journal of Pediatric Otorhinolaryngology*, 68: 1141–8.

Rochat, P. (1992) 'Self-sitting and reaching in 5- to 8-month-old infants: the impact of posture and its development on early eye-hand coordination', *Journal of Motor Behavior*, 24: 210–20.

Roncesvalles, M.N. and Jensen, J.L. (1993) 'The expression of weight-bearing ability in infants between four and seven months of age', paper presented at the annual meeting of the North American Society for Psychology of Sport and Physical Activity, Brainerd, MN, June.

Roncesvalles, M.N., Woollacott, M.H. and Jensen, J.L. (2000) 'The development of compensatory stepping skills in children', *Journal of Motor Behavior*, 32: 100–11.

Roncesvalles, M.N., Woollacott, M.H. and Jensen, J.L. (2001) 'Development of lower extremity kinetics for balance control in infants and young children', *Journal of Motor Behavior*, 33: 180–92.

Roncesvalles, M.N., Woollacott, M.H., Brown, N. and Jensen, J.L. (2004) 'An emerging postural response: is control of the hip possible in the newly walking child?', *Journal of Motor Behavior*, 36: 147–59.

Shumway-Cook, A., Hutchinson, S., Kartin, D., Price, R. and Woollacott, M. (2003) 'Effect of balance training on recovery of stability in children with cerebral palsy', *Developmental Medicine and Child Neurology*, 45: 591–602.

Stergiou, N. (2004) *Innovative Analyses of Human Movement*, Champaign, IL: Human Kinetics.

Sundermier, L., Woollacott, M., Roncesvalles, M.N. and Jensen, J. (2001) 'The development of balance control in children: comparisons of EMG and kinetic variables and chronological and developmental groupings', *Experimental Brain Research*, 136: 340–50.

Sveistrup, H. and Woollacott, M.H. (1996) 'Longitudinal development of the automatic postural response in infants', *Journal of Motor Behavior*, 28: 58–70.

Thelen, E. (1985) 'Developmental origins of motor coordination: leg movements in human infants', *Developmental Psychobiology*, 18: 1–22.

Thelen, E. (1986) 'Treadmill-elicited stepping in seven-month-old infants', *Child Development*, 57: 1498–506.

Thelen, E. and Spencer, J.P. (1998) 'Postural control during reaching in young infants: a dynamic systems approach', *Neuroscience and Biobehavioral Reviews*, 22: 507–14.

Thelen, E., Corbetta, D. and Spencer, J.P. (1996) 'Development of reaching during the first year: role of movement speed', *Journal of Experimental Psychology: Human Perception and Performance*, 22: 1059–76.

Thelen, E., Fisher, D.M., Ridley-Johnson, R. and Griffin, N.J. (1982) 'Effects of body build and arousal on newborn infant stepping', *Developmental Psychobiology*, 15: 447–53.

Usui, N., Maekawa, K. and Hirasawa, Y. (1995) 'Development of the upright postural sway of children', *Developmental Medicine and Child Neurology*, 37: 985–96.

van Dieën, J.H., Koppes, L.L. and Twisk, J.W. (2010) 'Postural sway parameters in seated balancing; their reliability and relationship with balancing performance', *Gait & Posture*, 31: 42–6.

van Zandwijk, R., Hill, P.M. and Jensen, J.L. (2010) 'The development of asymmetric stance in early childhood', *Journal of Sport and Exercise Psychology*, 32 (Suppl.): S51–2.

Winter, D.A. (1990) *Biomechanics and Motor Control of Human Movement*, New York: Wiley.

Winter, D.A. and Eng, P. (1995) 'Human balance and posture control during standing and walking', *Gait & Posture*, 3: 193–214.

Winter, D.A., Prince, F., Stergiou, P. and Powell, C. (1993) 'Medial-lateral and anterior-posterior motor-responses associated with center of pressure changes in quiet standing', *Neuroscience Research Communications*, 12: 141–8.

Winter, D.A., Patla, A.E., Prince, F., Ishac, M. and Gielo-Perczak, K. (1998) 'Stiffness control of balance in quiet standing', *Journal of Neurophysiology*, 80: 1211–21.

Winter, D.A., Prince, F., Frank, J.S., Powell, C. and Zabjek, K.F. (1996) 'Unified theory regarding A/P and M/L balance in quiet stance', *Journal of Neurophysiology*, 75: 2334–43.

Wong, S., Ada, L. and Butler, J. (1998) 'Differences in ankle range of motion between pre-walking and walking infants', *Australian Journal of Physiotherapy*, 44: 57–60.

Woollacott M., Shumway-Cook, A., Hutchinson, S., Ciol, M., Price, R. and Kartin, D. (2005) 'Effect of balance training on muscle activity used in recovery of stability in children with cerebral palsy: a pilot study', *Developmental Medicine and Child Neurology*, 47: 455–61.

Zelazo, P.R. (1983) 'The development of walking: new findings and old assumptions', *Journal of Motor Behavior*, 15: 99–137.

8 Biomechanical aspects of the development of walking

Beverly Ulrich and Masayoshi Kubo

Introduction

The goal of this chapter is to present four basic theoretical concepts that are core to understanding the development of neuromotor control, specifically applied to the development of walking. After stating each concept we provide a brief introduction to the unique biomechanics tools and principles that are used to study issues relevant to the concept. And we follow with numerous examples from our own and others' research to illustrate the most important points. At the end of the chapter we summarize key concepts and provide suggestions for application.

Basic theoretical concept 1

Well-practised and stable movements can be modelled because humans discover the relations among the mechanical properties of their systems by exploiting passive, non-muscular torques and adding muscular torques which are appropriate in magnitude and timing.

The human body is a physical system and therefore, all of our movements need to follow Newton's laws without exception. According to Newton's second law of motion, translation of a segment requires forces and rotation requires torques. Although there are multiple sources of forces and torques available, the sum of them determines the movement of a segment. For a more detailed description of the mechanical interactions between forces and movement, we refer the reader to Chapter 6.

Newton's third law states that for every action there is an equal and opposite reaction. Since human bodies consist of multiple segments connected by joints, any movement of one segment will affect the movement of adjacent segments. For a segment to rotate around a joint, a certain amount of torque needs to be applied to it (i.e. Newton's second law). According to Newton's third law, a reaction torque is applied to the adjacent segment that is equal in magnitude and opposite in direction.

Figure 8.1 illustrates a simple model of the arm on a table top, consisting of the forearm (A) and upper arm (B) segments connected to the trunk at the shoulder joint. The intended movement is an anticlockwise movement of segment A

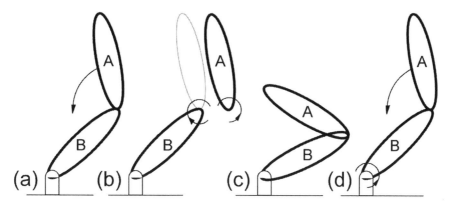

Figure 8.1 The effect of a reaction torque on the movement of multiple segments.

(Figure 8.1a). To accomplish that goal, an anticlockwise torque is generated at the proximal joint of segment A. At the same time, a clockwise reaction torque is generated at the distal joint of segment B (Figure 8.1b). If no additional torques are applied to the system, segment B will rotate in a clockwise direction (Figure 8.1c).

As we develop control, there are two ways we can respond to this intersegmental effect. If this is considered as a perturbation, an anticlockwise torque at the proximal joint of segment B is needed (Figure 8.1d) to avoid the *unintended* movement at the shoulder joint. Or, once we harness timing and the amount of multiple sources of interacting torques, we can utilize this motion-dependent effect to generate *intended* movement of multiple segments efficiently.

Biomechanical studies show that after healthy adults are given a few minutes to warm up, if asked to walk at their preferred speed over level ground and without any perturbations, they spontaneously select their own unique speed that minimizes their energy cost and maximizes the recovery of potential elastic energy (Brooks *et al.* 2000; Srinivasan 2009) (see Chapter 5). Further, adults perform this goal using remarkably similar control strategies. They alternate limb movements so that each leg initiates a new stride halfway through the cycle of the other leg. Furthermore, they add energy into the system (i.e. initiate push-off) just prior to the point at which the swing leg makes contact with the ground (Kuo and Donelan 2010). If we built a robot with similar physical properties as humans with the goal of optimizing energy efficiency and stability, the optimal limb kinematics and timing characteristics would be similar to what humans actually do. It is intuitive to conclude that humans evolved to have controllers (neural programmes) built into their genetic codes that achieve these optimal movement patterns. But what if, *in utero*, the foetus was exposed to a teratogen, such as thalidomide, resulting in asymmetric segmental growth or the absence of a joint or a segment? What if we were born on a space station, with zero gravity, and alternating leg movements would not be a sensible locomotor solution? How well would a prespecified programme designed for a symmetrical system with set properties and a set gravitational field work? A predesigned programme allows for very limited intrinsic or extrinsic variability. In reality, evolution has enabled us to be much smarter and more efficient than that. Instead of having predesigned solutions, we have developed more complexity within our systems. At the nervous system level, complexity means that even *in utero*, the billions of neurons in our brains already begin to establish a primary repertoire of interconnections. As we begin to move and receive sensory input, *in utero* and after birth, we gradually select and strengthen connections that reflect our experiences and meet our needs (see Chapter 2). However, this self-organizing characteristic of our nervous systems also means that we really have to 'work' to help our brains increase their functional organizational structure. We explore and practise to discover multimodal cooperative neural networks and reinforce or stabilize them as well as retain adaptivity among populations of neurons.

Fortunately, as babies we were pretty smart little explorers. We discovered quite rapidly the opportunities our bodies' mechanical and physiological properties held for us and we were eager to try them out. For example, one of the physiological properties our muscles and tendons have is viscoelasticity. That means that, when stretched, they recoil, like a spring (see Chapter 5). In a clever experiment, Goldfield *et al.* (1993) illustrated, using a novel toy that parents in the USA often provide for their babies, that infants not only discover easily neural control strategies, but also that they pick mechanically efficient ones. The toy, called a jolly jumper (Figure 8.2), consists of a soft fabric seat and harness suspended by a spring from a bar, which is usually fastened to a doorframe in the home. When placed in this seat, infants try a variety of leg pushes, taps, kicks, and so on until they gleefully figure out that, by pushing off with both legs, they can take advantage of their own 'leg springs' and the suspension spring. They control the system to produce quite vigorous and continuous bouncing. Goldfield *et al.* (1993) modelled the baby and jolly jumper system as a two-spring system (the baby's legs are one spring and the apparatus's spring is the second) and used each baby's own weight as the system's mass, to model this biomechanical system and predict the frequency at which the system would oscillate most effectively. Because babies' weights varied, so did their optimal frequency. It turns out that each baby selected a frequency that matched closely what the scientists' mathematical model predicted for producing the highest bounce amplitude and lowest variability from cycle to cycle. Each baby discovered how to control his system as a 'forced-mass-spring system operating at resonance' (p. 1128). In simple play settings like this, babies show us their fluency in testing their innate capacities and adapting them, such as exploring the amount of muscle force needed and the timing of it to fit the context – in this case the stiffness of the suspension spring – and ultimately achieve their goal: bouncing high! In fact, when Goldfield and colleagues tried to trick babies by changing the spring's stiffness, babies adapted their muscle force, up or down accordingly, to continue to maintain optimal 'bang for their buck' (Foo *et al.* 2004).

Bouncing up and down in an infant seat is a far cry from walking independently. Does that mean that bouncing is a simple action that might be discovered, but for

Figure 8.2 Baby supported in a jolly jumper will learn, all by herself, the optimal force to generate via parallel leg kicks to 'use' her leg springs cooperatively with the suspension spring and her body mass to bounce continuously.

something more universal and life-sustaining, like walking, that evolution must take hold of and build a template for us? We think the evidence stacks up against such a grand plan and towards a human neuromotor-control system that is complex, dynamic and opportunistic. Humans have genetic material that contributes to the assembly of an enormous number of neurons, muscle fibres and nascent interconnections. From this treasure-trove of resources infants build combinations of neurons, muscles and body segments to learn to walk, just as they learn to reach, to sit and to speak. They develop strength and control, bit by bit over time, via practice and goal-directed efforts. We can observe one example of an emergent adaptive solution to a simple goal that can be produced by babies long before they are able to walk, yet results in patterns of leg movement that we call 'stepping'. As Figure 8.3 illustrates, if we support infants who are developing typically, upright, so their feet rest on the belt of a paediatric treadmill, when the treadmill belt moves their legs backwards, they respond by producing leg movements that

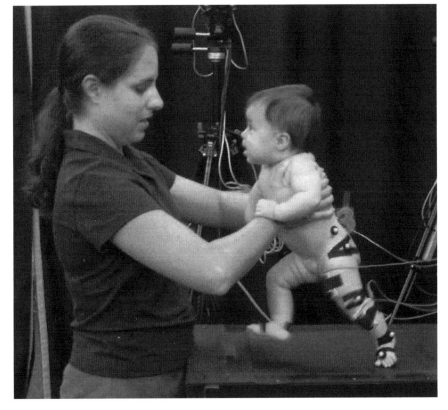

Figure 8.3 Infant with typical development stepping well when supported upright on a
       motorized treadmill, long before attempting to step or walk independently.

look very much like walking (Thelen 1986). In fact, they consistently alternate,
like adults do when they walk, with one leg swinging forwards halfway through
the stride cycle of the other leg, and they maintain this alternation even when
the treadmill belt is split down the middle and one half moves twice as fast as the
other (Thelen *et al.* 1987). But these patterns, that look so similar kinematically
to those of adults, are produced by quite different kinetics or combinations of
passive forces (gravity and motion-dependent forces) and active forces (muscle
force) (Ulrich *et al.* 1994). Adults consistently activate muscles to initiate and
terminate the forward motion of their legs but allow passive torques or momentum
to carry the leg through the middle of the swing phase. Infants, in contrast, display
many and varied combinations of active and passive torques during the forward
swing of the leg, even though kinematically speaking, the leg trajectory is similar
– swinging forwards through space and then back down to the surface. Most
commonly, the muscle torque is flexor throughout the swing phase, and hip-joint
reversal from flexion to extension emerges near the end of the swing phase when
the torque, due to gravity, becomes greater in magnitude than the flexor muscle

torque. From one step to the next, net muscle torque values did not reflect the smooth ramping up to or consistent timing of reversal; rather, there were multiple periods of rising and falling values, suggesting varied combinations of motor units firing during individual steps and over repeated cycles. Data such as these, collected in the very early stages of emergent limb control, suggest that babies have great flexibility in their neuromotor systems and many choices for harnessing the biomechanics of their body's levers and masses, in order to adapt to the contexts in which they find themselves. Their responses to treadmills and jolly jumpers emerge spontaneously; over time, nearly all healthy babies discover how to use these flexible properties to their advantage to achieve their own independent goals for locomotion.

During the later months of the first year, as infants continue their 'work' of moving, exploring, building muscle strength and learning to pull their bodies upright, they start to shift weight from one foot to the other as they hold on to coffee tables, couches or parents' hands. This may happen as they simply stretch and reach for a toy, but in the process they discover options for combining sufficient force and body positions to propel their bodies forwards without allowing their many loosely controlled body parts to simply topple over (Haehl *et al.* 2000). Toddlers' first independent steps, though unsteady, can be modelled as the behaviour of a mechanical system, because they must, like robotic walkers, deal with the forces of gravity, their own mass, axes of rotation (joints) and levers (body segments). Yet, the human neural organization that controls this set of linkages is unsteady, and thus new walkers fall frequently. One strategy new walkers use for improving their odds of staying upright while trying to move forwards is to position their little feet farther apart (Adolph *et al.* 2003). The energy cost of doing so is high, but initially staying upright is prioritized over, saving energy (Kubo and Ulrich 2006). With practice, the solution is adapted, bit by bit.

Kubo and Ulrich (2006) showed that the greatest challenge of 'letting go' is to learn to couple the oscillations of the centre of mass between the anterior–posterior direction (needed to move forwards) and the medial–lateral direction (needed to push off, shifting weight from the left foot to the right foot). Wider step widths and shorter steps aid in this process. And, within a few weeks and thousands and thousands of practice steps, infants begin to narrow their base, learn how to take advantage of gravity and motion-dependent torques, and insert muscle torques appropriately to take advantage of the pendulum and spring-like dynamics. Holt *et al.* (2006, 2007) argue that toddlers learn to control their walking in a two-phase process. By the end of one month of independent walking practice, toddlers made huge gains in timing the escapement pulse or insertion of muscle force during the step cycle, with less dramatic improvements over the next several months (Holt *et al.* 2006). Yet, the stiffness toddlers produced (think of this as the tension in their joints and muscles of their pendulum and spring system) remained much higher than a model would predict. Several more months of practice were needed before toddlers allowed this stiffness to reduce to a more mechanically efficient level (Holt *et al.* 2007).

Basic theoretical concept 2

Changes in systems' properties (mass, stiffness, length) lead to shifts in patterns of behaviour.

The behaviour of an oscillatory system is affected by the physical properties of components within the system and its surrounding environment. For the purpose of illustration, consider the behaviour of a pendulum. A simple pendulum, shown in Figure 8.4a, consists of a string with length l (m) and a mass M (kg). The period of oscillation T (s) of this pendulum is given by Equation 8.1:

$$T = \frac{2\pi}{\omega_0} \tag{8.1}$$

where ω_0 is angular velocity (rad/s). The angular velocity of a simple pendulum is a function of its length and the gravity constant g (m/s^2) and is given by Equation 8.2:

$$\omega_0 = \sqrt{\frac{g}{l}} \tag{8.2}$$

It should be noted that the period of oscillation is determined only by the length of string (l). Therefore, changes in mass will not change the oscillatory behaviour of this pendulum at all. If no other external forces are acting on this system, the pendulum will oscillate at this frequency once set into motion (i.e. at its 'natural frequency').

In order to make the pendulum a bit closer to a model of a human body segment, we can add a constant stiffness k (N/m) and a constant damping coefficient c (Ns/m) (Figure 8.4b). The damped natural frequency of the pendulum is given by $\omega_d = \omega\sqrt{1 - \xi^2}$. The damping factor ξ is a function of the damping coefficient c and other physical properties of the pendulum (i.e. k, l and M). Since the value of ξ is small in human body segments, the damped natural frequency (ω_d) of this pendulum is approximated by the natural frequency ω given in Equation 8.3:

$$\omega = \sqrt{\frac{kl + Mg}{Ml}} \tag{8.3}$$

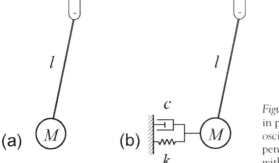

(a)   (b)

Figure 8.4 The effect of changes in physical property on oscillatory movement. A simple pendulum (a) and a pendulum with stiffness and damping (b).

This equation illustrates that any changes in length, mass or stiffness have the potential to affect the frequency of oscillatory movement. Over developmental time, changes in the body's mechanical properties such as weight and segment lengths (see Chapter 1) as well as the elastic properties of the musculoskeletal system (see Chapter 5) change dramatically. Therefore, humans must continually adapt their functional behaviours in response to these changing intrinsic characteristics of the system.

Modelling the biomechanics of human systems as inverted pendulum and spring systems assumes a minimum level of 'stiffness' across joints to avoid excessive instability of the system. Persons with Down syndrome have inherently less stiffness in their systems due to ligamentous laxity and low muscle tone. A medical treatment does not exist that can fix these problems, but persons with Down syndrome learn to adjust their body's stiffness, for example, by changing the levels of tension in relevant muscle groups. In a series of studies, Ulrich and colleagues (Black *et al.* 2009; Ulrich *et al.* 2004) examined gait in new walkers and highly experienced walkers, with and without Down syndrome. They showed that, in order to walk independently, toddlers with Down syndrome scale up on their body's stiffness to match that of toddlers with typical development, and they continue to do so, even after years of practice. The fact that the inherent stiffness (due to joint laxity and low tone) remains a constraint even after much experience is suggested by their responses when placed in particularly unstable gait situations. For example, when preadolescents with Down syndrome and peers with typical development walk on treadmills at a range of speeds, those with Down syndrome produce significantly higher levels of stiffness than their peers with typical development, at all speeds (Ulrich *et al.* 2004). We believe children with Down syndrome do this because they accurately perceive this context to have greater potential for them to lose control. Thus, they use a compensatory mechanism to guard against any potential imbalance. To put this into perspective, think about what you do, when you walk on an icy path: you may shorten your steps and stiffen your joints to ensure that your centre of mass remains upright and over your base of support. Children with Down syndrome change their bodies' properties in a similar way to adapt their behaviour and maintain functional, stable walking patterns.

While the previous example illustrates how intrinsic constraints can influence movement patterns, extrinsic (environmental) constraints are equally important. Jensen *et al.* (1994) studied how 3-month-old babies spontaneously changed the muscle forces they produced in order simply to kick their legs, when placed in three very different body positions relative to gravity: on their backs, held upright or supported at an angle of 45°, as they are in an infant seat. When placed on their backs, infants initiated kicks by creating a net flexor muscle torque at the hip joint. Extension was produced passively, by gravity and inertial forces, including the recoil of the extensor muscle-tendon units that had been passively stretched during the flexion phase. When moved to the more upright positions, gravity resisted leg flexion by four to ten times compared to the supine position, requiring infants to generate much more muscular torques to move their legs towards their trunks. Faced with these increased biomechanical demands, infants responded by adapting the limb trajectory. In particular, they moved the hip and knee through reduced ranges of motion and coupled them to work in greater synchrony.

When performing more functional behaviours such as walking, the challenges of centre of mass control and intersegmental coordination become a continuous process of adapting to the changing influences of passive forces. When infants learn to walk, they use a great deal of practice to discover the most efficient ways to move this mass forward on their little levers (short leg to trunk ratios compared to adults). When older children take on additional mass, such as by carrying a school bag on their shoulders and back, they spontaneously adapt by leaning forwards if the bag is positioned in the centre of their back and to the opposite side if it is asymmetrically positioned. But can we see such self-organizing principles at play in babies who have only begun to use their legs to walk?

One way to test this hypothesis is to change the extrinsic constraints, which can be achieved by adding mass to their bodies. Results from such experiments demonstrate that newly walking toddlers' responses to added weight (up to 25 per cent of total body weight) are remarkably consistent, immediate and effective solutions (Garciaguirre *et al.* 2007; Vereijken *et al.* 2009). Additional weight positioned on the shoulders caused toddlers to walk faster with longer steps, presumably to compensate for the increased instability introduced by the raised centre of mass and greater gravitational torques. Placing the weight closer to the feet resulted in slower speed, shorter steps and increased outward rotation of the feet. The shorter steps were a response to the requirement of lifting a greater mass with each step. The outward rotation of the feet was a means to increase the base of support needed to meet the increased stability requirements. Garciaguirre *et al.* (2007) further showed that intrinsic body mass influenced the solutions chosen. Chubbier toddlers made greater adjustments to the added weight than leaner toddlers, even when the weight added was proportionately identical. This suggests that their intrinsic body mass to size ratio created an inherently greater challenge for controlling balance for upright locomotion, which was made even more precarious when this type of novel perturbation must be dealt with.

In summary, during the first six to eight months after the onset of walking, toddlers improve their efficiency and stability as a result of practice. Furthermore, the adaptations produced by new walkers to changes in intrinsic or extrinsic constraints show significant differences in their solutions compared to experienced walkers. Nevertheless, new walkers demonstrate immediate, self-organized and sustained adaptations to changes of these constraints.

Basic theoretical concept 3

Stability of movement is adapted by changing mechanical constraints.

Static balance

Standing postural stability is often referred to as static balance. A vase on the table is in a state of static balance. In both cases, the projection of the centre of mass (COM) on to the support surface would be within the base of support (BOS), geometrically. However, unlike the vase on the table, the centre of pressure (COP) under the feet keeps moving when we are in a standing posture, reflecting our constant effort to maintain the posture against the influences of gravity. During quiet standing, the BOS is in static balance. However, the COP is constantly moving. Maintaining posture is therefore a dynamic effort.

During locomotion, we alter the position of the BOS. We can shift our BOS very slowly so that the vertical projection of COM is always within the BOS, and we can stop immediately any time we want without losing balance (Figure 8.5). Similarly, rock climbers use a static climbing technique, in which only one limb moves in the air at a time. The remaining three limbs are in contact with the environment, providing a solid BOS. This way, they can always be in static balance.

It may not be as precarious as rock climbing, but when babies crawl on the floor they use a variation of the same 'static balance-and-move' technique. When they cruise with one hand (or two) holding on to an object, the technique that they use is similar to that of crawling (or rock climbing, for that matter). Often, babies tend to move sideways and with a wider base of support. The BOS is most stable

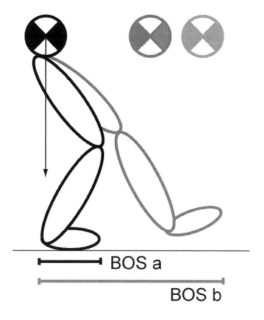

Figure 8.5 Static balance and transportation. The centre of mass moves forwards but does not leave the base of support (BOS) of the stance leg (a) until a new BOS (b) is created by the leading leg. This strategy for gait can be used only at slow speeds.

in the medio–lateral direction. When leaning or moving forwards, on the other hand, the stability offered in that direction by the BOS is low.

Dynamic balance in the early stage of walking

In the case of bipedal walking in healthy adults, the BOS is formed by one foot – the stance leg – in stance phase and two feet during the double-stance phase. The vertical projection of the COM momentarily passes through each of these BOSs during each stride cycle. Therefore, stability in the geometrical sense is not the constraining biomechanical variable. Instead, it is the ability to control the movement of the COM, so that its trajectory is dynamically aligned with the constantly progressing BOS.

In adults, the control of the COM during walking is mostly accomplished by the action of the lower limb muscles. While the stance leg is responsible for projecting the COM upwards, the trailing leg pushes the COM forwards. Once one stride is completed, the roles of stance and trailing legs are then reversed.

During the very early stage of development of biped walking infants do not only rely on their lower limb muscles because they cannot control them very well. Therefore, they explore their whole repertoire of movements (including movements which utilize upper body segments) and discover options that can be used to control the COM motion. New walkers employ a variety of strategies to cope with the increased demand for dynamic stability. Adults employ such upper-body control strategies when they walk on a narrow beam or a tightrope. Under these more challenging circumstances, upper-body control strategies are more effective than lower-limb strategies to control the COM.

Have you ever watched someone learning a challenging new motor skill who seemed to wobble and nearly lose control of the movement? For example, when a child makes their first attempts to ride a two-wheel bicycle, the bike and child seem to wobble from side to side erratically. Until they actually fall over, one could argue that they have not lost control, but that control is minimal. To assist the child, we might add a mechanical constraint to reduce the challenge, such as training wheels or the parent's hand on the bicycle (Figure 8.6). With such externally applied constraints the child is able to discover more easily how to piece together the different components required to perform the task successfully (i.e. alternating leg rotations to propel the bike, arm movements to move the handlebars, head and trunk posture to control the movement direction). Most of us are familiar with such a scenario and may even have experienced being the external force that changed the mechanical constraints and provided stability. Learning to ride a bike is a memorable childhood experience. In reality, we began adapting and changing our own personal mechanical constraints to improve stability at birth and, perhaps, pre-birth.

When infants take their first independent walking steps they often position their arms in what seems like very awkward locations compared to what skilled walkers do. Arms are held high, in what developmentalists call a 'high-guard' position (Figure 8.7). Traditional thinking was that they kept their arms high to

Figure 8.6 Child learning to ride a two-wheeled bicycle with the external support of an adult who stabilizes the movement of the system in order to help the child to discover workable solutions.

prepare for the inevitable fall, to be ready to catch themselves as they went down. However, Kubo and Ulrich (2006) analysed the trunk and arm motion and forces produced by new walkers and found that, in fact, this was a strategy that infants discovered to stabilize their movement. Early steps are marked by erratic timing and amounts of force produced with each step, making the control over their many body segments, from the feet up to the head, quite demanding. Positioning their arms above their centre of mass allowed them to use these segments to counter the over- or under-rotation of the trunk and head as the body progressed through space. Too much forward lean of the trunk is counterbalanced by the backward rotation of the upper limbs. Too little forward trunk and body movement is assisted by thrusting the arms forwards. This solution is, indeed, a dynamic one, varying across individuals and shifting over time as stability needs change. As neural control of the amplitude and timing of leg force becomes steadier and appropriate to the task demands, infants begin to move their arms downwards. Eventually, they reach the low carrying angle and alternation with leg action that provides the force needed to stabilize the trunk and limit its rotation in the transverse plane (Kubo and Ulrich 2006; Ledebt 2000).

The ability to discover without instruction how to adapt to their own instability by changing their body's mechanical constraints is not limited to infants with

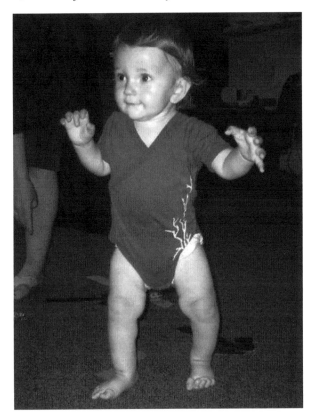

Figure 8.7 Newly walking toddler using arms in high-guard position to contribute to controlling the whole system while performing early, unstable, independent steps.

typical development. Children with disabilities, including those with cognitive disabilities, perceive fairly accurately their inherent level of control and the demands of the environment that challenge their stability. Some of the challenges that all new walkers face include obstacles in their path, transitions from smooth, paved surfaces to grass, uneven surfaces and so on. In a laboratory setting, when faced with a 5-cm-high barrier across their pathway, toddlers with Down syndrome who had only one month of walking experience demonstrated a variety of strategies to clear the obstacle (Mulvey *et al.* 2011). The most common, however, was to shift from bipedal to quadruped locomotion or from walking to crawling. And, like their peers with typical development, they began preparing for this challenge to their stability well in advance of reaching it, by slowing down, reducing step lengths and, unique to the group with Down syndrome, increasing their step width. The shift to crawling significantly increased stability for these novice walkers, averting all possibility of losing control and falling. In fact, at this point in developmental time, toddlers with Down syndrome maintained their stability, via this strategy, better than their peers with typical development. Infants with typical development seemed more daring and unconcerned if they fell in the process of crossing, which they did until they acquired a few more weeks of walking practice. Fortunately, new walkers are still in nappies that provide padding when they fall and they have relatively short legs, so they land not far from where they started!

Just as newly walking toddlers with typical development or Down syndrome prepare for stability challenges by adjusting their velocity and stride length before reaching an obstacle, so do older adults (Smith and Ulrich 2008). Again, comparisons of persons with typical development and ones with physical and cognitive differences show that people discover solutions that fit their own unique innate mechanics and control abilities. After preparing similarly in their approach to an obstacle, older adults with Down syndrome successfully cleared the barrier. But they positioned the obstacle further forwards of their COM as the lead foot moved over it, producing a lower, flatter trajectory, with less dorsiflexion at crossing. This is typical of much older persons with typical development and is believed to be a stability-enhancing strategy (Chen *et al.* 1991). In this posture the trunk does not tip forwards as early in the crossing, not until the lead foot is closer to touching down. People with Down syndrome have low tone and high instability in their joints due to ligamentous laxity. Thus, much earlier in life than their typically developing peers they perceive the very real need to adapt the mechanics of their gait in ways that provide greater protection from falling.

Basic theoretical concept 4

Forces needed to generate motor patterns must be appropriately timed and can be produced by many combinations of active and passive forces.

Mechanical resonance

Mechanical systems consisting of pendulums and mass-springs have their own preferred frequency of oscillation (natural frequency: ω_0). When a mechanical system is 'forced' (energy is added to cause movement) it oscillates periodically in the vicinity of its natural frequency and absorbs most energy. Consequently, the amplitude of oscillation increases rapidly. This phenomenon is referred to as 'mechanical resonance'.

Figure 8.8 shows an example of mechanical resonance, when a slightly damped system with natural frequency ω_0 is forced by a function $F(t) = F_0 \sin(\omega_f t)$. The x-axis of the graph shows the frequency ratio (r) expressed by Equation 8.4:

$$r = \frac{\omega_f}{\omega_0} \qquad (8.4)$$

while the y-axis shows the magnification factor (β), which is a ratio of the oscillation amplitude with repetitive forcing compared to that with only one-time forcing.

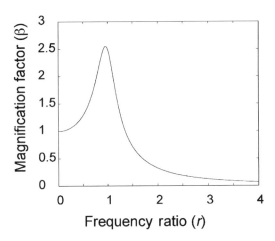

Figure 8.8 Mechanical resonance: when the frequency of forcing is close to the natural
frequency of the system, the amplitude of oscillation grows rapidly.

In our example the magnification factor is close to 2.5 when the ratio is close to
1. For a given mechanical system, the magnification factor depends on the amount
of damping in the system. If there was no friction at all, the amplitude would grow
infinitely. However, in reality, mechanical systems cannot sustain oscillations of
amplitudes very large and, and they would collapse if the amplitude exceeds a
certain threshold. Therefore, in a sense, the existence of damping (i.e. friction)
within a system serves as a 'fail-safe mechanism'.

One example of forcing a certain frequency on to a mechanical system is
pushing someone on a swing. In such a situation, we would not know the optimal
frequency in advance. Rather, we discover the best forcing frequency by observing
how the swing behaves in response to different 'forcing functions'.

Escapement

In the digital world in which we live, when we observe 'paced' behaviours
we naturally assume the existence of a prescribed forcing function such as
$F(t) = F_0 \sin \omega_j t$. However, there is a much more traditional way for rhythmicity
to emerge in behaviour. For example, most people eat three meals a day. This
happens because people choose to eat meals when they feel hungry, not because
a timer tells them to eat. The rhythmicity of eating meals emerges from the state
of people (i.e. hunger) and the availability of food.

Similarly, in walking, the right amount of energy needs to be provided at the
right time. The 'right' time is not prescribed beforehand but depends on which
part of the walking cycle the system displays. This state-dependent timing
mechanism can be compared to the escapement used in pendulum clocks. While
the source of energy is continuous in nature (i.e. elastic energy stored in and
released from a coil), the escapment provides energy to the pendulum in a

rhythmic manner. The timing of the energy release is dependent on the pendulum's movement and vice versa. This mutual relationship between the escapement and movement of the pendulum is key for the emergence of effective timing.

When you first learned how to propel your body high into the air on a swing, you faced the challenge of calculating how to control your body and time the energy you put into the system. You had to learn when to pump your legs and when to tilt your upper body forwards and backwards. When you 'got it', you had learned to take advantage of the pendular dynamics of the swing. You were able to insert just the right amount of energy, at the right time, to achieve the height you wanted.

When adults learn a new motor skill they explore their options for coupling their limbs and determining the optimal forcing, just as children do when learning to swing. Vereijken *et al.* (1992) demonstrated this in an experiment in which they asked their participants to learn to use an alpine ski simulator. The apparatus consisted of two curved parallel bars with a platform on top. The platform could move over the bars via rollers and springs were attached to the bars at the peak of the curve. When the springs were at their shortest, the platform was stable at the top of the arc. Performers stood on the platform and shifted their body weight from side to side, stretching the springs and causing the platform to move downwards. Once the springs were fully stretched, the platform would move back up to the top and down on the other side. The transition from demonstrating novice performance to producing highly skilled movement patterns was achieved through the discovery of the appropriate time to add energy into the system. The springs' elastic properties allowed the use of stored elastic potential energy, which assisted the upward movement of the platform. Early in the learning process, participants added muscular forces too soon resulting in a loss (waste) of energy. Later in the learning process, they discovered that they could maximize the recovery of elastic potential energy by adding muscular energy just after the platform had passed the summit of its curved path. Just as children discover when pushing a friend on a swing, the mechanical advantage for energy conservation occurs just after reversal of direction, when the system's mass has already begun to swing in the desired direction. In just one practice session, Vereijken *et al.*'s (1992) participants learned to optimize the use of active and passive forces to achieve the movement outcome efficiently.

During walking we take advantage of such interplay between active and passive forces in a similar fashion. Adult walking is characterized by many years of walking experience. Over time we learn to organize muscles into synergies in dynamic fashion that result in appropriately dosed and timed joint torques, which ultimately dictate the movement outcome. In Chapter 6, we have seen that net joint torques can be decomposed into muscular, gravitational and motion-dependent torques. Ulrich *et al.* (1994) showed that when adults walk on a treadmill, the net flexor torque used to initiate the forward motion of the leg in swing is predominantly generated by muscles. Lowering the leg back to the support surface requires a net extensor torque, which is dominated by gravitational contributions. The authors further demonstrated that 7-month-old infants show similar patterns of net joint torques (i.e. net flexor torques at step initiation and net extensor torques when lowering the leg to the ground) when supported on treadmills. However, when decomposing the net torques into muscular and non-muscular contributions, infants' strategies to achieve these net torques were much more variable than in adults. These findings suggest that infants explore multiple solutions to the task in order to discover optimal movement patterns. Variability in intermuscular coordination patterns during walking have also been observed in toddlers. Chang et al. (2006, 2009) examined the muscle activation patterns in toddlers during the first seven months after the onset of walking. The net extensor and flexor torques were appropriate in magnitude and timing. However, all four of the primary gait muscles – the gastrocnemius, tibialis anterior, quadriceps femoris and hamstrings – showed large variations in the timing and duration of muscle activity, with no dominant patterns emerging for several months. While over time, we become more and more consistent in the neuromuscular patterns that result in the required joint torques which produce the ultimate movement outcome, some variability remains, even during very skilful walking. In fact, this is one of the beauties of our system because the variability and the redundancy of the human system are the means by which human gait can be so highly flexible and adaptive to the many goals we create for ourselves and the challenges we face as we move through space.

Clinical and practical applications

Of the principles we described in this chapter, one of the most important applications for walking is that to facilitate the emergence of this skill or the subsequent stability of this behaviour in functional settings, we must provide new walkers with enormous amounts of practice. We use the adjective 'enormous' to make a point. Motor behaviour researchers who use simple laboratory tasks, like moving a joy stick to new locations or pointing to a target, and who work with healthy young adults, know they must provide hundreds of practice trials in order to allow participants to discover optimal movement trajectories. Yet, in clinical settings, therapists and teachers usually have limited time to work with individuals and can become frustrated with the lack of progress they see in their patients/students. To learn to walk, performers need many more repetitions than we generally provide to discover the match between their bodies, the context and the goal of the movement.

To maximize success rates, teachers/therapists must advocate for increased time and be creative in finding ways to keep their patients/students motivated to repeat their efforts. Further, they should provide safe and reasonable opportunities to experience the boundaries of limb control to highlight the contrast between too much, too little and just right with respect to both timing and magnitude of muscular forces. Teachers/therapists should consider using instructional cues that draw their attention to the specific desired functional outcomes. They should avoid giving specific instructions relating to the details about how to move individual body segments and in a particular sequence. Rather, cues should draw attention to the impact of the variations in relation to the desired movement. Furthermore, teachers/therapists should consider changing the environmental constraints to facilitate functional adaptations, which facilitate successful movement outcomes. Finally, they should encourage performers to perceive their own body and potential equipment (e.g. crutches, braces, the narrow width of a balance beam) as facilitators of the exploration of movement solutions, taking the constraints of the specific situation into consideration, rather than perceiving these items as objects simply to be manipulated or managed.

Key points

- Well-practised and stable walking patterns can be modelled because humans discover the relations among the mechanical properties of their systems and passive forces (gravity, motion-dependent torques) and add energy (muscle torque) into the system at efficient times and in efficient amounts, in order to acquire this skill.
- Changes in the systems' properties (mass, stiffness, length) can lead to shifts in walking patterns.
- Walking stability is adapted by changing the system's mechanical constraints.
- Net torques needed to generate walking patterns must be appropriate in timing and magnitude. They can be produced by many combinations of active and passive forces.
- When left to their own devices, infants and young children generate thousands of exploratory solutions as they discover how to map their own biomechanics on to the physical properties of the world around them in pursuit of ways to locomote.
- As a consequence, when trying to facilitate the development of walking, teachers/therapists should provide many opportunities to discover these important relationships, rather than focusing attention on a prescribed movement pattern.

Acknowledgements

We would like to thank Annette Pantall, Sandy Saavedra and Jennifer Sansom for their very helpful feedback on earlier versions of our chapter. Funds to support B. Ulrich's work on this chapter were provided by a grant from the National Institutes of Health (1 R01 HD047567-01A1).

References

Adolph, K.E., Vereijken, B. and Shrout, P.E. (2003) 'What changes in infant walking and why', *Child Development*, 74: 474–97.
Black, D., Chang, C.-L., Kubo, M., Holt, K. and Ulrich, B.D. (2009) 'Developmental trajectory of dynamic resource utilization during walking: toddlers with and without Down Syndrome', *Human Movement Science*, 28: 141–54.

Brooks, G.A., Fahey, T.D., White, R.P. and Baldwin, K.M. (2000) *Exercise Physiology: Human Bioenergetics and its Applications*, Mountain View, CA: Mayfield.

Chang, C.-L., Kubo, M. and Ulrich, B.D. (2009) 'Emergence of neuromuscular patterns during walking in toddlers with typical development and with Down syndrome', *Human Movement Science*, 28: 283–96.

Chang, C.-L., Kubo, M., Buzzi, U. and Ulrich, B. (2006) 'Early changes in muscle activation patterns of toddlers during walking', *Infant Behavior & Development*, 29: 175–88.

Chen, H.C., Ashton-Miller, J.A., Alexander, N.B. and Schultz, A.B. (1991) 'Stepping over obstacles: gait patterns of healthy young and old adults', *Journal of Gerontology*, 46: M196–203.

Foo, P., Goldfield, E.C., Kay, B. and Warren, W.H. (2004) 'Infant bouncing: spontaneous learning of task dynamics', paper presented at the International Conference on Infant Studies, Chicago, May.

Garciaguirre, J.S., Adolph, K.E. and Shrout, P.E. (2007) 'Baby carriage: infants walking with loads', *Child Development*, 78: 664–80.

Goldfield, E.C., Kay, B.A. and Warren, W.H. (1993) 'Infant bouncing: the assembly and tuning of action systems', *Child Development*, 64: 1128–42.

Haehl, V., Vardaxis, V. and Ulrich, B.D. (2000) 'Learning to cruise: Bernstein's theory applied to skill acquisition during infancy', *Human Movement Science*, 19: 685–715.

Holt, K.G., Saltzman, E., Ho, C.-L. and Ulrich, B.D. (2007) 'Scaling of dynamics in the earliest stages of walking', *Physical Therapy*, 87: 1458–67.

Holt, K.G., Saltzman, E., Ho, C.-L., Kubo, M. and Ulrich, B.D. (2006) 'Discovery of the pendulum and spring dynamics in the early stages of walking', *Journal of Motor Behavior*, 38: 206–18.

Jensen, J.L., Ulrich, B.D., Thelen, E., Schneider, K. and Zernicke, R.F. (1994) 'Adaptive dynamics of the leg movement patterns of human infants: I. The effects of posture on spontaneous kicking', *Journal of Motor Behavior*, 26: 303–12.

Kubo, M. and Ulrich, B. (2006) 'A biomechanical analysis of the "high guard" position of arms during walking in toddlers', *Infant Behavior and Development*, 29: 509–17.

Kuo, A.D. and Donelan, J.M. (2010) 'Dynamic principles of gait and their clinical implications', *Physical Therapy*, 90: 157–74.

Ledebt, A. (2000) 'Changes in arm posture during the early acquisition of walking', *Infant Behavior and Development*, 23: 79–89.

Mulvey, G.M., Kubo, M., Chang, C.-L. and Ulrich, B.D. (2011) 'New walkers with Down syndrome use cautious but effective strategies for crossing obstacles', *Research Quarterly for Exercise and Sport*, 82: 210–19.

Smith, B.A. and Ulrich, B.D. (2008) 'Early onset stabilizing strategies for gait and obstacles: older adults with Down syndrome', *Gait & Posture*, 28: 448–55.

Srinivasan, M. (2009) 'Optimal speeds for walking and running and walking on a moving walkway' *Chaos*, 19: 026112.

Thelen, E. (1986) 'Treadmill-elicited stepping in seven-month-old infants', *Child Development*, 57: 1498–1506.

Thelen, E., Ulrich, B.D. and Niles, D. (1987) 'Bilateral coordination in human infants: stepping on a split-belt treadmill', *Journal of Experimental Psychology: Human Perception and Performance*, 13: 405–10.

Ulrich, B.D., Haehl, V., Buzzi, U., Kubo, M. and Holt, K.G. (2004) 'Modeling dynamic resource utilization in populations with unique constraints: preadolescents with and without Down syndrome', *Human Movement Science*, 23: 133–56.

Ulrich, B.D., Jensen, J.L., Thelen, E., Schneider, K. and Zernicke, R.F. (1994) 'Adaptive dynamics of the leg movement patterns of human infants: II. Treadmill stepping in infants and adults', *Journal of Motor Behavior*, 26: 313–24.

Vereijken, B., Whiting, H.T.A. and Beek, W.J. (1992) 'A dynamical systems approach to skill acquisition', *Quarterly Journal of Experimental Psychology*, 45A: 323–44.

Vereijken, B., Pedersen, A.V. and Storksen, J.H. (2009) 'Early independent walking: a longitudinal study of load perturbation effects', *Developmental Psychobiology*, 51: 374–83.

9 Biomechanical aspects of the development of object projection skills

Stephen Langendorfer, Mary Ann Roberton and David Stodden

Introduction

In this chapter we examine the biomechanical principles that unify forceful projection tasks, the characteristics of advanced forceful throwing, striking and kicking, and how each of these exemplars changes developmentally. Skilled performance in object projection tasks is generally related both to high-effort force production as well as to projectile accuracy. Our focus in this chapter, however, is to discuss characteristics/mechanisms related to the development of movement patterns that produce maximal projectile speed. Inherent in these discussions is the understanding that accuracy will constrain the behaviours we describe.

Basic theoretical concepts

While advanced throwing, striking and kicking differ in their respective mechanics, they all integrate the following biomechanical and neuromuscular properties during the production of their specific movement patterns: (a) generation and transfer of linear and rotational energy within an open kinetic chain, using proximal-to-distal movement sequencing of body segments; (b) effective use of segmental inertial characteristics; and (c) exploitation of elastic tissue characteristics within the context of eccentric and concentric muscle actions.

Kinetic chain

High distal segment velocities and projectile speeds are promoted via proximal-to-distal sequencing of body segments through a kinetic chain mechanism. In a human kinetic chain, the movement of proximal segments sequentially influences distal segments in the chain. A more massive proximal segment transfers its linear and rotational energy to its adjacent, less massive distal segment via complex interactions among muscles, tendons, ligaments, cartilage and bony structures. As a proximal segment accelerates near its maximum velocity, its more distal neighbour spatially and temporally lags behind due to its own inertial characteristics. This lag results in an eccentric lengthening (i.e. stretching) of the associated musculature and connective tissue that crosses the joint. The proximal segment subsequently slows down due to the motion-dependent effect of the distal segment on the proximal segment (Putnam 1993). If optimally timed, the distal segment recovers the stored elastic potential energy and accelerates, demonstrating higher linear and/or angular velocity than the proximal segment. Figure 9.1 illustrates two hypothetical graphs demonstrating similar proximal-to-distal sequences in kinetic chains with different relative timing. Compared to Figure 9.1a, Figure 9.1b represents more favourable temporal sequencing of shoulder, elbow and wrist angular velocities, with relative timing of peak angular velocities promoted closer in time to take better advantage of elastic potential energy and to optimize energy transfer.

Segmental inertial characteristics

Energy transfer within a kinetic chain is augmented by the effective use of segmental inertial characteristics. High levels of energy contributed by both linear centre of mass translations and angular rotations are promoted by segmental configurations that optimize joint moment inertial characteristics (Stodden *et al.* 2006a, 2006b). Optimal preparatory configurations of segments distal to a specific joint may initially reduce the moment of inertia in the more proximal segment. For example, increasing knee flexion during upper leg acceleration in kicking decreases the moment of inertia about the hip and increases hip angular velocity. In contrast, increasing the moment of inertia may promote more advantageous energy generation and transfer. In throwing, a 90° elbow angle maximizes the inertial characteristics of the forearm which, in turn, promotes external rotation (Stodden *et al.* 2006b). Thus, optimal preparatory positions of distal segments in the kinetic chain may either reduce or increase inertial characteristics to maximize energy transfer capability.

Neuromuscular contributions

Optimizing segmental inertia can promote an eccentric stretch of musculature crossing a joint (e.g. rapid elbow flexion eccentrically stretches elbow extensors) and briefly store elastic potential energy. Optimally timed eccentric muscle contraction stimulates muscle spindles which promote a subsequent stretch reflex, leading to a concentric contraction of that specific musculature. The evoked stretch reflex promotes high motor-unit activation and can enhance voluntary muscle force contributions (Fleisig *et al.* 1996a; Roberts and Metcalfe 1968). Enhanced muscle force produces higher torque about a joint and is supplemented by recovery of elastic potential energy from the preceding eccentric stretch. The result is increased joint angular velocities and resultant energy transfer to distal segments.

While these mechanisms have been described individually, they are collectively integrated within the entire movement pattern. If optimally integrated, they serve to increase energy generation and transfer within the system and

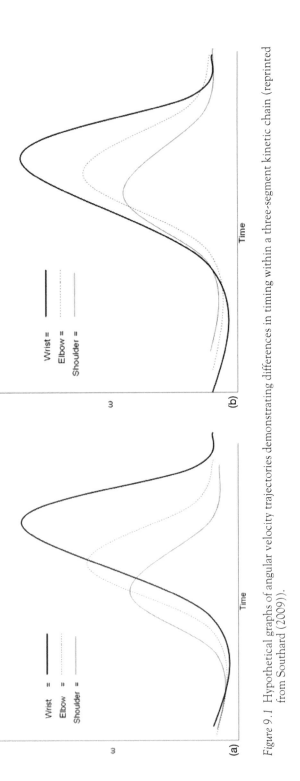

Figure 9.1 Hypothetical graphs of angular velocity trajectories demonstrating differences in timing within a three-segment kinetic chain (reprinted from Southard (2009)).

promote maximal projection speeds. 'Optimal' integration refers to absolute and relative timing of segmental interactions within the system. For instance, the absolute time from stride foot contact to ball release in highly skilled throwers is consistently around 150 ms (Fleisig *et al.* 1999). In that time, the ball can be accelerated from 0 m/s to over 44 m/s (100 mph). Such optimal timing is needed not only to maximize ball speed but also to minimize the high joint forces and torques generated at the spine, gleno-humeral, elbow and wrist joints (Fleisig *et al.* 1996a; Stodden *et al.* 2005).

Moreover, in throwing, decreased time from stride foot contact (SFC) to ball release or maximum shoulder internal rotation (which is near the time of ball release) results in higher ball velocities among different level throwers (Chu *et al.* 2009; Matsuo *et al.* 2001; Stodden *et al.* 2005, 2006a). This finding suggests that maximal energy transfer is facilitated by decreasing the overall time across multiple segment interactions within the kinetic chain. In contrast, Stodden *et al.* (2005) demonstrated that increased shoulder horizontal adduction time during arm cocking and arm acceleration was associated with increased ball velocities within highly skilled pitchers. This increased horizontal adduction time was associated with increased horizontal abduction at SFC, indicating that pitchers were able to generate shoulder adduction force for a longer period of time over a greater range of motion. Thus, the term 'optimize' speaks to the complexity of absolute and relative timing interactions within the kinetic chain.

Biomechanics and mechanisms of advanced overarm throwing

Throwing mechanics have generally been classified into three different patterns: (a) overarm; (b) underhand; and (c) sidearm. Fleisig *et al.* (1996a) suggested that the mechanics of most 'overhand' throwing activities are similar, especially with regard to baseball pitching and throwing. Similar biomechanics and mechanisms are demonstrated in other sports such as football (Fleisig *et al.* 1996b; Rash and Shapiro 1995), javelin (Mero *et al.* 1994) and cricket (Bartlett *et al.* 1996). Differences in the size and mass of implements, projected objects and task-specific goals promote specific differences in their respective biomechanics. In the following section we focus on the overarm throw used in baseball pitching since it has received the most biomechanical and developmental study.

Phases of advanced overarm throwing

As illustrated in Figure 9.2, joint movements during skilled throwing interact across phases of throwing (e.g. pelvis and upper torso rotation occur during multiple phases of the throw). These phases are separated by convenient positional points within the throwing motion. Below, we describe phase descriptions of the kinematics, kinetics and timing of advanced throwing.

Figure 9.2 Sequence of movements during the baseball pitch (reprinted from Chu *et al.* (2009)).

Wind-up

Preparatory motions in the wind-up vary considerably in skilled throwing, taking approximately two seconds in a baseball pitch (Dillman *et al.* 1993; Fleisig *et al.* 1996a). They include backward movement of the centre of mass (COM)and trunk rotation away from the intended target. The leg contralateral to the throwing arm generally flexes at the hip and knee, while the trunk rotates away from the intended target. This combination of movements aligns body segments to facilitate sequential derotation and linear translation of the COM towards the target.

Stride

The initial phase in which motion starts in the direction of throwing takes approximately 0.5 s (Fleisig *et al.* 1996a) and comprises two important movements: the initiation of linear movement towards the intended target with a contralateral step, and bilateral, pendulum-like arm movements that include elbow extension followed by flexion and shoulder abduction and external rotation. At SFC the shoulder has abducted approximately 100° with 60° of external rotation (Chu *et al.* 2009; Fleisig *et al.* 1996a). Elbow flexion is generally between 75° and 100° at SFC (Chu *et al.* 2009; Fleisig *et al.* 1996a, 1999). Near the end of the stride phase the pelvis begins to open towards the target prior to SFC (27°±13°) while the upper torso remains in a more closed position (−19°±15°) demonstrating differentiated spinal rotation of 40° to 56° (Chu *et al.* 2009; Stodden *et al.* 2001). Delayed rotation of the upper torso is facilitated by high angular velocities of the pelvis (400–600°/s) and the inertial characteristics of the upper torso and throwing arm positions relative to the pelvis.

The placement of the stride foot is aligned with toes pointed towards the intended target, helping the lead leg act as a stable brace over which the upper body can rotate (Dillman *et al.* 1993). Stride length in highly skilled throwers is generally between 70 per cent and 94 per cent of standing height (Atwater 1979; Chu *et al.* 2009; Fleisig *et al.* 1996a) with a knee angle of approximately 38° (Fleisig *et al.* 2006a). At the end of the stride phase the throwing arm is abducted at the shoulder (93°±2°), externally rotated (67°±24°), with the elbow flexed (98°±18°) and horizontally abducted behind the head (−11°±24°) (Fleisig *et al.* 1996a). The forearm is pronated (~100°), pointing the ball away from the intended target. The wrist maintains a neutral position (Sakurai *et al.* 1993).

Arm cocking

Arm cocking, the phase interval from SFC to maximum external rotation of the throwing arm, lasts between 100 and 150 ms (Fleisig *et al.* 1996a). During arm cocking, the pelvis and upper torso continue to 'open up' or derotate towards the target. By the end of arm cocking, the upper torso rotation (88°±10°) has slightly surpassed pelvic rotation (85°±7°) (Stodden *et al.* 2001). As the trunk rotates to face the target, maximum passive shoulder external rotation reaches high values

(175°±11°) that cannot be actively attained (Fleisig *et al*. 1996a). From SFC until the instant of maximal external rotation, trunk and shoulder horizontal adduction musculature are the main factors responsible for upper extremity movements (Feltner 1989). The net horizontal abduction acceleration of the upper arm approaches zero during arm cocking and becomes horizontal adduction acceleration due to both linear and angular acceleration of the throwing shoulder (Feltner 1989). The upper arm and forearm lag spatially behind the upper trunk due to the acceleration of the upper trunk and inertial characteristics of the throwing arm.

Shortly before maximal external rotation is reached, the arm begins to extend at the elbow from a right angle (~90°) position. To extend the elbow, the triceps muscle applies an active extension torque to the forearm. This active extension torque is not large enough to produce the high elbow angular velocities demonstrated in highly skilled throwers (Werner *et al*. 1993). Rather, the combination of decreased biceps activity, which lessens elbow flexion torque and upper torso rotation, may be a better explanation.

Arm acceleration

Arm acceleration, lasting approximately 30–50 ms, is the most dynamic and explosive phase of the throw (Atwater 1979; Fleisig *et al*. 1996a; Pappas *et al*. 1985). Arm acceleration begins when the humerus starts to rotate internally at the shoulder and is initially produced by active torques generated by the subscapularis, latissimus dorsi, pectoralis major, teres major and by passive torques exerted by the joint capsule near the limits of the shoulder joint's range of motion (i.e. close packed position). Internal rotation velocities during arm acceleration can approach 8000°/s, even in 10–12-year-old children (Fleisig *et al*. 2006b). These high internal rotation velocities are augmented by a decrease in external rotation torque at the shoulder as the elbow extends (Feltner and Dapena 1986). Elbow extension serves to increase the linear aspect of the hand or ball movement as it deviates from the angular movements generally produced by the trunk and shoulder musculature (Atwater 1979; Feltner 1988). Elbow extension also serves to decrease the moment of inertia about the longitudinal axis of the upper arm, favouring large internal rotation velocities. Stride knee extension during arm acceleration facilitates energy transfer by continuing to promote a stable base of support over which the upper torso and arm rotate (Matsuo *et al*. 2001). Increasing trunk flexion corresponds to increased linear velocity during arm acceleration and augments ball speeds (Matsuo *et al*. 2001). From an approximate 40° hyperextended position, the wrist begins flexing about 20 ms before ball release, ending in a neutral position at release (Pappas *et al*. 1985; Sakurai *et al*. 1993).

Arm deceleration and follow-through

During arm deceleration and follow-through, stride leg musculature maintains a dynamic, single leg stance as the upper torso and upper extremities rotate and linearly translate about the fixed stride leg. Strong eccentric contractions are necessary in lower extremity posterior musculature (i.e. gluteals, hamstrings and plantar flexors), shoulder decelerators (i.e. rotator cuff and scapular stabilizers) and elbow flexors to dissipate the forces generated during the follow-through (Fleisig *et al.* 1996a; Campbell *et al.* 2010).

Developmental considerations for overarm throwing

Qualitative analyses

Developmental sequences for the overarm throw have been validated for children and adolescents from both cross-sectional and longitudinal data (Table 9.1) (Roberton 1977, 1978; Roberton and Halverson 1984; Roberton and Langendorfer 1980). In the following sections we examine those sequences and their intercomponent relationships for biomechanical implications. Developmental sequences are typically divided into qualitatively different categories known as developmental levels or steps.

Table 9.1 Developmental sequences for the overarm throw for force

Preparatory backswing component

Step 1 No backswing action. Ball is thrown from whatever initial position it is grasped.

Step 2 Shoulder/elbow flexion. Both shoulder and elbow are flexed in sagittal plane.

Step 3 Upward circular. Shoulder is flexed while elbow remains extended during upward circular path.

Step 4 Lateral circular with supination. Ball-in-the-hand moves lateral to the trunk, usually below level of the shoulder. At point of greatest backswing, hand and forearm are supinated so ball faces laterally or forwards.

Step 5 Lateral circular with pronation. Ball-in-the-hand moves lateral to the trunk, usually downwards and backwards from the shoulder. At point of greatest backswing, hand and forearm are pronated so ball faces backwards.

Trunk action component during forward swing

Step 1 No trunk action or forward–backward movements. Only the arm is active in force production. Sometimes, the forward thrust of the arm pulls the trunk into a passive left rotation (assuming a right-handed throw), but no twist-up precedes that action. If trunk action occurs, it accompanies the forward thrust of the arm by flexing forwards at the hips. Preparatory extension sometimes precedes forward hip flexion.

Step 2 Upper trunk rotation or total trunk 'block' rotation. Both the spine and pelvis rotate away from the intended line of flight and then simultaneously begin forward rotation, acting as a unit or 'block'. Occasionally, only the upper spine twists away, then towards the direction of force. The pelvis then remains fixed, facing the line of flight, or joins the rotary movement after forward spinal rotation has begun.

Step 3 Differentiated rotation. The pelvis precedes the upper spine in initiating forward rotation. The child twists away from the intended line of ball flight and then begins forward rotation with the pelvis while the upper spine is still twisting away.

Table 9.1 continued

Humerus (upper arm) action component during forward swing

Step 1 Humerus oblique. The humerus moves forwards to ball release in a plane that inter-sects the trunk obliquely above or below the horizontal line of the shoulders. Occasionally, during the backswing, the humerus is placed at a right angle to the trunk, with the elbow pointing towards the target. It maintains this fixed position during the throw.

Step 2 Humerus aligned but independent. The humerus moves forwards to ball release in a plane horizontally aligned with the shoulder, forming a right angle between humerus and trunk. By the time the shoulders (upper spine) reach front facing, the humerus (elbow) has moved independently ahead of the outline of the body (as seen from the side) via hori-zontal adduction at the shoulder.

Step 3 Humerus lags. The humerus moves forwards to ball release horizontally aligned, but at the moment the shoulders (upper spine) reach front facing, the humerus remains within the outline of the body (as seen from the side). No horizontal adduction of the humerus occurs before front facing.

Forearm action component during forward swing

Step 1 No forearm lag. The forearm and ball move steadily forwards to ball release throughout the throwing action.

Step 2 Forearm lag. The forearm and ball appear to 'lag', i.e. to remain stationary behind the child or to move downwards or backwards in relation to him/her. The lagging forearm reaches its furthest point back, deepest point down or last stationary point *before* the shoulders (upper spine) reach front facing.

Step 3 Delayed forearm lag. The lagging forearm delays reaching its final point of lag until the moment of front facing.

Foot (stepping) action

Step 1 No movement. Thrower throws from whatever position feet happen to be in.

Step 2 Homolateral step. Final forward step is with the same foot as the throwing hand.

Step 3 Contralateral step. Final forward step is with the opposite foot from the throwing hand.

Step 4 Long contralateral step. Length of the final forward step with the opposite foot is over half of the thrower's standing height.

Note: Modified from Roberton and Halverson (1984). Hypothesized backswing sequence, adapted from Langendorfer (1999).

Backswing

The backswing is important in the development of throwing because it constrains the arm movements possible in the forward swing. Primitive backswing levels assume final positions that do not allow shoulder external rotation or forearm pronation during the forward swing (Langendorfer and Roberton 1999). These developmentally early levels are also characterized by minimal arm movement. As the backswing becomes more circular, either to the side or down and back, the greater range of movement gives the hand or ball more time and space in which to accelerate forwards.

Trunk action

At the most primitive level, the trunk either remains stationary, presumably to brace against forces generated by the moving arm, or it exhibits flexion in the sagittal plane. We speculate that flexion precedes rotation developmentally in the throw because it perturbs balance less and enables the thrower to keep their head and shoulders aligned to face the target. Trunk rotation requires the head to face the target while the shoulders twist beneath it (Figure 9.3).

Figure 9.3 Visual descriptions of differences in trunk action levels 1–3. Level 1: no rotation (A–C); level 2: block rotation, simultaneous rotation of pelvis and upper torso (D–F); level 3: differentiated rotation (G–I).

The final step in the trunk sequence, differentiated rotation, is also the last aspect of the throw to develop, typically appearing after the forearm component has reached its advanced level. Langendorfer and Roberton (2002a), like Fleisig *et al.* (1996a) and Stodden *et al.* (2001) previously, hypothesized that the inertia of the arms and shoulders in the advanced arm action causes the upper trunk to lag behind when forward trunk rotation begins, producing the pelvic lead associated with differentiated rotation. We conclude that the neural command to rotate results either in trunk action level 2, block rotation, or trunk action level 3, differentiated rotation, depending on the biomechanical context. Since instances of level 2 humerus and forearm action have been observed with differentiated rotation, Langendorfer and Roberton (2002a) suggested three other conditions that may potentially produce differentiated rotation: (1) a relatively large range and high velocity of forward trunk rotation; (2) a position of the arm segments at the end of the backswing that optimizes the inertial moment; and/or (3) optimal timing relationships among the backswing, the long contralateral step and the initiation of forward trunk rotation. Barton and French (2004) favoured this last 'timing' hypothesis in their study of American Little League baseball players.

Humerus action

The position oblique to the horizontal line of the shoulders that is characteristic of the earliest developmental level places the gleno-humeral joint in what the therapeutic literature refers to as a 'loose-packed' position (Langendorfer and Roberton 2002b). Loose-packed positions reduce the tension on the joint, allowing the tendons and ligaments to be fairly lax, presumably giving more comfort as the humerus moves forwards primarily in the sagittal plane. In level 2 the humerus moves closer to a 'close-packed' starting position in that it is held close to 90° with the shoulders; however, during the forward phase of the throw horizontal adduction moves the humerus into a more loose-packed position. In level 3, the humerus remains in line with or lags behind the shoulders until the thrower rotates to front facing. Since research on advanced throwers has shown that the moment for horizontal adduction of the humerus remains positive until front facing (Hong *et al.* 2001), inertia of the upper extremity caused by trunk rotation is greater than this adduction moment, constraining the humerus to 'lag'. Thus, the same horizontal adduction moment is operating in both developmental levels 2 and 3, but in level 3 the thrower has positioned their arm in a way that takes advantage of the segmental energy transferred from forward trunk rotation (i.e. promoting lag; Figure 9.4).

Most likely, optimal arm positioning also involves the elbow angle. Many intermediate level throwers collapse (flex) their elbow before or during the forward movement, pulling their hand/ball in towards their head. This flexion action serves to move the mass of their arm segments closer to the centre of rotation, thereby reducing the inertial moment (Cross 2004). In contrast, advanced throwers maintain a right angle at their elbow, which increases the rotational

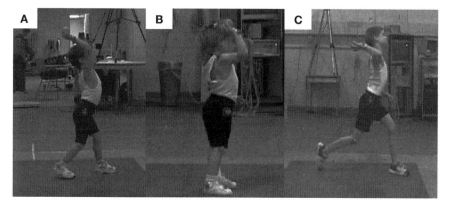

Figure 9.4 Humerus action levels 1–3. Level 1: humerus oblique – shoulder abduction >90° (A); level 2: humerus aligned but independent – shoulder abduction >90°, horizontal adduction at front facing >0° (B); level 3: humerus lag – shoulder abduction >90°, shoulder horizontal adduction at front facing ≤0° (C).

inertia that enables the humerus to 'lag' despite the adductor moment. The elbow right angle greatly increases the tension on the gleno-humeral joint by causing external rotation of the humerus, a true 'close-packed' position, which passively generates a large amount of elastic potential energy from shoulder internal rotation and horizontal adduction musculature (i.e. latissimus dorsi and pectoralis major).

Forearm action

The developmental levels of forearm action actually describe the degree and timing of external rotation of the humerus during the forward movement of the throw. Thus, this component focuses on changes in the behaviour of the distal aspects of the kinetic chain. The initial developmental level of forearm action shows no lag (i.e. no external rotation). Indeed, the triceps may even initiate the forward movement of the throw via active elbow extension rather than receiving energy transferred from the trunk and humerus. In the second level of forearm action, external humeral rotation concludes before the arm has received full energy transfer; thus demonstrating an ineffective kinetic chain mechanism. In the advanced throw, the point of deepest lag coincides with front facing, so the hand/ball receives energy transfer from the other segments in the chain (Figure 9.5).

Foot action

Although other component developmental sequences in Table 9.1 have been validated through longitudinal study, the foot action sequence was hypothesized by Roberton in 1984, but has never been validated. Indeed, Langendorfer (1987b)

Figure 9.5 Forearm action levels 1–3. Level 1, no lag: two consecutive frames of the same individual with no passive external rotation and only active elbow extension during the throw (A–B); level 2, forearm lag: two consecutive frames of the same individual showing maximum external rotation occurring before front facing (C), and decreased external rotation at front facing (D); level 3, delayed forearm lag: two consecutive frames of the same individual showing maximum external rotation occurs at front facing (E–F).

questioned its reliability and validity: not all children showed developmental level 2, step with the foot homolateral to the throwing arm; also across a series of throwing trials, children frequently varied their foot action among categories (Figure 9.6). What is certain, however, is that linear momentum generated by the presence and length of stepping action increases with development (Stodden *et al.* 2006a). Mastering the trunk rotation that occurs over this moving base of support is a difficult developmental task, presumably constrained by the thrower's degree of dynamic balance.

Figure 9.6 Visual descriptions of differences in step action levels 1–4. Level 1: no step (A); level 2: homolateral step (B); level 3: contralateral short step (C); level 4: contralateral long step (D).

Developmental relationships among components

Several investigators have begun to study the developmental order between levels across throwing components (Barton and French 2004; Langendorfer and Roberton 2002a; Robinson *et al.* 2007). In a study of 39 children followed longitudinally from 5.7 to 8.0 years of age, Langendorfer and Roberton (2002b) found that children first began to block rotate their trunks while their humerus and forearm actions remained at level 1. Block rotators then showed various combinations of the first and second levels of humerus and forearm action. Eventually, 25 of the children arrived at the second levels of trunk, humerus and forearm action. Those who developed further primarily changed to level 3 in their humerus action. Langendorfer and Roberton (2002b) speculated that changes in the trunk kinematics (particularly angular forward acceleration) associated with block rotation may enable the changes in arm action. As indicated earlier, after the humerus reached its most advanced level, the forearm tended to progress to level 3. Lastly, the trunk attained differentiated rotation. While some individual differences occurred in this common developmental pathway, Barton and French (2004) verified these results in their American Little League players.

Quantitative analyses

Descriptions of throwing kinematics, kinetics and temporal measures have focused on elite performers, with limited developmental considerations (Southard 2002, 2009; Stodden 2006; Stodden and Rudisill 2006; Stodden *et al.* 2006a, 2006b); however, reported kinematic differences within and between highly skilled and less skilled throwers seem to parallel the verbal descriptions of the developmental sequences in Table 9.1. For instance, Stodden *et al.* (2006a, 2006b) demonstrated that maximal linear and angular velocities of the pelvis and upper torso increased significantly from low- to high-skilled throwers. Dramatic increases in linear and angular velocities of the trunk paralleled increases in step lengths and forward trunk tilt angle, and were related to higher ball velocities (Matsuo *et al.* 2001; Stodden *et al.* 2006a). In essence, the increased COM translations and trunk angular velocities provided increased energy to the kinetic chain system.

Maximal energy transfer in throwing development is augmented by optimal absolute and relative timing of segmental interactions. For example, the time from SFC to ball release decreases as ball velocity increases in low- to high-skilled children (Stodden *et al.* 2006a). Southard (2002) argued that skill level increased with the number of joints experiencing distal lag, which he defined as the time difference between peak velocity of a distal joint and peak velocity of the proximal joint. In addition, Southard (2009) found increased elbow lag and decreased wrist lag in higher skilled throwers. Thus, developmentally advanced movement patterns optimize timing and segmental inertial characteristics to increase initial energy generation and maximize energy transfer. Less advanced individuals do not effectively exploit these mechanisms. Other quantitative differences between low- and high-skilled individuals occur in the range of motion, step length, pelvis rotation, upper torso rotation, shoulder external/internal rotation, elbow flexion and shoulder abduction deviations from 90° (Stodden *et al.* 2006a, 2006b).

One additional aspect of skill development is variability in movement patterns. Fleisig *et al.* (2009) documented variability in kinematic parameters among youth, high school, college, minor League and Major League pitchers. As skill level increased, throwers demonstrated decreased variability in 6 of 11 kinematic parameters, but no differences in temporal or kinetic parameter variability. Similarly, Southard (2002, 2009) found that variability in the timing of shoulder, elbow and wrist joint lag was greater in low-skilled throwers.

Differences in joint kinetics among different throwing levels are the product of differences in a multitude of factors, including strength, absolute and relative timing, anthropometric characteristics, range of motion and joint moments of inertia. Higher kinetic values, in turn, are generally associated with differences in ball speeds (Fleisig *et al.* 1999). It is important to note that, while many of the aforementioned factors will lead to increased kinetics, developing optimal timing may lead to increased ball speeds without necessarily increasing joint torques to produce that speed (Herring and Chapman 1992; Stodden *et al.* 2005). The process of optimizing kinematics and absolute and relative timing during development may improve performance while minimizing the increase in joint torques associated with injury.

Biomechanics and mechanisms of advanced striking

Like throwing, striking occurs in many different sports (e.g. volleyball, tennis, baseball, softball, cricket, badminton, golf and handball). One main difference among these activities is whether they require an implement. Various masses, shapes and lengths of implements alter the kinetics and their resulting kinematic patterns. Differences in (or lack of) implements lead to dramatic differences in distal endpoint velocities. In volleyball, hand speeds reach upwards of 22 m/s (Ciapponi and Hudson 2000), whereas baseball batting (~33 m/s) (Welch *et al.* 1995), tennis serving (~55 m/s) (Fleisig *et al.* 2003) and golf driving (~52 m/s) (Nesbit 2005) can achieve significantly higher implement speeds.

Phases of advanced striking

The movement pattern of skilled striking can be defined with three temporally sequenced phases: a preparatory backswing, an acceleration phase and a follow-through phase. Below are phase descriptions of the biomechanics of advanced striking.

Backswing

The backswing in striking skills generally includes a brief lateral shift of the COM away from the intended target. Differences in specific movement patterns (underarm vs. sidearm vs. overarm) and the specific sport context (i.e. golf swing vs. baseball/tennis stroke) promote variations in how much COM linear translation occurs (Hume *et al.* 2005) and the extent to which subsequent forward COM translation correlates with implement/distal segment speed. An exception to this pattern is displayed in the volleyball strike in which the moving approach is similar to kicking. A general similarity between striking and throwing backswings is initial trunk/upper torso rotation away from the target which eccentrically loads the trunk musculature (Escamilla *et al.* 2009). In most striking skills, preparatory shoulder movements and elbow extension/flexion (either bilateral or unilateral, depending on the skill) also promote distal segment/implement trajectories away from the intended target and the consequent eccentric loading of the associated musculature.

Acceleration

Kinematic trajectories of the distal segment/implement in the forward acceleration phase occur in the horizontal (e.g. baseball/softball batting, tennis forehand/backhand) or sagittal (tennis serve, cricket batting, golf swing) planes. High pelvis and upper torso rotation angular velocities are comparable to those in elite throwing (Escamilla *et al.* 2009; Fleisig *et al.* 2003; Hume *et al.* 2005). Presumably, trunk energy transfer within the kinetic chain, in conjunction with the added arc distance to accelerate an implement, promotes higher terminal end velocity in striking with implements as compared to either the volleyball strike or throwing. An advanced wrist flexion/extension mechanism also has been strongly correlated with high implement velocities (Elliott *et al.* 1995; Hume *et al.* 2005).

Interestingly, Fleisig *et al.* (2003) indicated that advanced tennis serving did not follow the typical proximal-to-distal sequence of movements. Maximum upper torso angular velocity occurred before maximum pelvis angular velocity, while maximum shoulder internal rotation velocity followed maximum elbow and wrist angular velocities. There was no explanation of why this might be the case. It is therefore unclear why tennis serving does not follow a proximal-to-distal sequence.

Follow-through

The striking follow-through decelerates body segments and the implement. In a review of shoulder muscle recruitment patterns of upper extremity sports, eccentric contractions generally ranged from 30 to 74 per cent of maximal voluntary isometric contractions (Escamilla and Andrews 2009), which is less than demonstrated in throwing. While the range of motion of upper extremity joints in the follow-through is similar to throwing, the decreased muscle activity during follow-through is not well understood. Clearly, more research on muscle activation patterns in striking skills is needed to better understand the mechanical nature of these types of skills.

Developmental considerations for striking

Qualitative analyses

The development of overarm striking and the tennis serve have been studied pre-longitudinally in childhood and adolescence by Langendorfer (1987a, 1987b) and Messick (1991), respectively. Langendorfer filmed a cross-sectional sample of 58 boys, aged 2–12 years, striking a suspended tennis ball with a squash racquet that was lightweight and had a shortened shaft. He documented that developmental changes in the action of the trunk and humerus in overarm striking emerged similarly to the changes in overarm throwing for force (Table 9.1). Messick (1991) added cross-sectional observations on 9–19-year-olds performing overarm tennis serving with tennis racquets. In her preparatory trunk action component she found that younger tennis players held their trunks stationary or extended during the backswing while older tennis players employed minimal or maximal trunk rotation. Langendorfer (1987b) also observed that his participants employed one of the four stepping categories that paralleled those used in overarm throwing (Table 9.1), although the long, contralateral step was not observed as frequently in overarm striking. Apparently, the accuracy demand of contacting the suspended tennis ball constrained the strikers to limit their step. Interestingly, developmental changes in racquet action both in striking (Langendorfer 1987a, 1987b) and in tennis serving (Messick 1991) parallel forearm action development in throwing. From a biomechanical perspective, this result should not be surprising because the racquet becomes the distal segment in the kinetic chain. Further research on quantitative aspects of striking development is needed to assess the biomechanical parameters associated with qualitative developmental change in striking.

Co-development of throwing and striking

Langendorfer (1987a) also compared his participants' developmental levels in both throwing and striking. He found that they displayed either the same level of trunk and humerus action in both tasks or their overarm striking lagged behind throwing by one developmental level. For example, although some children used a non-rotating trunk action (level 1) in both throwing and striking, others demonstrated block rotation (level 2) in throwing while still striking with level 1. Some older boys used block rotation in both tasks while others used differentiated rotation (level 3) in throwing, but block rotation during striking. Langendorfer (1987a) inferred from these cross-sectional results that one or more constraints were acting systematically to delay striking compared to throwing. He hypothesized that possible constraints could be the double accuracy demand within the striking task (before directing the path of the object, the striker first has to make contact with the object) or the decreased moment of inertia resulting from increased wrist and elbow flexion.

Biomechanics and mechanisms of advanced kicking

Forceful kicking is constrained by the same 'double accuracy' demand encountered in striking: the kicker must first contact the ball with some part of the foot as well as send the ball in a desired direction. The constraints associated with a moving ball accentuate these accuracy demands. Of the many kinds of kicks, we will focus on place kicking.

Compared to overarm throwing, fewer research studies have examined place kicking (Roberts and Metcalfe 1968). Those available studies of elite place kickers describe a series of open kinetic chain movements focused primarily around actions of the lower extremities (Lees *et al.* 2010). The biomechanical principles that apply to forceful soccer instep kicking are again similar to those previously described (Lees *et al.* 2010). For example, the stretch-shortening cycle harnesses elastic energy when the kicking knee flexes during the beginning of the forward swing. The proximal-to-distal sequence transfers limb segment energy from the thigh to the shank just prior to ball contact in forceful kicking.

Phases of advanced kicking

Advanced kicking occurs across three temporal phases: the approach, acceleration and follow-through. In each phase, the trunk, the kicking and non-kicking legs and upper extremities display typical movement patterns.

Approach

A highly skilled kick begins with an angled, curved running approach to a stationary ball. Advanced soccer kickers prefer an approach at an angle of 45° using two to four steps. The final stride taken by the support leg involves an extended leap with the foot landing lateral to the ball (Lees *et al.* 2010). During the forward approach, both arms swing in opposition to the legs as they do in running. The trunk is upright with a minimal forward lean. At the point of support leg contact ending the approach, the pelvis has rotated away from the ball while the trunk has inclined backwards and laterally (Roberts and Metcalfe 1968).

Acceleration

From the point of support leg contact until ball contact, rapid angular rotation of the pelvis accompanies forward acceleration of the thigh via hip and knee flexion (Roberts and Metcalfe 1968). Prior to ball contact, hip flexion angular velocity decreases while energy from the thigh transfers to the shank, resulting in rapid knee extension (Roberts and Metcalfe 1968; Wickstrom 1983). The arm opposite the kicking leg simultaneously flexes at the shoulder while the trunk flexes forwards.

Follow-through

As the foot contacts the ball, hip flexion and knee extension continue to carry the kicking leg and then the entire body forwards and upwards. With a sufficiently forceful kick, the support foot is pulled off the ground. The flexion of the arm opposite the kicking leg continues as does trunk flexion (Wickstrom 1983). At the end of the follow-through the kicking leg contacts the ground prior to the support foot.

Developmental considerations for kicking

Kicking has rarely been studied developmentally across childhood and adolescence, although the few available studies present a fairly consistent picture (Bloomfield *et al.* 1979; Deach 1950; Lisy 2002). Early kicking shows limited or no approach steps, minimal range of motion in the leg backswing and limited involvement of the trunk and arms (Bloomfield *et al.* 1979). As children advance developmentally, they increase the number of steps in approaching the ball, enlarge the range of motion at the hip and knees during both back and forward leg swings and move at least one arm in opposition to the kicking leg. As the approach steps increase in number and length, the backswing is incorporated into the approach (Mally *et al.* 2011). Dynamic balance is clearly a rate-limiter for kicking development, as it is for throwing and striking. In early kicks, the child must balance on one foot while swinging the other leg backwards and forwards in a pendulum-like manner. Later, as the approach is incorporated into the kick, the child must land on one foot from a leap, then control their balance as they swing the kicking leg forwards.

Clinical and practical applications

These descriptions of the ordered changes occurring within throwing, striking and kicking presume that performers were attempting to produce each skill with high levels of effort and minimal emphasis on hitting a target. Several studies of throwing and kicking have actually examined the developmental impact of demanding accuracy through use of a target. Results suggested that the target effect is dependent on the distance or effort with which the person is required to project the object (Button *et al.* 2005; Langendorfer 1990; Manoel and Oliveira 2000; Teixeira 1999). Hamilton and Tate (2002) found that target size did not affect the throws of 8-year-olds, but the farther the children were from the target, the more advanced were their stepping, trunk and humerus actions. Barrett and Burton (2002) observed that college baseball players frequently used alternatives to the advanced throw at short distances. Southard (2006) found that asking throwers to throw with maximal velocity was the most effective instructional method for producing distal temporal lag.

These results show that situations demanding increased kinetic chain energy elicit more developmentally advanced movements. This fact supports Roberton's (1996) contention that ballistic skills, such as throwing, striking and kicking, should be introduced to children by stressing 'force' (i.e. high effort) goals with minimal emphasis on accuracy. There are a number of ways to augment the feedback that children receive when they throw, kick or strike 'hard'. These range from urging loud-sounding hits against a wall to measuring ball velocity with a radar gun. Herkowitz (1984) offered additional clever feedback techniques and equipment in this vein.

Results from the studies just described also imply that the distance children are asked to throw, strike or kick should challenge them to use close-to-maximal effort. Roberton (1996) further argued that, as children reach intermediate levels within the component developmental sequences, accuracy challenges can then be added to the force goals. At first these should be quite broad: 'Can you make the ball go over (or under) the horizontal line?' 'Can you make the ball hit to the left (right) of the vertical line?' Gradually, the accuracy goals can become more specific until traditional targets are introduced.

Key points

When performed at an advanced level, the object projection skills of throwing, striking and kicking demonstrate and integrate multiple principles of bio-mechanical and neuromuscular function.

- First, these skills are exemplars of how linear and rotational energy is generated and transferred across body segments in the form of an open kinetic chain. This transfer is accomplished by proximal-to-distal movement sequencing of those segments.
- Second, skilful object projection takes advantage of segmental inertial characteristics.
- Third, skilful object projection exploits elastic muscle/connective tissue contributions within the context of eccentric/concentric muscle actions.
- Fourth, advanced striking, kicking and throwing are similar in that energy is initially generated during preparatory approach/backswing movements, transferred from more massive proximal segments to less massive distal segments and, ultimately, to the projectile.
- Overall, highly skilled individuals, as compared to less skilled individuals, effectively incorporate more segments into their movement patterns and utilize kinetic chain principles more effectively to maximize energy transfer.

Developmental changes in kicking, throwing and striking also reveal similarities.

- As with the initial 'arm-dominated' patterns of throwing and striking, the early developmental levels of kicking might be called 'foot dominated' because the kicking leg punches or pushes at the ball in isolation from trunk movements.
- Early kickers, throwers and strikers use a minimal backswing/approach, thereby storing little elastic potential energy to convert into kinetic energy

and limiting the range over which they can accelerate their leg or arm/ implement prior to ball contact/release.

- During development, more body segments are effectively incorporated to promote energy generation and transfer through the kinetic chain system.
- Of great importance is the incorporation of trunk rotation into all these skills. This rotary action allows the body to take greater advantage of the inertia of the more distal segments; however, dynamic balance may constrain the development of trunk rotation. Acquiring the ability to rotate the trunk over a moving base of support presents a developmental challenge to the less skilled mover.
- Finally, viewing all these occurrences from a biomechanical perspective suggests that motor development can at least be partially described as acquiring the ability to take advantage of the mechanical characteristics inherent in our human biology.

References

Atwater, A.E. (1979) 'Biomechanics of overarm throwing movements and of throwing injuries', *Exercise and Sport Sciences Reviews*, 7: 43–85.

Barrett, D.D. and Burton, A.W. (2002) 'Throwing patterns used by collegiate baseball players in actual games', *Research Quarterly for Exercise and Sport*, 73: 19–27.

Bartlett, R.M., Stockill, N.P., Elliott, B.C. and Burnett, A.F. (1996) 'The biomechanics of fast bowling in men's cricket: a review', *Journal of Sports Sciences*, 14: 403–24.

Barton, G. and French, K. (2004) 'Throwing profiles of 7-, 8-, 9-, and 10-year-old Little League baseball players', *Research Quarterly for Exercise and Sport*, 75 (Suppl. A): 44.

Bloomfield, J., Elliott, B. and Davies, C. (1979) 'Development of the soccer kick: a cinematographical analysis', *Journal of Human Movement Studies*, 5: 152–9.

Button, C., Smith, J. and Pepping, G.J. (2005) 'The influential role of task constraints in acquiring football skills', in T. Reilly, J. Cabri and D. Araujo (eds) *Science and Football V*, London: Routledge, pp. 500–8.

Campbell, B.M., Stodden, D.F. and Nixon, M.K. (2010) 'Lower extremity muscle activation during baseball pitching', *Journal of Strength and Conditioning Research*, 24: 964–71.

Chu, Y., Fleisig, G.S., Simpson, K.J. and Andrews, J.R. (2009) 'Biomechanical comparison between elite female and male baseball pitchers', *Journal of Applied Biomechanics*, 25: 22–31.

Ciapponi, T. and Hudson, J. (2000) 'The volleyball approach: an exploration of balance', *International Symposium on Biomechanics in Sports, Conference Proceedings Archive*, 18.

Cross, R. (2004) 'Physics of overarm throwing', *American Journal of Physics*, 72: 305–12.

Deach, D. (1950) 'Genetic development of motor skills in children two through six years of age', unpublished doctoral dissertation, University of Michigan.

Dillman, C.J., Fleisig, G.S. and Andrews, J.R. (1993) 'Biomechanics of pitching with emphasis upon shoulder kinematics', *Journal of Orthopedic and Sports Physical Therapy*, 18: 402–8.

Elliott, B.C., Marshall, R.N. and Noffal, G.J. (1995) 'Contributions of upper limb segment rotations during the power serve in tennis', *Journal of Applied Biomechanics*, 11: 433–42.

Escamilla, R.F. and Andrews, J.R. (2009) 'Shoulder muscle recruitment patterns and related biomechanics during upper extremity sports', *Sports Medicine*, 39: 569–90.

Escamilla, R.F., Fleisig, G.S., DeRenne, C., Taylor, M.K., Moorman, C.T., Imamura, R., Barakatt, E. and Andrews, J.R. (2009) 'A comparison of age level on baseball hitting kinematics', *Journal of Applied Biomechanics*, 25: 210–18.

Feltner, M.E. (1988) 'Three-dimensional segment interactions of the throwing arm during overarm fastball pitching in baseball', unpublished doctoral dissertation, Indiana University.

Feltner, M.E. (1989) 'Three-dimensional interactions in a two-segment kinetic chain. Part II: application to the throwing arm in baseball pitching', *International Journal of Sport Biomechanics*, 5: 420–50.

Feltner, M.E. and Dapena, J. (1986) 'Dynamics of the shoulder and elbow joints of the throwing arm during baseball pitch', *International Journal of Sport Biomechanics*, 2: 235–59.

Fleisig, G.S., Escamilla, R.F. and Andrews, J.R. (1996a) 'Biomechanics of throwing', in J.E. Zachazewski, D.J. Magee and W.S. Quillen (eds) *Athletic Injuries and Rehabilitation*, Philadelphia, PA: W.B. Saunders Company.

Fleisig, G.S., Chu, Y., Weber, A. and Andrews, J.R. (2009) 'Variability in baseball pitching biomechanics among various levels of competition', *Sports Biomechanics*, 8: 10–21.

Fleisig, G.S., Nicholls, R.L., Elliott, B.C. and Escamilla, R.F. (2003) 'Kinematics used by world class tennis players to produce high-velocity serves', *Sports Biomechanics*, 2: 17–30.

Fleisig, G.S., Barrentine, S.W., Zheng, N., Escamilla, R.F. and Andrews, J.R. (1999) 'Kinematic and kinetic comparison of baseball pitching among various levels of development', *Journal of Biomechanics*, 32: 1371–5.

Fleisig, G.S., Escamilla, R.F., Andrews, J.R., Matsuo, T.M., Satterwhite, Y. and Barrentine, S.W. (1996b) 'Kinematic and kinetic comparison between baseball pitching and football passing', *Journal of Applied Biomechanics*, 12: 207–24.

Fleisig, G.S., Kingsley, D.S., Loftice, J.W., Dinnen, K.P., Ranganathan, R. and Andrews, J.R. (2006a) 'Kinetic comparison among the fastball, curveball, change-up, and slider in collegiate baseball pitchers', *American Journal of Sports Medicine*, 34: 423–30.

Fleisig, G.S., Phillips, R., Shatley, A., Loftice, J., Dun, S. and Andrews, J.R. (2006b) 'Kinematics and kinetics of youth baseball pitching with standard and lightweight balls', *Sports Engineering*, 9: 155–63.

Hamilton, M. and Tate, A. (2002) 'Constraints on throwing behavior of children', in J. Clark and J. Humphrey (eds) *Motor Development: Research and Reviews*, Reston, VA: NASPE.

Herkowitz, J. (1984) 'Developmentally engineered equipment and playgrounds', in J. Thomas (ed.) *Motor Development During Childhood and Adolescence*, Minneapolis, MN: Burgess.

Herring, R.M. and Chapman, A.E. (1992) 'Effects of changes in segmental values and timing of both torque and torque reversal in simulated throws', *Journal of Biomechanics*, 25: 1173–84.

Hong, D., Cheung, T. and Roberts, E. (2001) 'A three-dimensional, six-segment chain analysis of forceful overarm throwing', *Journal of Electromyography and Kinesiology*, 11: 95–112.

Hume, P.A., Keogh, J. and Reid, D. (2005) 'The role of biomechanics in maximizing distance and accuracy of golf shots', *Sports Medicine*, 35: 429–49.

Langendorfer, S.J. (1987a) 'A prelongitudinal test of motor stage theory', *Research Quarterly for Exercise and Sport*, 58: 21–9.

Langendorfer, S.J. (1987b) 'Prelongitudinal screening of overarm striking development performed under two environmental conditions', in J. Clark and J. Humphrey (eds) *Advances in Motor Development Research (Vol. 1)*, New York: AMS Press, pp. 17–47.

Langendorfer, S.J. (1990) 'Motor-task goal as a constraint on developmental status', in J. Clark and J. Humphrey (eds) *Advances in Motor Development Research (Vol. 3)*, New York: AMS Press, pp. 16–28.

Langendorfer, S.J. (1999) 'Developmental sequences for the preparatory phase of the overarm throw', paper presented at the annual conference of the American Alliance for Health Physical Education Recreation and Dance, Reston, VA, April.

Langendorfer, S.J. and Roberton, M.A. (1999) 'Developmental pathways between overarm throwing profiles: a longitudinal picture of commonalities and individual differences', *Journal of Sport and Exercise Psychology*, 21: S73.

Langendorfer, S.J. and Roberton, M.A. (2002a) 'Developmental profiles in overarm throwing: searching for "attractors," "stages," and "constraints"', in J. Clark and J. Humphrey (eds) *Motor Development: Research and Reviews (Vol. 2)*, Reston, VA: NASPE, pp. 1–25.

Langendorfer, S.J. and Roberton, M.A. (2002b) 'Individual pathways in the development of forceful throwing', *Research Quarterly for Exercise and Sport*, 73: 245–58.

Lees, A., Asai, T., Andersen, T.B., Nunome, H. and Sterzing, T. (2010) 'The biomechanics of kicking in soccer: a review', *Journal of Sports Sciences*, 28: 805–17.

Lisy, M. (2002) 'Testing developmental sequences for the forceful kick', unpublished masters thesis, Bowling Green State University, OH.

Mally, K., Battista, R. and Roberton, M.A. (2011) 'Distance as a control parameter for kicking', *Journal of Human Sport Exercise*, 6(1): 122–34.

Manoel, E.J. and Oliveira, J.A. (2000) 'Motor developmental status and task constraint in overarm throwing', *Journal of Human Movement Studies*, 39: 359–78.

Matsuo, T., Escamilla, R.F., Fleisig, G.S., Barrentine, S.W. and Andrews, J.R. (2001) 'Comparison of kinematic and temporal parameters between different pitch velocity groups', *Journal of Applied Biomechanics*, 17: 1–13.

Mero, A., Komi, P.V., Korjus, T., Navarro, E. and Gregor, R.J. (1994) 'Body segment contributions to javelin throwing during final thrust phases', *Journal of Applied Biomechanics*, 10: 166–77.

Messick, J.A. (1991) 'Prelongitudinal screening of hypothesized developmental sequences for the overhead tennis serve in experienced tennis players 9–19 years of age', *Research Quarterly for Exercise and Sport*, 62: 249–56.

Nesbit, S.M. (2005) 'A three dimensional kinematic and kinetic study of the golf swing', *Journal of Sports Science and Medicine*, 4: 499–519.

Pappas, A.M., Zawacki, R.M. and Sullivan, T.J. (1985) 'Biomechanics of baseball pitching: a preliminary report', *The American Journal of Sports Medicine*, 13: 216–22.

Putnam, C.A. (1993) 'Sequential motions of body segments in striking and throwing skills: descriptions and explanations', *Journal of Biomechanics*, 26 (Suppl. 1): 125–35.

Rash, G.S. and Shapiro, R. (1995) 'A three-dimensional dynamic analysis of the quarterback's throwing motion in American football', *Journal of Applied Biomechanics*, 11: 443–59.

Roberton, M.A. (1977) 'Stability of stage categorizations across trials: implications for the "stage theory" of over-arm throw development', *Journal of Human Movement Studies*, 3: 49–59.

Roberton, M.A. (1978) 'Longitudinal evidence for developmental stages in the forceful overarm throw', *Journal of Human Movement Studies*, 4: 167–75.

Roberton, M.A. (1984) 'Changing motor patterns during childhood', in J. Thomas (ed.) *Motor Development during Preschool and Elementary Years*, Minneapolis, MN: Burgess, pp. 48–90.

Roberton, M.A. (1996) 'Put that target away until later: developing skill in object projection', *Future Focus*, 17: 6–8.

Roberton, M.A. and Halverson, L.E. (1984) *Developing Children – Their Changing Movement*, Philadelphia, PA: Lea & Febiger.

Roberton, M.A. and Langendorfer, S.J. (1980) 'Testing motor development sequences across 9–14 years', in C. Nadeau, W. Halliwell, K. Newell and G. Roberts (eds) *Psychology of Motor Behavior and Sport*, Champaign, IL: Human Kinetics, pp. 269–79.

Roberts, E.M. and Metcalfe, A. (1968) 'Mechanical analysis of kicking', in J. Wartenweiler, E. Jokl and M. Hebbelink (eds) *Biomechanics I*, New York: Karger, pp. 315–19.

Robinson, L., Goodway, J. and Williams, E.J. (2007) 'Developmental trends of overarm throwing performance in young children', *Research Quarterly for Exercise and Sport*, 78 (Suppl.): A48.

Sakurai, S., Ikegami, Y., Okamoto, A., Yabe, K. and Toyoshima, S. (1993) 'A three-dimensional cinematographic analysis of upper limb movement during fastball and curveball baseball pitches', *Journal of Applied Biomechanics*, 9: 47–65.

Southard, D. (2002) 'Changes in throwing pattern: critical values for control parameter of velocity', *Research Quarterly for Exercise and Sport*, 73: 396–407.

Southard, D. (2006) 'Changing throwing pattern: instruction and control parameter', *Research Quarterly for Exercise and Sport*, 77: 316–25.

Southard, D. (2009) 'Throwing pattern: changes in timing of joint lag according to age between and within skill level', *Research Quarterly for Exercise and Sport*, 80: 213–22.

Stodden, D.F. (2006) 'Facilitating the acquisition of complex ballistic motor skills: promoting proximal or distal system perturbations', *Journal of Human Movement Studies*, 51: 197–220.

Stodden, D.F. and Rudisill, M.E. (2006) 'Integration of biomechanical and developmental concepts in the acquisition of throwing: effects on developmental characteristics and gender differences', *Journal of Human Movement Studies*, 51: 117–41.

Stodden, D.F., Fleisig, G.S., McLean, S.P. and Andrews, J.R. (2005) 'Relationship of biomechanical factors to baseball pitching velocity: within pitcher variation', *Journal of Applied Biomechanics*, 21: 44–56.

Stodden, D.F., Langendorfer, S.J., Fleisig, G.S. and Andrews, J.R. (2006a) 'Kinematic constraints associated with the acquisition of overarm throwing Part I: step and trunk actions', *Research Quarterly for Exercise and Sport*, 77: 417–27.

Stodden, D.F., Langendorfer, S.J., Fleisig, G.S. and Andrews, J.R. (2006b) 'Kinematic constraints associated with the acquisition of overarm throwing Part II: upper extremity actions', *Research Quarterly for Exercise and Sport*, 77: 428–36.

Stodden, D.S., Fleisig, G.S., McLean, S.P., Lyman, S.L. and Andrews, J. R. (2001) 'Relationship of trunk kinematics to pitched ball velocity', *Journal of Applied Biomechanics*, 17: 164–72.

Teixeira, L.A. (1999) 'Kinematics of kicking as a function of different sources of constraint on accuracy', *Perceptual and Motor Skills*, 88: 785–89.

Welch, C.M., Banks, S.A., Cook, F.F. and Draovitch, P. (1995) 'Hitting a baseball: a biomechanical description', *Journal of Orthopedic & Sports Physical Therapy*, 22: 193–201.

Werner, S.L., Fleisig, G.S., Dillman, C.J. and Andrews, J.R. (1993) 'Biomechanics of the elbow during baseball pitching', *Journal of Orthopedic & Sports Physical Therapy*, 17: 274–8.
Wickstrom, R. (1983) *Fundamental Motor Patterns*, Philadelphia, PA: Lea & Febiger.

Part IV

Selected clinical applications

10 The biomechanical basis of injury during childhood

Caroline F. Finch and Dara Twomey

Introduction

Childhood is a time of rapid growth and development, during which children learn to explore and interact with their environments in new ways. While this is both critical and necessary for healthy child motor development, such interactions with the physical environment can be associated with exposure to hazards that increase injury risk. Other chapters in this book have described different aspects of the relationship between neuromuscular development and chronological age. This chapter explains how these developmental factors can be associated with a risk of injury during childhood sport and physical activity, and demonstrates how biomechanically based interventions can be developed and implemented to reduce this risk. This chapter also includes specific examples of studies that have either identified child sport-/activity-related injury risks that are biomechanical in nature or which have devised solutions based on an understanding of the biomechanical or neuromuscular nature of this risk.

The objectives of this chapter are to: (1) provide an overview of the risk of musculoskeletal and impact-related injuries in children associated with active settings; (2) describe the intrinsic and extrinsic biomechanical basis of injury risk in children; (3) summarize the evidence for efficacious biomechanically based interventions to reduce injury risk in children; and (4) discuss the interaction between biomechanical and other considerations in injury prevention.

An understanding of the stages of development and associated mechanical tissue changes is critical to comprehend fully the biomechanical basis of injury risk in children. The following tutorial briefly summarizes the key musculoskeletal developmental changes experienced by children, through various growth stages and chronological ages (see Chapter 1 for more detail), to enhance the understanding of the biomechanical risk factors presented throughout this chapter.

Stages of growth, development and maturation are commonly divided into three age periods: infancy (the first year of life), childhood that extends from infancy to adolescence (approximately 1–10 years of age) and adolescence (10–18 years of age). Bone growth and modelling take place rapidly throughout infancy and childhood. Growing bone is more porous and less dense than adult bone, and therefore is less resistant to tensile and compressive stresses. During adolescence,

---

Example box 1: Age-related game delivery modifications

Lesson: Age-related modifications to a game (such as reduced field size, fewer players on the field at any given time, smaller and softer ball design, reduced game length and restriction of body contact) are a safer way to introduce children to game skills and techniques (e.g. tackling) through reducing the presence of biomechanical hazards.

Reference: Romiti *et al.* 2008.

Sport/s: Junior club Australian football.

Players and age group: 51 teams of players across the under 9 years to under 18 years competitions.

Nature of biomechanical injury risk: Australian football is a team ball sport with a high degree of both intentional and unintentional body contact and hence impact collision forces through the tackling of players and contesting for the ball either in flight or on the ground. Children are introduced to the skills of the game (particularly the tackling aspect) gradually, with a heavily modified version of the game delivered for young children. The level of modification is reduced until the adult form of the game, with full tackling, is introduced in the U18 (16+ years) competition.

Study design: Prospective cohort study over 40,208 hours of player exposure over one playing season. Information collected on the number of players who attended each game and training session, and injury details.

Results: Injury rates were much higher in the junior level closest to the adult form of the game. Modified rules for all other age groups appear to be effective in reducing injury risk. While fierce competition for the ball, faster game speeds, etc. might be more common in a competition game setting, injuries also occurred during training sessions and interventions should also be implemented for that context of play/practice.

Outcome: Decreasing the size of the playing field may increase the likelihood of collisions. Modifications to equipment need to be graduated to minimize injuries due to the transition between the modified and full version of a sport. Coach development programmes now encourage the gradual introduction of tackling skills to junior players.

---

bone continues to grow and remodel according to the amount of exposure to mechanical loading. Skeletal maturity is achieved with the fusion of the epiphyses which begin in childhood and are usually completed by 25 years of age (Cech and Martin 2002).

Muscle maturation also occurs in childhood and the onset of puberty marks an increase in growth of the musculoskeletal system, as the muscles have to lengthen to re-establish their length–tension relationship. At puberty, the complex inter-action between growth and biological maturity determines the muscular strength of the child. From infancy through to adolescence, the tensile strength of ligaments and tendons increases and, prior to skeletal maturity, ligament and tendons may be more viscous and compliant.

Importantly, as explained in detail in Chapter 1, chronological age is not nec-essarily the same as physical or musculoskeletal maturity as children develop and grow at different rates. This poses some challenges when delivering safe sport and other activities to children in a way to minimize their injury risk.

Basic theoretical concepts

Injuries in childhood

It is well recognized that childhood (through the early years to late adolescence) can be a particular period of increased injury risk for some children (Ozanne-Smith 1995). Indeed, this has been acknowledged as a global problem by the World Health Organization (WHO) (Peden *et al.* 2008). The relationship between child injury risk and developmental stages has been described elsewhere (Berger and Mohan 1996; Ozanne-Smith 1995) and the major external causes of injuries according to age group are summarized in Table 10.1.

Table 10.1 Relationship between child development stage and injury risk

Age-group (years)	Mechanism of impact and/or musculoskeletal injury	Settings/contexts of child activity
0–1	Most associated with a lack of supervision in particularly hazardous environments (e.g. baths, swimming pools, etc.) and/or falls from nursery furniture, prams, etc.	Beginning of movement patterns and child exploration of surroundings.
1–4	Falls are the major mechanism of injury; drownings, poisonings and burns are also common.	Hazards become within reach as children develop gross motor skills and the ability to crawl, walk and climb; improved fine motor skills mean children are able to open containers, cupboards, etc.
5–9	Falls are the major cause, particularly from playground equipment and trampolines.	Active play and further development of gross motor skills; immature bone development increases risk of fractures.
10–14	Sport and active recreational injuries; active transportation such as bicycling, inline skating and skate boarding.	Children commence participation in formal sport either recreationally or at school; further development of gross motor skills manifest in use of

Table 10.1 continued

Age-group (years)	Mechanism of impact and/or musculoskeletal injury	Settings/contexts of child activity
		wheeled recreational devices and other 'movement toys' that move at speed; immature bone development/ musculoskeletal system increases risk of fractures; children with 'sporting talent' often play more than one sport.
15–19	Sport and recreational injuries; active transportation such as bicycling, inline skating and skate boarding; road trauma.	Children participate in more competitive forms of sport both at high standards of play and with increased duration and frequency; increased use of active transportation devices used at faster speeds, including snow-sport equipment; exposure to more hazardous environments such as as riding/skating/blading on roads rather than backyards; children with 'sporting talent' often play more than one sport and train/play for increasingly more hours, leading to tissue overload.

As shown in Table 10.1, the ability of children to engage in the types of activities that are most commonly associated with injuries to others of their same age is directly related to their stage of biomechanical and neuromuscular control and development. Most evidence for the types of injuries sustained by children comes from routine data collections such as presentations for medical treatment at emergency departments or population surveys (e.g. of school children). Injuries during sport and active recreation (also sometimes referred to as leisure injuries and including activities such as formal and informal sport, use of playground equipment, use of wheeled recreational devices such as bicycles/skateboards/ skates) are the number one reason for injury-related hospital emergency department presentations and hospital admissions in children and adolescents in developed countries (Bijur *et al.* 1995; Cheng *et al.* 2000; Conn *et al.* 2003; Finch *et al.* 1998). A survey of non-fatal injuries in 11-, 13- and 15-year-olds from 11 different countries also found sport activity to be the most common cause of injury in those ages (Molcho *et al.* 2006). These data are supported by a Dutch prospective cohort study of physical activity-related injuries in 10–12-year-olds (Verhagen *et al.* 2009).

In the WHO Child Injury Report (Peden *et al.* 2008), head injuries are globally the single most common injury, with intracranial injury (mild traumatic brain injury, concussion) accounting for 17.1 per cent of all unintentional injuries in children. Fractures of bones, joints and other tissue injuries collectively account for almost one-third of all childhood non-intentional injuries, with the single

Example box 2: Exposure to injury risk varies by chronological age and position played

Lesson: Injury risk tends to increase as children grow and as they become exposed to more adult forms of the game. In some sports, the nature of the injury risks are related to the playing position, so that not all children are at the same risk of injury, particularly where specialization of sporting position/team role also develops as the child ages.

Reference: Finch *et al.* 2010.

Sport/s: Junior club cricket.

Players and age group: 88 under 12 years, 203 under 14 and 120 under 16 players.

Nature of biomechanical injury risk: When a cricket team is batting, only 2/11 players are on the field at any given time, both in batting positions. At the same time, the opposing fielding team has a wicket keeper (to catch/field balls close behind where the batter is), a bowler (who delivers a limited number of balls – this role may be shared among a number of players) and the remainder are in fielding positions requiring ball handlings and some running skills. Despite the plethora of literature in adult cricket showing that bowlers are most at risk of injury due to the repetitive nature of their movement patterns, in junior cricket, players are vulnerable to injury in any position. The nature of this risk varies with age, especially as players adopt more specialized positions (i.e. bowler or batter) as they get older.

Study design: Prospective cohort study with player attendance at all training sessions as well as the game positions participated in during games recorded. All injuries associated with the specific activities of batting, bowling and fielding in both game and training sessions were reported.

Results: Junior cricketers were at risk of injury irrespective of what playing position they adopt. Injury risk varied by age and player position and there was a significant interaction between these two factors. There was a trend towards more batting and fielding injuries in matches, and more bowling and batting injuries during training sessions.

Outcome: Just drawing on the adult literature for this sport would ignore the injury risks for children while batting and fielding, as well as bowling. Therefore, different prevention strategies will be needed to address these joint age and playing position effects in junior cricket players.

Example box 3: Variety and variation is needed when delivering exercise interventions to children

Lesson: In many instances, it is assumed that interventions developed for adults to address their injury concerns (e.g. exercise training programmes) would also be suitable for preventing similar injuries in children. However, variety and variation are needed when delivering some interventions to children, while maintaining biomechanical fidelity.

References: Kilding *et al.* 2008; Saunders *et al.* 2010.

Sport/s: Junior girls netball; young male football.

Players and age group: Teams of girls under 11, under 13 or under 15 years; 24 male football players average age 10.4 years.

Nature of biomechanical injury risk: In sports involving a large amount of running, changing direction, acceleration and sudden deceleration and/ or landing, a relatively high rate of lower limb injuries have been reported. It has been suggested that this is largely due to poor neuromuscular control of muscles, joints and ligaments during these movements. Accordingly, it has been suggested that having players participate in specifically designed exercise training programmes to improve neuromuscular control will prevent these injuries.

Study design: Survey feedback from coach about the challenges of delivering a safe landing technique to young netballers; questionnaire feedback from young football players' experience of an exercise training programme 'The FIFA 11' in relation to enjoyment, frequency of execution, perceived benefits, intention to continue/adhere, feedback on specific exercises.

Results: Both studies implemented exercise programmes based largely on evidence obtained from studies of adults. In the football study, players queried the number of times the programme should be performed as well as the benefits/value of some of the exercises. A number of the exercises were also considered challenging. In the netball study, coaches also raised the need for exercises to be more varied to keep the interest of the child athletes.

Outcome: The results from these studies can be used to inform the implementation and adoption of injury prevention strategies in the future.

most common category being to the ulna or radius, often resulting from a fall on to an outstretched arm. These examples illustrate the importance of impact mechanisms in causing traumatic injury in children of all ages. The sports medicine literature also shows an increasing trend towards more biomechanically or neuromuscular-based non-impact and/or overuse-related injuries in children as they age (Finch *et al.* 2010; Meyers 1993; Micheli 1983). It is beyond the scope of this chapter to describe the nature and rate of sport and active recreational injuries in children, but reviews across several sports can be found elsewhere (Caine and Maffulli 2005; Caine *et al.* 2008; Hootman *et al.* 2007; Maffulli and Caine 2005; McGuine 2006; Radelet *et al.* 2002).

Aetiological mechanistic models of injury causation and prevention

From a biomechanical viewpoint, most injury results from a failure of the body and its tissues to withstand the transfer of an applied energy or force (McIntosh 2004b). While there are varying definitions used in the literature to describe sports injury, for the purpose of this chapter, an injury is defined according to the tissue damage dimension only (i.e. irrespective of whether or not it also leads to lost-time from the activity).

Injury or tissue damage can be considered to result from the logical consequence of a series of actions, precipitating factors (including both intrinsic or personal characteristics and exposure to environmental hazards) and inciting events. This was first conceptualized by Haddon in the 1960s (Haddon 1972) and has since formed the basis of many prevention approaches. Once injury was recognized to be the result of a chain of events, it was then possible to conceive that breaking links in that chain would prevent injury from occurring (Berger and Mohan 1996; McIntosh 2004a; Robertson 2007). This approach forms the basis of current well-recognized sports injury aetiology models (Bahr and Krosshaug 2005; McIntosh 2005; Meeuwisse *et al.* 2007; Micheli 1983) that have been developed for all sporting populations but which also apply to children. These 'sequence of events' models are summarized in the upper part of Figure 10.1. The lower part of the figure indicates how injury causation can largely be considered to result from biomechanical responses to risks and hazards, both in the pre-event and event phases.

Based on the well-accepted energy exchange theory of injury, Haddon (1973) proposed a set of ten countermeasures (or injury-prevention approaches) that could be used to reduce injury risk by removing or reducing specific links in this chain. The ten countermeasure strategies are also indicated against the risk factor issues in the lower part of Figure 10.1.

According to the mechanistic model shown in Figure 10.1, a range of risk factors contributes to the risk of injury. Emery (2005) summarized and categorized the risk factors for paediatric sports injury into those that are extrinsic or intrinsic to the child and into those that are potentially modifiable:

Extrinsic, non-modifiable	Sport played (contact/no contact); level of play (recreational/elite); position played; weather; time of season/time of day.
Extrinsic, potentially modifiable	Rules; playing time; playing surface (type/condition); equipment (protective/footwear).
Intrinsic, non-modifiable	Previous injury, age, sex.

Intrinsic, potentially modifiable Fitness level pre-participation;
sport-specific training flexibility; strength;
joint stability; biomechanics; balance/
proprioception; psychological/social factors.

Unfortunately, this categorization only specifically considers biomechanics as a potentially modifiable intrinsic risk factor. It ignores the fact that biomechanical influences clearly have a significant role either as effect modifiers of the other risk factors or as a key consideration in the development and design of solutions

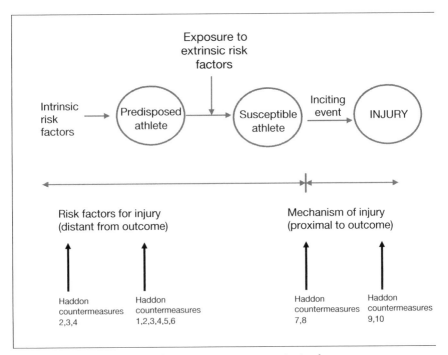

Haddon countermeasure strategies

1 Prevent the creation of the hazard.

2 Reduce the amount of the hazard created.

3 Prevent the release of a hazard that already exists.

4 Modify the rate or spatial distribution of the hazard from its source.

5 Separate in time or space the hazard from that which is to be protected.

6 Separate the hazard and what is to be protected by a material barrier.

7 Modify relevant basic qualities of the hazard.

8 Make what is to be protected more resistant to damage from the hazard.

9 Move rapidly to detect and evaluate damage that has occurred and counter its continuation and extension.

10 Stabilize, repair and rehabilitate the damage or injured person.

Figure 10.1 Mechanistic model of the chain of events leading to sports injury and opportunities for intervention according to Haddon principles. (This draws on the Bahr and Holme model (Bahr and Holme 2003) and Haddon's ten countermeasure strategies (Haddon 1973).)

to address other risk factors. Moreover, the listing omits several important risk factors that are also related to child growth and development, and suitable biomechanical controls for addressing these factors, including: extrinsic factors of playing equipment and coach education, and intrinsic factors such as risk perceptions and risk-taking behaviours, sports/playing experience and skill level. A more comprehensive listing of potential biomechanically related risk factors for child injury in sport and recommended solutions is given in Table 10.2.

Example box 4: The value of modifying environmental factors

Lesson: Environmental factors, such as low-impact absorbing surfaces or inappropriately designed fixed equipment, can increase the risk of injury in children's playgrounds and sporting environments.

Reference: Chalmers *et al.* 1996.

Sport/s: Playground activities.

Players and age group: School children under 14 years.

Nature of biomechanical injury risk: Surfaces with low-impact absorbing properties can increase the risk of a child sustaining a traumatic injury. Coupled with this is the height of the fall from playground equipment, which will determine the severity of injury. Higher fall heights result in increased acceleration and if the surface is not built to absorb some of the force on landing, an injury is likely to occur.

Study design: Data collected from 300 children who had fallen from playground equipment were split into two groups: those who had sustained an injury and received medical attention and those who sustained an injury but did not require medical attention.

Results: Children who fell from equipment that did not satisfy the height and surfacing requirements of the New Zealand playground standard had an increased injury risk. There was a significant interaction between equipment height and shock absorbency of surfaces, i.e. the combination of a harder surface and higher equipment greatly increased the risk of injury.

Outcome: Safety standards are critical to ensuring the reduction in injury risk in children inactive play areas.

Table 10.2 Biomechanical basis of sports injury risk factors and solutions that could be implemented to reduce the risk of sports injury in children

Risk factor	Biomechanical nature of risk	Biomechanical modifying strategy*	Example of solution	Linked example and related references
Age	Variable developmental changes	Indirect	Developmental matching in activities, especially when settings teams for team sports	Example boxes 1, 2: Finch et al. 2010; McIntosh et al. 2010; Romiti et al. 2008
Bone plasticity	Lower energy and force absorption	Indirect	Reduce numbers of players or increase size of playing area to limit large impact forces	Bernard et al. 1993; McMahon et al. 1993; Romiti et al. 2008
Coach/player lack of knowledge and attitudes	Mis- or non-understanding of the nature of risks in chosen sport	Indirect	Include information drawing on biomechanical evidence into education programmes	Cook et al. 2003; Finch et al. 2001
Excessive body contact	Impact forces lead to traumatic injury	Direct	Introduce rules to reduce or prevent body contact; introduce body contact into game development programmes gradually with age levels	Example box 1: Froholdt et al. 2009; Hootman et al. 2007; Romiti et al. 2008; Veigel and Pleacher 2008
Gender	Strength peaks at different ages in boys and girls and diverges thereafter	Indirect	Gender-specific activities at key developmental stages	Chau et al. 2007
Growth spurt	Tissue strength imbalance	Indirect	Screening for asynchronous tissue development and case-specific modifications	Caine and Maffulli 2005; Maffulli and Caine 2005; Meyers 1993
Ill-conditioned body soft tissues (e.g. ligaments)	Joints have difficulty in withstanding impact forces	Direct	Conditioning and other (e.g. balance) exercise training programmes	Example box 3: Abernethy and Bleakley 2007; Kilding et al. 2008; McHugh et al. 2007; Veigel and Pleacher 2008

Risk factor	Mechanism	Direct/Indirect	Intervention	Examples
Immature musculature	Decreased capacity for force production and absorption	Indirect	Developmental matching in activities and design of maturity stage-appropriate game activities and skills	Example box 1: Romiti et al. 2008
Inappropriate environmental design	Poorly designed facilities and sports ground can lead to increased impact forces and risk of injury; can also increase risk of falling	Direct	Better design of grounds for children; protective padding on surfaces such as goal posts; firm anchoring of devices such as goal posts; risk management checks of sports fields prior to play	Example box 4: Chalmers et al. 1996; Laforest et al. 2001
Inappropriate or poorly designed equipment	Overload	Direct	Modify equipment; ensure children use equipment appropriate to their size	Finch et al. 2010
Inappropriate personal protective equipment	Inappropriate impact absorption and energy transfer	Direct	Adopt age-appropriate personal protective equipment	Example boxes 2, 5: Abernethy and Bleakley 2007; Finch et al. 2010; Veigel and Pleacher 2008
Incorrect technique	Decreased mechanical efficiency	Direct	Teach optimal technique	Example box 2
Lack of, or non-enforcement of, rules	Exposure to inappropriate forces	Indirect	Modify rules on the basis of biomechanical principles to eliminate inappropriate conditions or to enforce appropriate protective equipment use	Froholdt et al. 2009; McIntosh et al. 2010; Shaw and Finch 2008
Maturity stage	Underdeveloped musculoskeletal system	Indirect	Developmental matching in activities and design of maturity stage-appropriate game activities and skills	Example boxes 1, 2, 3: Finch et al. 2010; Froholdt et al. 2009; Kilding et al. 2008; Romiti et al. 2008
Mismatching of body masses	Forces exceed tolerance levels	Indirect	Determine participation in contact activities by weight categorization	Brust et al. 1992; McMahon et al. 1993
Nature of the activity at the time of injury	Different playing positions and actions within a given sport require different movement patterns and hence exposure to forces	Indirect	Require position-specific protective equipment or modification of training loads	Finch et al. 2010; Gregory et al. 2002; Shaw and Finch 2008

Table 10.2 continued

Risk factor	Biomechanical nature of risk	Biomechanical modifying strategy*	Example of solution	Linked example and related references
Playing experience/skill level	Decreased mechanical efficiency	Direct	Establish and teach optimal technique in key essential skills	Example boxes 1, 2: Finch et al. 2010; Froholdt et al. 2009; Romiti et al. 2008
Previous injury	Damaged tissue	Direct	Encourage the use of suitable prophylactic or protective devices; design appropriate exercise training programmes	Abernethy and Bleakley 2007; McHugh et al. 2007
Too much exposure time	Overload	Direct	Adjust the duration of activities	Example box 6: Hootman et al. 2007
Unsafe/ unsuitable surfaces	Insufficient impact absorption	Direct	Change to surfaces with approved force reduction and energy restitution	Example box 4: Chalmers et al. 1996; Laforest et al. 2001
Workload issues – inappropriate training regimes/ match loads	Overload/insufficient rest periods	Direct	Assess workloads and adopt policies	Example box 6: Dennis et al. 2005; Fecteau et al. 2008; Gregory et al. 2004; Hootman et al. 2007

* Direct refers to strategies that specifically target the biomechanical nature of the injury risk and indirect refers to one where a biomechanical solution could be adopted to reduce the impact of that particular risk factor.

Understanding the age-related biomechanical basis of sports injury risk in children

It is evident from previous chapters that children are neither uniform nor 'mini-adults' and develop physically at different rates at the same chronological age. Accordingly, their injuries are generally influenced by age-related stages of musculoskeletal development and the increased intensity and competitive nature of sport as game delivery builds towards their adult forms (Hass *et al.* 2003; Magra *et al.* 2007). Although the growth spurt occurs at different ages in different children, both the changes that they undergo and how rapidly they occur can significantly increase injury risk (Caine *et al.* 1989; Micheli 1983).

An injury primarily occurs when a tissue (bone, tendon or ligament), experiences an excessive mechanical load or a regular load with insufficient recovery (DiFiori 2002). The biomechanical basis of child injuries includes both intrinsic and extrinsic risk factors that can change as the child moves through their various stages of growth and maturation. Intrinsic injury risk factors include elements such as age, sex, immature musculature, muscle group imbalance, timing of the growth spurt and asynchronous tissue development. On the other hand, extrinsic risk factors include such aspects as technique, training regimes, nature of the game or activity, equipment design and playing surfaces. Knowing and recognizing these factors and the interactions between them, as well as consideration of the extent to which they are modifiable, is critical in injury risk reduction in children of all ages and developmental stages. Table 10.2 lists potential child sports injury risk factors, their biomechanical nature and whether they can be modified directly or indirectly by a biomechanical intervention or strategy.

Throughout childhood, changes occur in musculoskeletal tissues that make them more or less susceptible to injury. Chronological age is not a good indicator of skeletal maturity (Malina *et al.* 2007), and bone modelling and growth occur up to the pubertal stage. However, peak gain in bone mineral content occurs approximately one year after the peak rate of longitudinal growth in girls and boys and the period between these two stages is a period of relative bone weakness which can expose the child to an increased risk of fracture (Whiting and Zernicke 2008). Fractures occur when the bone experiences a force greater than it can tolerate and are usually associated with a traumatic force that usually occurs during acute impacts (e.g. a fall from height, collision with another player, being struck by a ball or bat). Since children are small, such traumatic forces are dissipated over relatively less body mass, making them also more vulnerable to multi-system injury (Kerr 2008).

Many overuse injuries in children are also influenced by the age-related stages of musculoskeletal development. During childhood, the growth cartilage is less resistant to repetitive microtraumas than the surrounding musculoskeletal tissue, making it more vulnerable to injury. With immaturity, avulsion fractures are more common due to the viscoelastic and relatively more compliant nature of ligaments and tendons; as skeletal maturity is reached, ligament and tendon injuries are more prevalent (Klingele and Kocher 2002). Asynchronous changes, which can occur during the growth spurt, combined with repetitive loading characterize risk factors for overuse injuries that are unique to adolescents (DiFiori 2002), such as the weakness of the growth cartilage compared to the tendon associated with conditions such as Severs or Osgood-Schlatter disease.

Different stages of ossification at different joints can also contribute to increased injury risk in children. For example, skeletal maturity is sequential at the elbow, from the capitulum to the external epicondyle, but is not complete until the mid-teens in most children (Benjamin and Briner 2005). Sports and activities that have high biomechanical demands on the elbow (e.g. throwing actions in ball sports or loading movements in gymnasts) increase the risk of elbow injuries in children. Differential bone growth in relation to muscle length can result in a decrease in flexibility and strength. The lack of flexibility can lead to the inability of many paediatric athletes to generate sufficient power or strength in activities such as tennis, cricket or baseball.

Due to many of the musculoskeletal immaturities and developmental stages in children discussed in previous chapters, the equipment used and worn by children when playing sport or participating in active recreation must be appropriate to the biomechanical capacities of their immature bodies. In many instances, equipment designed for adults is not suitable and will increase the risk of injury in the younger age group by transferring the application of the force to a more vulnerable part of the body or altering the energy transfer between the equipment and body part. This has been evident in sports such as downhill skiing where the most frequent injuries reported in children are the result of excessive stresses on the lower leg (Meyers *et al.* 2007) or to snowboarders in relation to wristguard use (Hagel *et al.* 2005). The nature of stiff ski boots results in the concentration of forces at the tibia and as bone diameter is smaller in children it increases their likelihood of tibial fracture (Meyers *et al.* 2007). Similar issues occur where a racquet or bat is too heavy for a child (Marx *et al.* 2001), thereby resulting in alterations to biomechanically sound techniques. Proper mechanical progressions and age-appropriateness in the teaching of skills at an early age or adjusting equipment to suit the musculoskeletal capacity of the child is critical in the reduction of overuse injuries.

Training regimes directly adopted from the adult context can also contribute to an increased injury risk if they are not modified for children. In addition to training regimes, the biomechanical basis of injury can be directly linked to the amount of practice and playing time experienced by child athletes. Chronic overuse injuries occur at the muscle-bone-tendon unit (Magra *et al.* 2007) due to repetitive microtraumatic overload. For example, young high-performance cricket fast-bowlers have been shown to be at particularly high risk of chronic back injuries and conditions due to repetitive tissue overload (Dennis *et al.* 2005; Elliott *et al.* 1993; Gregory *et al.* 2002, 2004).

Frequently, children are participating in many different sports or leisure activities or participate in numerous teams or contexts within the same sport, for example, a child might play and train for a school team and concurrently play for a few teams within a local club or at representative level. There is concern that

Example box 5: Protective equipment is not always the only answer

Lesson: Sometimes it may be quite easy to come up with a biomechanically driven intervention to an injury risk, e.g. protective equipment to guard against impact injury. However, the final choice of intervention needs to be made within a holistic approach, as sometimes another approach may be more effective, either because it addresses more injury risks or because one particular solution may have low uptake/adoption.

References: McIntosh *et al.* 2009, 2010.

Sport/s: Youth club/school rugby union.

Players and age group: Male players from schoolboy under 13, under 15 and under 18 years and club under 20 years.

Nature of biomechanical injury risk: Rugby union is a team-based collision/contact ball sport where there is a high level of both incidental and intentional body contact as part of competition for the ball. The nature of tackling allowed varies with age-level of play. Given the nature of the impact injuries and other factors involved, it has been suggested that players should wear protective headgear to prevent head/neck/face injuries.

Study design: Prospective collection of injury, exposure and risk-factor data as part of a group-clustered randomized controlled trial of the efficacy of headgear.

Results: The overall rate of head/neck/face injury was comparable to that in other studies. The main incident types leading to head, neck and facial injury were impacts. Tackling another player or being tackled, as injury mechanisms, were associated with the highest percentage of injuries (22–30 per cent) across all age levels.

Outcome: While it may be tempting to think first of protective headgear as the only solution to this problem as a means of providing a physical barrier for mitigating the transfer of energy to the head, another trial found no evidence for any protective effect for the soft shell headgear commonly worn in this sport. This indicates that other strategies for prevention need to be developed and should probably focus on reducing the transfer of energy during game phases/actions such as the tackle and scrum. This would be consistent with the hierarchy of countermeasures suggested in Haddon's ten strategies.

too much participation in sport can lead to increased risk of injury, especially among high-performing junior athletes (Finch *et al.* 2002). The combined training requirements across excessive workload of one sport or accumulated across several sports, together with the reduced capacity for adequate rest could increase the risk of overuse repetitive microtraumatic injuries in children.

In sport, child injury rates have been consistently found to be higher in formal competition than during training or practice sessions (Caine and Maffulli 2005; Caine *et al.* 2008; Durie and Munroe 2000; Finch *et al.* 2010; Junge *et al.* 2004; Maffulli and Caine 2005; McIntosh *et al.* 2010; Romiti *et al.* 2008). This is likely to be related to the increased nature of the competitive setting and faster speeds of games, which all lead to the generation of high forces that could lead to increased injury risk in some children.

In terms of safe performance of work tasks or safe participation in sport/activity, it could be argued that physical maturity and size are more important than chronological age (Kidd *et al.* 2000). In a study of Bantam ice-hockey players (aged 13–15 years), it was found that the body mass and stature differences between the smallest and largest were 53 kg and 53 cm, respectively (Brust *et al.* 1992). In a contact sport like ice hockey, the resultant differences in momentum and hence impact collision forces would place smaller and lighter children at increased injury risk. Indeed, in a similar Bantam ice-hockey competition, a 357 per cent difference in the force of impact between the smallest and largest player was reported (Bernard *et al.* 1993).

An Australian prospective cohort study reported injuries in junior cricketers according to the field position at the time of injury, age levels of play (under 12 years, under 14 years and under 16 years) and match versus training settings (Finch *et al.* 2010). As the junior version of this game in Australia has a gradual progression according to age level of play and no injuries were reported in the U12 players, the authors concluded that the age-appropriateness of the game and its gradual introduction to more complex game demands as children grow has been a success. As in a similar study of junior Australian football players (Romiti *et al.* 2008), this cricket study showed that injury rates increase with playing age, as players progress towards the more adult form of the game. In sports such as these with heavily modified rules for children, rates increase with age because some of the external protections put in place for younger children are gradually removed as players enter older level competitions. For example, compared to younger ones, children in older age competitions may (a) play with larger, heavier or harder equipment, (b) play on larger grounds and so run around more, (c) are involved in teams with more players of varying size and maturity, (d) are introduced to body contact elements of the game such as tackling skills and (e) have longer games and more training sessions. If the young players are not well skilled in technique, rules of the game or safety measures, then their risk of injury can be increased as they move to older levels of play because they do not have the full skills to handle these changes in external game factors in a safe manner.

Due to the exploratory and often carefree nature of children, they will always be prone to falls and associated injuries, therefore making the condition and impact-absorbing properties of the surfaces they land on essential. Compared to adults, children have a stiffer technique when landing, with more hip and knee extension, leading to a high impulse and short time to peak vertical ground-reaction forces (Swartz *et al.* 2005), resulting in the reduced ability to absorb impact forces on landing. To counteract this and reduce the risk of injury, better impact-absorbing surfaces are becoming more common in playing areas for children, particularly in playground settings (Gunatilaka *et al.* 2004; Mack *et al.* 2000). Equipment height has also been implicated as a contributing factor to child injury risk in playgrounds (Fiissel *et al.* 2005) and hence safety standards for playground surfaces and equipment specifications have been introduced in many countries (e.g. NZS 5828, BS EN 1176-7) (Sherker and Ozanne-Smith 2004). Differences in landing mechanics, coactivation during landing and dynamic stability of the knee between children and adults also place children at a greater risk of injury during these movements. These predisposing factors to the proposed increase in the relative risk of landing and pivoting injury during childhood and potential prevention strategies will be discussed in more detail in Chapter 11.

Clinical and practical applications

Throughout this chapter example boxes have been used to describe the application of biomechanical principles to specific examples of child injury risk. While biomechanical principles can be employed to develop a targeted and appropriate intervention, its introduction to players and subsequent uptake require other prevention strategies to also be adopted. Just having a well-designed protective helmet available for a game does not ensure players use it unless there is also wide usage of it by players (McIntosh *et al.* 2009). Accompanying strategies such as mandating protective equipment use turns a well-designed biomechanical device into a successful prevention measure (Cameron *et al.* 1994; Shaw and Finch 2008).

Of course, the value of biomechanically based injury-prevention measures is directly related to their uptake and adoption, and hence is heavily influenced by social and behavioural factors, including player/coach knowledge levels and attitudes. A study of high-school rugby players found that while they generally understood what caused concussion and what its symptoms were, they were unaware of return to play guidelines after a concussion as the major means of preventing further physical injury damage (Sye *et al.* 2006). Other studies have questioned whether protective equipment use increases risk compensatory behaviour in young athletes, with players using protective equipment potentially more likely to take greater physical risks because they perceive themselves to be safer (Finch *et al.* 2001; McIntosh 2005). It is also possible that players modify their on-field behaviours if they perceive significant injury hazards to be present (White *et al.* 2011).

Recent studies describing the implementation of exercise training programmes by coaches to prevent lower limb injuries in young football (Kilding *et al.* 2008) and netball (Saunders *et al.* 2010) players have not been fully adopted because the

---

Example box 6: Excessive workload and inadequate rest periods

Lesson: Excessive workloads relating to a specific game-related activity can lead to increased risk of overuse injury through overloading of tissues.

Reference: Dennis *et al.* 2005.

Sport/s: Club and district cricket.

Players and age group: 44 fast bowlers aged 12–17 years.

Nature of biomechanical injury risk: Overuse injuries resulting from repetitive microtrauma. Fast bowling in cricket requires repetitive hyper-extension of the lumbar spine. Young fast bowlers are at increased risk of back injury, e.g. pars interarticularis. Predisposing factors for overuse injury to fast bowlers include poor technique, poor physical preparation and overuse.

Study design: Prospective cohort study over one playing season. Junior players completed daily workload diaries to record bowling workloads in all games and training sessions, and to report injuries. All injury reports were validated by a physiotherapist. After the season, workloads were compared for those who were and were not injured.

Results: 25 per cent of bowlers sustained an overuse injury. Injured bowlers had significantly higher bowling workloads than uninjured bowlers in terms of the length of rest periods between bowling episodes.

Outcome: This study led to national workload policy change for junior cricket players but its success has not yet been measured. Coaches should ensure adequate rest periods for bowlers.

---

programmes, which were largely direct applications of programmes developed for adults, were not fully appropriate for application with children. These examples show that particular consideration needs to be given towards the development of interventions that are specifically targeted at the physical maturation level of players to ensure that they are still enjoyable and relevant for this group, while maintaining the biomechanical fidelity of the underlying exercise intervention.

Another factor influencing the value of biomechanical countermeasures is their widespread availability, appropriateness and acceptability. Just because protective equipment could be available, it does not necessarily mean that this is appropriate for use by child athletes, especially if the only biomechanical development and testing has been undertaken for adult populations. This poses a challenge when there is a community need, often driven by parents, for prevention measures such as protective equipment to be mandated for use by their children. For this reason, biomechanical injury solutions are most effective when paired with other prevention approaches such as policy setting, manufacturing standards, education, rules and regulations (Christoffel and Gallagher 2006; McClure *et al.* 2004).

Conclusion

This chapter summarizes some of the best evidence about the nature of biomechanical-based injury risk in children across their life span and some of the strategies that could be adopted to reduce their chance of significant injury. Such strategies are relevant at different points in the chain of events that lead to injury and generally target either intrinsic or extrinsic risk factors. Unfortunately, there is still a major knowledge gap with regard to many sport and active recreational injuries in children and the potential solutions that could be developed or adopted to prevent them. Until there are further properly designed large-scale prospective studies with appropriate and detailed mechanistic information about the true nature of injury risk in children (Finch *et al.* 2011), we run the risk of only being able to best guess what interventions would be most appropriate for children. By drawing only on information from adult studies, there is the chance that the need to optimize intervention effectiveness in accordance with child growth and development will be lost, and our children will remain at sustained risk of injury.

Key points

- Injury occurs when there is a transfer of energy beyond the threshold that the body can withstand.
- This energy exchange can be as a single acute traumatic episode (e.g. a collision in sport, a fall from playground, being struck by a ball) or by an accumulation of microtrauma, as is the case in overuse injuries.
- Children are not 'mini adults' and their capacity to withstand these energy insults is different from that of adults and also changes as their musculo-skeletal systems develop.
- The fact that chronological age is not the same as maturational or developmental age poses additional considerations when prescribing injury prevention strategies in children.
- Current biomechanical interventions protect child athletes by modifying their exposure to inherent hazards (e.g. through modified rules), directly protecting them from injury risks (e.g. through protective equipment use) or by appropriate physical preparation to allow them to meet the physical requirements of the activity (e.g. reduced training loads).

Acknowledgements

Caroline Finch was supported by a National Health and Medical Research Council (NHMRC, of Australia) Principal Research Fellowship. The Australian Centre for Research into Injury in Sport and its Prevention (ACRISP) is one of the International Research Centres for Prevention of Injury and Protection of Athlete Health supported by the International Olympic Committee (IOC).

References

Abernethy, L. and Bleakley, C. (2007) 'Strategies to prevent injury in adolescent sport: a systematic review', *British Journal of Sports Medicine*, 41: 627–38.

Bahr, R. and Holme, I. (2003) 'Risk factors for sports injuries – a methodological approach', *British Journal of Sports Medicine*, 37: 384–92.

Bahr, R. and Krosshaug, T. (2005) 'Understanding injury mechanisms: a key component of preventing injuries in sport', *British Journal of Sports Medicine*, 39: 32–49.

Benjamin, H. and Briner, W. (2005) 'Little league elbow', *Clinical Journal of Sport Medicine*, 15: 37–40.

Berger, L.R. and Mohan, D. (1996) *Injury Control. A Global View*, Delhi: Oxford University Press.

Bernard, D., Trudel, P., Marcotte, G. and Boileau, R. (1993) 'The incidence, types, and circumstances of injuries to ice hockey players at the Bantam level (14–15 years)', in C. Castaldi, P. Bishop and E. Hoerner (eds) *Safety in Ice Hockey*, Philadelphia, PA: American Society for Testing and Materials, pp. 44–55.

Bijur, P., Trumble, A., Harel, Y., Overpeck, M., Jones, D. and Scheidt, P. (1995) 'Sports and recreation injuries in US children and adolescents', *Archives of Pediatrics & Adolescent Medicine*, 149: 1009–16.

Brust, J., Leonard, B., Pheley, A. and Roberts, W. (1992) 'Children's ice hockey injuries', *American Journal of Diseases in Childhood*, 146: 741–7.

Caine, D. and Maffulli, N. (eds) (2005) *Epidemiology of Pediatric Sports Injuries: Individual Sports*, Basel, Switzerland: Karger.

Caine, D., Maffulli, N. and Caine, C. (2008) 'Epidemiology of injury in child and adolescent sports: injury rates, risk factors and prevention', *Clinics in Sports Medicine*, 27: 19–50.

Caine, D., Cochrane, B., Caine, C. and Zemper, E. (1989) 'An epidemiological investigation of injuries affecting young gymnasts', *American Journal of Sports Medicine*, 17: 811–20.

Cameron, M.H., Vulcan, A.P., Finch, C.F. and Newstead, S.V. (1994) 'Mandatory bicycle helmet use following a decade of helmet promotion in Victoria, Australia – an evaluation', *Accident Analysis and Prevention*, 26: 325–37.

Cech, D.J. and Martin, S.T. (2002) *Functional Movement Development*, Philadelphia, PA: Elsevier.

Chalmers, D., Marshall, S., Langley, J., Evans, M., Brunton, C., Kelly, A. and Pickering, A. (1996) 'Height and surfacing as risk factors for injury in falls from playground equipment: a case-control study', *Injury Prevention*, 2: 98–104.

Chau, N., Predine, R., Aptel, E., D'houtaud, A. and Choquet, M. (2007) 'School injury and gender differentials: a prospective cohort study', *European Journal of Epidemiology*, 22: 327–34.

Cheng, T., Fields, C., Brenner, R., Wright, J., Lomax, T. and Scheidt, P. (2000) 'Sports injuries: an important cause of morbidity in urban youth', *Pediatrics*, 105: e32.

Christoffel, T. and Gallagher, S. (2006) *Injury Prevention and Public Health: Practical Knowledge, Skills, and Strategies*, Sudbury, UK: Jones and Bartlett Publishers.

Conn, M., Annest, J. and Gilchrist, J. (2003) 'Sports and recreation related injury episodes in the US population, 1997–99', *Injury Prevention*, 9: 117–23.

Cook, D., Cusimano, M., Tator, C. and Chipman, M. (2003) 'Evaluation of the ThinkFirst Canada, Smart Hockey, brain and spinal cord injury prevention video', *Injury Prevention*, 9: 361–6.

Dennis, R., Finch, C. and Farhart, P. (2005) 'Is bowling workload a risk factor for injury to Australian junior cricket fast bowlers?', *British Journal of Sports Medicine*, 39: 843–6.

DiFiori, J.P. (2002) 'Overuse injuries in young athletes: an overview', *Athletic Therapy Today*, 7: 25–9.

Durie, R. and Munroe, A. (2000) 'A prospective survey of injuries in a New Zealand schoolboy rugby population', *New Zealand Journal of Sports Medicine*, 28: 84–90.

Elliott, B.C., David, J.W., Khangure, M.S., Hardcastle, P. and Foster, D. (1993) 'Disc degeneration and the young fast bowler in cricket', *Clinical Biomechanics*, 8: 227–34.

Emery, C. (2005) 'Injury prevention and future research', in D. Caine and N. Maffulli (eds) *Epidemiology of Pediatric Sports Injuries: Individual Sports*, Basel, Switzerland: Karger, pp. 179–200.

Fecteau, D., Gravel, J., D'angelo, A., Martin, E. and Amre, D. (2008) 'The effect of concentrating periods of physical activity on the risk of injury in organized sports in a pediatric population', *Clinical Journal of Sport Medicine*, 18: 410–14.

Fiissel, D., Pattison, G. and Howard, A. (2005) 'Severity of playground fractures: play equipment versus standing height falls', *Injury Prevention*, 11: 337–9.

Finch, C., McIntosh, A. and Mccrory, P. (2001) 'What do under 15 year old schoolboy rugby union players think about protective headgear?', *British Journal of Sports Medicine*, 35: 89–94.

Finch, C., Ullah, S. and McIntosh, A. (2011) 'Combining epidemiology and biomechanics in sports injury prevention research – a new approach for selecting suitable controls', *Sports Medicine*, 41: 59–72.

Finch, C., Valuri, G. and Ozanne-Smith, J. (1998) 'Sport and active recreation injuries in Australia: evidence from emergency department presentations', *British Journal of Sports Medicine*, 32: 220–5.

Finch, C., Donohue, S., Garnham, A. and Seward, H. (2002) 'The playing habits and other commitments of elite junior Australian football players', *Journal of Science and Medicine in Sport*, 5: 266–73.

Finch, C., White, P., Dennis, R., Twomey, D. and Hayen, A. (2010) 'Fielders and batters are injured too: a prospective cohort study of injuries in junior club cricket', *Journal of Science and Medicine in Sport*, 13: 489–95.

Froholdt, A., Olsen, O. and Bahr, R. (2009) 'Low risk of injuries among children playing organized soccer', *American Journal of Sports Medicine*, 37: 1155–60.

Gregory, P., Batt, M. and Wallace, W. (2002) 'Comparing injuries of spin bowling with fast bowling in young cricketers', *Clinical Journal of Sport Medicine*, 12: 107–12.

Gregory, P.L., Batt, M.E. and Wallace, W.A. (2004) 'Is risk of fast bowling injury in cricketers greatest in those who bowl most? A cohort of young English fast bowlers', *British Journal of Sports Medicine*, 38: 125–8.

Gunatilaka, A., Sherker, S. and Ozanne-Smith, S. (2004) 'Comparative performance of playground surfacing materials including conditions of extreme non-compliance', *Injury Prevention*, 10: 174–9.

Haddon, W. (1972) 'A logical framework for categorizing highway safety phenomena and activity', *Journal of Trauma*, 12: 197–207.

Haddon, W.J. (1973) 'Energy damage and the 10 countermeasure strategies', *Journal of Trauma*, 13: 321–31.

Hagel, B., Pless, I.B. and Goulet, C. (2005) 'The effect of wrist guard use on upper-extremity injuries in snowboarders', *American Journal of Epidemiology*, 162: 149–56.

Hass, C.J., Schick, E.A., Chow, J.W., Tillman, M.D., Brunbt, D. and Cauraugh, J.H. (2003) 'Lower extremity biomechanics differ in prepubescent and postpubescent female athletes during stride jump landings', *Journal of Applied Biomechanics*, 19: 139–52.

Hootman, J., Dick, R. and Age, J. (2007) 'Epidemiology of collegiate injuries for 15 sports: summary and recommendations for injury prevention initiatives', *Journal of Athletic Training*, 42: 311–19.

Junge, A., Cheung, K., Edwards, T. and Dvorak, J. (2004) 'Injuries in youth amateur soccer and rugby players – comparison of incidence and characteristics', *British Journal of Sports Medicine*, 38: 168–72.

Kerr, M. (2008) 'Paediatric chest trauma (part 1) – initial lethal injuries', *Trauma*, 10: 183–94.

Kidd, P.S., Mccoy, C. and Steenbergen, L. (2000) 'Repetitive strain injuries in youth', *Journal of the American Academy of Nurse Practitioners*, 12: 413–26.

Kilding, A., Tunstall, H. and Kuzmic, D. (2008) 'Suitability of FIFA's "the 11" training programme for young football players – impact on physical performance', *Journal of Sports Science and Medicine*, 7: 320–6.

Klingele, K.E. and Kocher, M.S. (2002) 'Little league elbow: valgus overload injury in the paediatric athlete', *Sports Medicine*, 32: 1005–15.

Laforest, S., Robitallie, Y., Lesage, D. and Dorval, D. (2001) 'Surface characteristics, equipment height, and the occurrence and severity of playground injuries', *Injury Prevention*, 7: 35–40.

Mack, M., Sacks, J. and Thompson, D. (2000) 'Testing the impact attenuation of loose-fill playground surfaces', *Injury Prevention*, 6: 141–4.

Maffulli, N. and Caine, D. (eds) (2005) *Epidemiology of Pediatric Sports Injuries: Team Sports*, Basel, Switzerland: Karger.

Magra, M., Caine, D. and Maffulli, N. (2007) 'A review of epidemiology of paediatric elbow injuries in sports', *Sports Medicine*, 37: 717–35.

Malina, R., Dampier, T., Powell, J., Barron, M. and Moore, M. (2007) 'Validation of a noninvasive maturity estimate relative to skeletal age in youth football players', *Clinical Journal of Sport Medicine*, 17: 362–8.

Marx, R., Sperling, J. and Cordasco, F. (2001) 'Overuse injuries of the upper extremities in tennis players', *Clinics in Sports Medicine*, 20: 439–51.

McClure, R., Stevenson, M. and McEvoy, S. (eds) (2004) *The Scientific Basis of Injury Prevention and Control*, Melbourne: IP Communications.

McGuine, T. (2006) 'Sports injuries in high school athletes: a review of injury-risk and injury-prevention research', *Clinical Journal of Sport Medicine*, 16: 488–99.

McHugh, M., Tyler, T., Mirabella, M., Mullaney, M. and Nicholas, S. (2007) 'The effectiveness of a balance training intervention in reducing the incidence of noncontact ankle sprains in high school football players', *American Journal of Sports Medicine*, 35: 1289–94.

McIntosh, A. (2004a) 'Developing injury interventions: the role of biomechanics', in R. McClure, M. Stevenson and S. McEvoy (eds) *The Scientific Basis of Injury Prevention and Control*, Melbourne: IP Communications, pp. 202–13.

McIntosh, A. (2004b) 'Risk factor identification: the role of biomechanics', in R. McClure, M. Stevenson and S. McEvoy (eds) *The Scientific Basis of Injury Prevention and Control*, Melbourne: IP Communications, pp. 144–53.

McIntosh, A. (2005) 'Risk compensation, motivation, injuries, and biomechanics in competitive sport', *British Journal of Sports Medicine*, 39: 2–3.

McIntosh, A.S., McCrory, P., Finch, C.F. and Wolfe, R. (2010) 'Head, face and neck injury in youth rugby: incidence and risk factors', *British Journal of Sports Medicine*, 44: 188–93.

McIntosh, A.S., McCrory, P., Finch, C.F., Chalmers, D., Best, J.P. and Wolfe, R. (2009) 'Efficacy of padded headgear in rugby union football: cluster randomised trial', *Medicine and Science in Sports and Exercise*, 41: 306–13.

McMahon, K.A., Nolan, T., Bennett, C.M., and Carlin, J.B. (1993) 'Australian Rules football injuries in children and adolescents', *Medical Journal of Australia*, 159: 301–6.

Meeuwisse, W.H., Tyreman, H., Hagel, B. and Emery, C. (2007) 'A dynamic model of etiology in sport injury: the recursive nature of risk and causation', *Clinical Journal of Sport Medicine*, 17: 215–19.

Meyers, J. (1993) 'The growing athlete', in P. Renstrom (ed.) *Sports Injuries: Basic Principles of Prevention and Care*, Oxford: Blackwell Scientific Publications.

Meyers, M.C., Laurent, C.M., Higgins, R.W. and Skelly, W.A. (2007) 'Downhill ski injuries in children and adolescents', *Sports Medicine*, 37: 485–99.

Micheli, L. (1983) 'Overuse injuries in children's sport: the growth factor', *Orthopedic Clinics of North America*, 14: 337–60.

Molcho, M., Harel, Y., Pickett, W., Scheidt, P., Mazur, J. and Overpeck, M. (2006) 'The epidemiology of non-fatal injuries among 11-, 13- and 15-year old youth in 11 countries: findings from the 1998 WHO-HBSC cross national survey', *International Journal of Injury Control and Safety Promotion*, 13: 205–11.

Ozanne-Smith, J. (1995) 'Child injury prevention', in J. Ozanne-Smith and F. Williams (eds) *Injury Research and Prevention: A Text*, Melbourne: Monash University Accident Research Centre.

Peden, M., Oyegbite, K., Ozanne-Smith, J., Hyder, A., Branche, C., Rahman, A., Rivara, F. and Bartolomeos, K. (eds) (2008) *World Report on Injury Prevention*, Geneva: World Health Organization.

Radelet, M., Lephart, S., Rubinstein, E. and Myers, J. (2002) 'Survey of the injury rate for children in community sports', *Pediatrics*, 110, e28.

Robertson, L. (2007) *Injury Epidemiology: Research and Control Strategies*, New York: Oxford University Press.

Romiti, M., Finch, C. and Gabbe, B. (2008) 'A prospective cohort study of the incidence of injuries among junior Australian football players – evidence for a playing age level effect', *British Journal of Sports Medicine*, 42: 441–6.

Saunders, N., Otago, L., Romiti, M., Donaldson, A., White, P. and Finch, C. (2010) 'Coaches' perspectives on implementing an evidence-informed injury prevention program in junior community netball', *British Journal of Sports Medicine*, 44: 1128–32.

Shaw, L. and Finch, C. (2008) 'Injuries to junior club cricketers – the impact of helmet regulations', *British Journal of Sports Medicine*, 42: 437–40.

Sherker, S. and Ozanne-Smith, J. (2004) 'Are current playground safety standards adequate for preventing arm fracture?', *Medical Journal of Australia*, 180: 562–5.

Swartz, E.E., Decoster, L.C., Russell, P.J. and Croce, R.V. (2005) 'Effects of developmental stage and sex on lower extremity kinematics and vertical ground reaction forces during landing', *Journal of Athletic Training*, 40: 9–14.

Sye, G., Sullivan, S. and McCrory, P. (2006) 'High school rugby players' understanding of concussion and return to play guidelines', *British Journal of Sports Medicine*, 40: 1003–5.

Veigel, J. and Pleacher, M. (2008) 'Injury prevention in youth sports', *Current Sports Medicine Reports*, 7: 348–52.

Verhagen, E., Collard, D., Paw, M. and Van Mechelen, W. (2009) 'A prospective cohort study on physical activity and sports-related injuries in 10–12-year-old children', *British Journal of Sports Medicine*, 43: 1031–5.

White, P.E., Finch, C.F., Dennis, R. and Siesmaa, E. (2011) 'Understanding perceptions of injury risk associated with playing junior cricket', *Journal of Science and Medicine in Sport*, 14: 115–20.

Whiting, W. and Zernicke, R. (2008) *Biomechanics of Musculoskeletal Injury*, Champaign, IL: Human Kinetics.

11 Dynamic knee stability during childhood

Mark De Ste Croix and Martine Deighan

Introduction

Participation in sport and exercise places the participant at an increased risk of injury, independent of age and sex. This, of course, includes the period of childhood, and it has been proposed that children may be at a greater risk than adults due to developing musculoskeletal systems (see Chapter 1) that may not be considered fully mature (Funasaki 2009). Collard *et al.* (2009) suggest that all forms of physical activity place the child at greater risk of injury than sedentary individuals due to a high level of exposure to situations where muscles and joints are placed under stress. Due to the high prevalence of physical activity- and sport-related injuries, and the associated negative short- and long-term consequences, prevention of these injuries in children is important. With an estimated 45 million children involved in organized sports in the United States alone, it is hardly surprising that sport contributes to the major risk of injury during childhood, estimated in the USA to be about 3 million per year (Fecteau *et al.* 2008; Veigel and Pleacher 2008). These injuries influence health and fitness and have a socioeconomic impact, but most of them can be prevented. There are a number of well-recognized paediatric conditions associated with the mechanical stresses placed on joints, tendons and ligaments during sporting activities. These include osteochondral injury of the elbow joint, Sever's disease of the ankle, avulsion fracture of the pelvis and Osgood-Schlatter disease (Funasaki 2009). Some of these injuries may lead to joint dysfunction such as deformity, pain and limited range of motion. Although some of these conditions may progress into adulthood, early detection can lead to complete recovery without complications. More importantly, a clear understanding of the possible causes of such injuries means that prevention can be put in place during childhood to avoid such consequences. It is beyond the scope of this chapter to explore all of these paediatric conditions, and therefore the focus of this chapter is on dynamic knee stability, which is important in preventing non-contact anterior cruciate ligament (ACL) injury. Dynamic knee stability is also mostly influenced by muscular, neuromuscular, hormonal and biomechanical factors. The incidence, severity, costliness and long-term consequences from ACL tears make prevention a high priority in the medical and research communities (Bonci 1999). There is good evidence that

ACL injury at a young age increases the relative likelihood of suffering from osteoarthritis in older age (Palmieri-Smith and Thomas 2009). The focus of this chapter is to explore the role that dynamic stability of the knee plays in reducing the relative risk of injury from a muscular and neuromuscular perspective.

Basic theoretical concepts

Many previous papers have described the anatomy and biomechanics of the knee joint, and the reader is directed to the review of Goldblatt and Richmond (2003). Functional stability is provided by static and dynamic restraints within the joint and relies on feedforward and feedback systems (Wikstrom *et al.* 2006). Feedforward systems include anticipatory muscle actions occurring before the sensory detection of a disruption (Bonci 1999; Kaminski *et al.* 1998; Riemann and Lephart 2002). Conversely, feedback is the stimulation of a corrective response of the neuromuscular system after sensory detection. Riemann and Lephart (2002) found that joint stability and protection against injury relied on adequate feedback and feedforward systems to improve muscular stiffness during functional tasks. Therefore, it is useful for the physician involved in injury prevention or rehabilitation to be able to accurately measure the amount of joint stiffness induced by the surrounding muscles. The use of neuromuscular biomechanical modelling to understand knee ligament loading and subsequent knee joint stability has emphasized the importance and effectiveness of the muscles in providing this stabilization (Lloyd *et al.* 2005). During landing, when potential to injure the ACL exists, the response of the neuromuscular system is critical. The ACL can provide up to 86 per cent of the resistance to anterior tibial translation. However, it is well recognized that the internal and external forces incurred at the knee during landing can stress the passive ligament structures beyond their capacity. Muscular forces are crucial in maintaining joint stability predominantly by increasing joint stiffness through co-contraction of antagonistic muscles. *In vivo* exploration of hamstring or quadriceps co-contraction suggests that it can double or even triple joint stiffness and decreases joint laxity by up to 50 per cent (Russell *et al.* 2007). Data from adults indicate that sidestepping and cutting movements are most likely to load the ACL and subsequently increase the relative risk of injury (Lloyd *et al.* 2005). There are no comparative data on children. However, it is likely that such movements in children demonstrate a similar pattern of ligament loading to that of adults. It may be that due to the development of kinematic motor control during childhood the growing child may be predisposed to additional ligament loads compared to adults. However, this suggestion requires further investigation, utilizing biomechanical modelling to explore ligament loading during childhood and during various dynamic tasks. It is beyond the scope of this chapter to explore the effect of altering movement mechanics on ACL loading, but adult studies seem to demonstrate that reduction in peak tibial shear force is possible through changes in knee flexion angle, shank angle and foot location during landing (Myers and Hawkins 2010).

Relative risk of non-contact ACL injury during childhood

Participation in sports with a high frequency of decelerating manoeuvres increases the risk of injury (Griffin *et al.* 2006), and in those sports, females participating at the same level as males have a three to six times higher injury incidence than males (Hewett *et al.* 2001). Some studies have suggested that sex differences in injury rate may be as great as 9.7 times higher in females compared with males (Russell *et al.* 2007). Observations of high-risk situations have been used to formulate three theories as potential reasons for this bias: (a) the ligament dominance theory (Hewett *et al.* 2001), which relates to deficiencies in neuromuscular control and strength (Ford *et al.* 2003); (b) the quadriceps dominance theory (Hewett *et al.* 2001) where inadequate quadriceps recruitment patterns create excessive anterior translation of the tibia (Ford *et al.* 2003; Hewett *et al.* 1996; Huston and Wojtys 1996); and (c) the straight knee landing theory (Huston *et al.* 2001), which addresses angle-specific neuromuscular characteristics to explain the occurrence of injuries near full knee extension (Decker *et al.* 2003; Huston *et al.* 2001).

The incidence of injury in children is discussed in some detail in Chapter 10. However, the relative risk of acquiring a non-contact ACL injury during childhood is not well understood due to the difficulty in collecting incidence data. There appear to be few sets of age-specific data available on non-contact ACL injury that cover the period of childhood (de Loës *et al.* 2000; Emery and Meeuwisse 2010; Shea *et al.* 2004; Yu *et al.* 2002). These data need to be viewed with a degree of caution, as Yu *et al.* (2002) derived information from ACL reconstruction data and Shea *et al.* (2004) obtained their information from insurance claims relating to ACL injury, which do not adequately reflect incidence rates.

Current evidence suggests that the knee is especially susceptible to injury during childhood, with around 63 per cent of sports-related injuries in 6–12-year-olds occurring at the knee (Gallagher *et al.* 1984). Emery and Meeuwisse's (2010) study on 13–18-year-old male and female soccer players (n = 744) reported a higher injury rate in older age groups but found no significant sex difference in the rate of all injuries. However, although not statistically significant, they did note a trend towards a greater risk of ankle and knee injuries in females. Myer *et al.* (2004) note that most of the published literature demonstrates no evidence of a sex difference in ACL injury rates in prepubertal children. Limited incidence data appear to suggest an increased risk in girls compared to boys as they enter puberty and subsequently linked to periods when growth spurts occur. How this increased incidence in girls is related to differences in peak height velocity, gains in muscle size and issues relating to proportionality remain to be investigated. Interestingly, a recent retrospective observational study indicated that females of all ages are more likely to injure the ACL of their supporting leg during a kicking action compared to males who appear to injure their kicking leg (Brophy *et al.* 2010).

Etiology of non-contact ACL injury in children

Many factors, both intrinsic and extrinsic, have been proposed as predisposing factors for non-contact ACL injury. These include: lower extremity malalignment, ligamentous laxity, lower extremity muscular strength, neuromuscular control and proprioception, hormonal influences, intercondylar notch width and biomechanics of player technique (including landing and pivoting biomechanics) (Bonci 1999; Huston 2007). Although no definitive etiology for the discrepancy in the occurrence of ACL injuries between the sexes has been established, structural, hormonal and neuromuscular factors have all been proposed (Hewett *et al.* 2007). All of these factors are of particular relevance for the developing child when anatomical structures are being formed and developed, and as changes in hormonal and neuromuscular functioning occur (see Chapters 1 and 3 for more detail). Dynamic muscular control of knee-joint alignment, specifically differences in muscle recruitment, firing patterns and strength may be partly responsible for the sex differences in the incidence of ACL injury. Results from Myer *et al.* (2010) suggest that increased relative quadriceps recruitment, decreased knee flexion range of movement, concomitant with increased tibia length and mass normalized to stature are all related to increased ACL injury risk in both children and adults. Many of these factors are rarely stable during childhood due to issues of growth and maturation, and change in individuals at differing rates. How these factors are related to the age- and sex-associated risk of injury remains to be identified.

Adult studies have clearly indicated significant sex difference in landing and pivoting mechanics, with females exhibiting higher knee abduction moments (KAM) than males. The few studies that are available from paediatric populations have focused on high-school athletes (Ford *et al.* 2003; Myer *et al.* 2010), and they have also demonstrated similar findings of altered landing mechanics in females compared to males. Young girls appear to land in a more erect position, with less knee flexion, causing higher forces to be transferred through the knee joint when compared to boys (Decker *et al.* 2003; Yu *et al.* 2005). There do not appear to be any longitudinal studies examining age- and sex-associated changes in landing kinematics, especially in younger age groups, and further investigation is required. Nevertheless, current evidence suggests a sex difference in landing and pivoting kinematics during childhood.

The dynamic stabilizers

It is well recognized that varus and valgus moments, and internal and external rotation moments subject the knee ligaments to high forces. Therefore, stabilization of the knee during these movements is crucial in reducing the injury risk to the knee ligaments. The muscles spanning the knee joint are the key dynamic stabilizers and have the potential to support varus and valgus moments. Lloyd *et al.* (2005) suggest that the knee muscles are the main defence against knee-ligament injury during flexion and extension movements, as well as sidestepping movements. Anatomically, the quadriceps have both valgus and varus moment arms by virtue of being indirectly connected to the patella ligament; however, these moment arms are smaller than those of other knee muscles (e.g. the hamstring muscles). Due to the high forces incurred at the joint during sporting activity that are beyond the capacity of the passive ligamentous constraints, the addition of active muscular force is essential to maintain joint equilibrium and stability (Shultz and Perrin 1999). It is well recognized that near full extension coactivation of the quadriceps and hamstrings serves primarily to increase joint stiffness and effectively reduce anterior–posterior, varus–valgus and rotational laxity through increased tibiofemoral joint compression irrespective of age or sex (Shultz 2007).

Lloyd *et al.* (2005) raise an interesting point when they suggest that although the knee muscles have the potential for providing stability, it does not necessarily mean that people activate their muscles effectively to produce that stability. The neuromuscular functioning of muscles during childhood may be as critical, if not more so, than simple force production itself. As we will see later in this chapter, the development of both concentric and eccentric moments of the knee muscles during growth and maturation has important implications for the effectiveness of the dynamic stabilizers and the potential to reduce the relative risk of injury.

Conventional and functional hamstring/quadriceps ratio

The ratio of eccentric hamstring to concentric quadriceps torque is an important measure as it describes the dynamic stability of the knee and an improved ratio has been shown to reduce ACL load. However, most of the available literature on the age- and sex-related changes in both functional hamstring to quadriceps ratio (H/Q_{func}) and neuromuscular functioning for examining joint stability have focused on adults. Few studies have examined the age- and sex-associated changes in factors that may be associated with dynamic knee stability and relative risk of injury during childhood. This is somewhat surprising given the epidemiological data indicating that adolescent females appear to be the most 'at-risk' group for non-contact ACL injury. Recent studies on adults clearly link the relative risk of injury to muscle imbalance of the H/Q_{func} ratio (Yeung *et al.* 2009).

To quantify intra-limb dynamic muscle balance, reciprocal muscle group ratios are calculated from torque measurements (Figure 11.1). Torque measurements are routinely determined using isokinetic dynamometry, and previous studies have demonstrated the reliability and reproducibility of such measurements in children (Deighan *et al.* 2003; Iga *et al.* 2006).

The 'conventional' ratio is the most widely reported ratio in the literature and is calculated by dividing the concentric hamstrings peak torque by the concentric quadriceps peak torque (H/Q_{con}). A H/Q_{con} ratio of 0.6 at an angular velocity of 1.05 rad·s^{-1} is considered to represent normal knee function (Aagaard *et al.* 1995). The conventional ratio has been examined by numerous authors, but it has been suggested to lack functional relevance. In fact, during knee extension, antagonistic eccentric, not concentric, hamstrings coactivation decreases the anterior

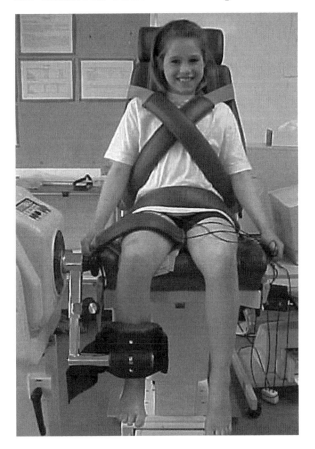

Figure 11.1 Isokinetic set-up for assessment of eccentric and concentric torque to determine the functional hamstring/quadriceps ratio.

shear forces induced by the concentric quadriceps muscle group action (Senter and Hame 2006). Therefore, the H/Q_{func} ratio has been suggested as more relevant (Aagaard *et al.* 1995). To calculate the H/Q_{func} ratio the peak eccentric hamstrings torque is divided by the peak concentric quadriceps torque to evaluate the relative ability of the hamstrings to act eccentrically and stabilize the knee. A H/Q_{func} ratio of less than 0.6 has been linked with a 17-fold increase in the relative risk of hamstring injury (Yeung *et al.* 2009). A 1:1 value is accepted as the reference value for the point of equality (Coombs and Garbutt 2002), and a value within the 0.7–1 range is accepted to represent adequate dynamic stability (Aagaard *et al.* 1995). Values below 0.7 illustrate muscle imbalance and as such may be a risk factor due to hamstring weakness compared to the quadriceps. One previous study that examined functional ratios of the knee demonstrated that an H/Q_{func} ratio of less than 0.6 represents a 77.5 per cent probability of knee injury in elite soccer players (Dauty *et al.* 2003).

Epidemiological findings highlight the importance of joint angle, angular velocity and action-specificity when calculating the H/Q_{func} ratio. It has been found that ACL rupture is most likely to occur near full knee extension during a high velocity movement. The H/Q_{func} ratio, outside the 0.7–1 range, when calculated near full knee extension (0°) suggests an increase in the injury risk. If it is to be accepted that injury occurrence is due to a specific hamstring weakness, the H/Q_{func} ratio should decrease when approaching full knee extension and with increasing angular velocity or in the presence of eccentric fatigue of the hamstrings. This would represent the inability of the hamstrings to absorb the anterior tibial forces induced by the concentric quadriceps action. However, Aagaard *et al.* (1995) found that the H/Q_{func} ratio increased with decreased relative knee angle and increased angular velocity. The antagonist muscle group function is angle-specific because it has to provide adequate anti-shear stabilizing torque at the same joint angle as the concentric action of the agonist muscle group. The joint angle specific H/Q_{func} ratio has been briefly considered (Aagaard *et al.* 1998) but has remained largely ignored in the literature, despite being a potentially confounding variable when describing functional deficiencies. The calculation of the H/Q_{func} should be joint angle- and angular velocity-specific, as hamstrings eccentric peak torque and quadriceps concentric peak torque do not occur at the same joint angle, and the angular specificities of the H/Q_{func} remain to be elucidated during childhood, especially near full knee extension and after fatigue.

Age- and sex-associated development in the H/Q_{con} ratio

There are a number of paediatric studies that have examined the age- and sex-associated changes in the H/Q_{con} ratio, and, despite their limited relevance to dynamic knee control, it is important to acknowledge the findings. Forbes *et al.* (2009a, 2009b), Holm and Vøllestad (2008) and Croce *et al.* (1996) have all reported no significant age or maturational effects on H/Q_{con} ratio in 12–16-year-old youth footballers, 7–12-year-old boys and girls, and between young boys and men over a range of velocities. However, Holm and Vøllestad (2008) did report significant sex differences in 7–12-year-olds, with girls demonstrating lower ratios than boys in all age groups and at both velocities. These results are in contrast to findings from Hamstra-Wright *et al.* (2006) who demonstrated no significant sex difference in isometric determined H/Q_{con} ratio in 9-year-old children. Likewise Barber-Westin (2005) also found no significant sex difference in the H/Q_{con} ratio in prepubertal children. It would appear that the H/Q_{con} ratio may be similar in children compared with adults, but it provides us with limited evidence to understand the age- and sex-associated development in dynamic stability of the knee.

Age- and sex-associated development in the H/Q_{func} ratio

There are few studies examining age- and sex-associated changes in H/Q_{func} ratio. This may in part be due to the small number of studies that have determined eccentric forces in children. At least for the adult population there is an abundance of literature on isokinetic eccentric (ECC) torque production but fewer studies are available on children, especially including females (De Ste Croix *et al.* 2003). It is possible that the limited information on ECC strength capabilities of children may have arisen from the concern that ECC testing, with its potential for high muscle-force production, might predispose children to higher risk of muscle injury. However, there is no reason to expect greater muscle injury with ECC actions in children compared to adults, provided they are given sufficient warm-up and familiarization (Blimkie and Macauley 2001).

As dictated by the force-velocity relationship, the H/Q_{func} ratio has been shown to increase as angular velocity increases in prepubertal children, teenagers and adults for both sexes (De Ste Croix *et al.* 2007; Kellis and Katis 2007). The H/Q_{func} ratio was significantly higher at 3.14 rad·s^{-1} (1.12) compared to 0.52 rad·s^{-1} (0.8). This increase in the ratio at higher velocities is due to the significant anterior tibial translation or shear at high quadriceps forces, and increase in internal rotation of the tibia in relation to the femur (Gerodimos *et al.* 2003). Irrespective of age or sex, the increase in coactivation of the hamstrings during high-velocity movements significantly contributes to counterbalance this tibial shear or rotation. These data reinforce the notion that the H/Q_{func} ratio is a more relevant estimate of the capacity for muscular knee joint stabilization than conventional ratios in children. It is important, therefore, that when making age- and sex-associated comparisons of the H/Q_{func} ratio that movement velocity is taken into account. Meaningful interpretation of the H/Q_{func} ratio in relation to age and sex has been problematic due to the order of action cycling in the isokinetic protocol. For example, during eccentric/concentric cycles, theoretically the eccentric action may potentiate the following concentric action. Only one study appears to have examined angle-specific H/Q_{func} ratio in a small group of 13-year-old boys (Kellis and Katis 2007). Kellis and Katis (2007) reported that the H/Q_{func} ratio increases near full knee extension due to the knee flexors' greater relative eccentric torque capability compared with the extensors' concentric torque. This suggests that young boys have reduced injury risk during fast velocity movements near full knee extension.

Age differences in H/Q_{func} ratio

It cannot be assumed that the relationship between concentric and eccentric actions is the same across ages during childhood and puberty because children have immature neuromuscular systems as evidenced by the incomplete myelination of nerve fibres during childhood (Brooks and Fahey 1987). The few studies that have measured the H/Q_{func} ratio in children have generally found significantly higher eccentric compared to concentric strength, depending upon movement velocity with ratios ranging between 0.84 at the slowest velocity to 1.47 at the highest velocity (De Ste Croix *et al.* 2007; Forbes *et al.* 2009b; Iga *et al.* 2009; Kellis and Katis 2007). As can be seen in Table 11.1, very few studies have directly measured the H/Q_{func} ratio in children and there appears to be only one study that has included girls (De Ste Croix *et al.* 2007).

Table 11.1 Paediatric functional hamstring/quadriceps ratio findings

Authors	n	Sex	Age (y)	Velocity (°/s)	Key findings
De Ste Croix *et al.* (2007)	121	M (n = 49) F (n = 72)	9, 17, 24y	30°/s, 180°/s	Ratio ↑ as velocity increases
					No significant sex difference
					Significant age effect (children ↓ ratio than adults)
Forbes *et al.* (2009)	157	M	11–18y	60°/s	Significant age effect (↓ ratio in 18-year-olds)
Gerodimos *et al.* (2003)	180	M (30 per age group)	12–17y	60°/s	No significant age effect
					Ratio ↑ as velocity increases
Iga *et al.* (2009)	45	M	15y	60°/s, 250°/s	↓ Ratio in conventionally trained footballers compared with resistance-trained and controls
Kellis and Katis (2007)	17	M	13y	60°/s, 180°/s	Ratio ↑ as knee extends
					Ratio ↑ as velocity increases

Gerodimos *et al.* (2003) reported a non-significant age effect on H/Q$_{func}$ ratios between 12- and 17-year-old trained male basketball players. Conflictingly, a cross- sectional study of prepubertal children, teenagers and adults has reported a significant age effect for the H/Q$_{func}$ ratio at a relatively high angular velocity of 3.14 rad·s^{-1} with 9–10-year-olds producing a significantly lower ratio (0.97) than teenagers (1.23) and adults (1.19) (De Ste Croix *et al.* 2007). Interestingly, the H/Q$_{func}$ ratio at the slower velocity (0.52 rad·s^{-1}) did not demonstrate a significant age effect. The functional ratio found below 1.0 for prepubertal children may be attributed to the inability of 9–10-year-old children to recruit their entire motor unit pool during eccentric actions (De Ste Croix *et al.* 2007). Kawakami *et al.* (1993) demonstrated a flattening of the force-velocity curve in the eccentric condition and hypothesized that boys might be too immature to recruit their entire motor unit pool compared to adults. This seems plausible as Weltman *et al.* (1986) reported that gains from resistance training in boys must be due to factors other than hypertrophy, such as reduced neuromuscular inhibition. The specula-tion that there might be a more pronounced force suppression in prepubertal children to protect their immature musculoskeletal systems is supported by the findings of the study of De Ste Croix *et al.* (2007).

Evidence to support the fact that neuromuscular inhibition is present not only in eccentric actions but also in slow-velocity concentric actions in children is provided by Seger and Thorstenson (2000) who found that EMG amplitude decreased with decreasing velocity during concentric muscle actions, especially in prepubertal boys. Although there was no significant age difference in this measure, there was a trend for the largest decline in EMG activity with decreasing velocity to occur in the boys (21 per cent) and the smallest for the men (7 per cent). No such trend was found in females. These findings suggest that at a velocity of 0.52 rad·s^{-1} there would also be inhibition during concentric actions as well as eccentric actions which could account for the non-significant difference in H/Q$_{func}$ ratios from prepuberty to adulthood during slow-velocity movements.

Conflicting data are available, and a large cross-sectional study of elite male footballers from 12 to 18 years has demonstrated significant age effects in the H/Q$_{func}$ ratio with significantly lower ratios in 18-year-olds (0.84) compared with 12-year-olds (1.01) (Forbes *et al.* 2009a, 2009b). This reduction in H/Q$_{func}$ ratio with age is attributed to a relatively greater increase in concentric quadriceps torque compared with the relative increase in eccentric hamstring torque. This age-related difference in the studies of De Ste Croix *et al.* (2007) and Forbes *et al.* (2009a, 2009b) may be reflective of a limited focus on eccentric hamstring training in youth footballers. Recent work has confirmed that loading patterns experienced during soccer asymmetrically strengthen the muscles about the knee, altering the balance towards quadriceps dominance (Iga *et al.* 2009). Iga *et al.* (2009) demonstrated a training effect in 15-year-old football players on H/Q$_{func}$ ratio with lower ratios in conventionally trained footballers compared with resistance-trained footballers and controls. These data must be viewed with a degree of caution as the ratio is calculated using eccentric testing at 2.16 rad·s^{-1} and concentric testing at 1.08 and 4.32 rad·s^{-1}. The current conflicting findings on the age-related changes in H/Q$_{func}$ ratio demonstrate how training status may influence the findings. However, a wider exploration of whether training status influences the H/Q$_{func}$ ratio or if ratios from joints other than the knee show different age- and sex-associated patterns is required.

Significant age and sex effects have also been observed for the functional quadriceps/hamstrings (Q/H$_{func}$) ratio during knee flexion (e.g. eccentric quadriceps/concentric hamstrings) at slow- and fast-velocity movements with adults demonstrating significantly lower ratios than prepubertal children and teenagers (De Ste Croix *et al.* 2007). These data suggest that adults have a reduced capacity for dynamic knee-joint stabilization during forceful knee flexion movements with accompanying eccentric muscle actions. This may be attributed to a greater ability of adults to create large forces during eccentric actions due to fully mature musculoskeletal systems. It may also be that adults have different neural mechanisms that control eccentric muscle actions compared to children, but direct evidence to support this notion remains to be established. Although these data are of interest, the knee is rarely injured during fast flexion movements when the quadriceps muscles are working eccentrically.

Sex differences in H/Q$_{func}$ ratio

It appears that adult females demonstrate a quadriceps dominance, which in turn makes them more susceptible to injury. A high level of quadriceps strength compared to hamstring strength reduces the H/Q$_{func}$ ratio and a ratio of less than 0.55 may represent a quadriceps dominant athlete. However, there are limited data investigating sex differences in the H/Q$_{func}$ ratio in children.

It has been suggested that sex differences in adults in the H/Q$_{func}$ ratio of the knee joint are due to differences in neuromuscular recruitment during maximal voluntary actions, with women having a lower percentage recruitment than men during concentric actions (Westing and Seger 1989) but not eccentric actions (Enoka 2002).

Another factor that might explain these sex differences is the capacity to store and release elastic energy in the muscle-tendon unit during eccentric actions. There is evidence suggesting a superior ability of females compared to males in utilizing stored elastic energy in the muscle-tendon unit (Komi and Bosco 1978). Komi and Bosco (1978) discovered that female participants were able to utilize significantly more of the energy produced in the pre-stretching phase compared with males.

The cross-sectional study of De Ste Croix *et al.* (2007) appears to be the only study that has examined sex differences in the H/Q$_{func}$ ratio in children. These authors reported no significant sex differences in H/Q$_{func}$ ratio at either test velocity in prepubertal, teenage or adults. These findings may in part be attributed to that fact that isokinetic concentric/eccentric tests do not allow a great deal of elastic energy storage and thus any potential benefits that females may have in utilizing stored elastic energy is not evident. More longitudinal data are required to reinforce whether or not there are sex differences in the H/Q$_{func}$ ratio throughout childhood. However, based on very limited data concerning sex differences in the H/Q$_{func}$ ratio we might speculate that sex differences in the relative risk of non-contact ACL injury may be attributed to factors other than those that are muscular in nature. These data would suggest that neuromuscular recruitment plays a crucial role in the sex difference related to relative risk of injury during childhood.

Interestingly, the same study reported significant sex differences for the Q/H$_{func}$ ratio at slow and fast movement velocities, irrespective of age, with males demonstrating significantly higher ratios than females (De Ste Croix *et al.* 2007). Females had similar quadriceps eccentric muscle torque compared to males and the lower ratio can be attributed to a lower capacity for hamstring concentric actions. It is generally accepted that females have a greater capacity for generating eccentric torque compared to males (Seger and Thorstenson 2000). However, females, irrespective of age or movement velocity, appear to have a lower capacity for generating concentric hamstring torque than males. It is speculated that this trend is due to lower motor-unit activation during concentric actions in females (Westing and Seger 1989). In an applied sense, these data suggest that stabilization of the knee joint during flexion movements, with high eccentric knee extension actions, is lower in females compared to males during childhood and into adulthood.

One of the biggest issues with the current literature examining the H/Q$_{func}$ ratio during childhood is that no studies appear to have determined the functional ratio at the point where knee injury is most likely to occur (0–30° of full knee extension). All of the studies reported in this chapter have used a peak torque value for concentric quadriceps torque and divided by an eccentric peak torque value for eccentric hamstring actions. This may not accurately represent the functional ratio as it is likely that the angle at which peak torque occurs for each of these actions and muscle groups is different. The recent work of Forbes *et al.* (2009b) highlights this issue where angles of peak torque for concentric quadriceps ranged from 72° to 78° in 12–18-year-olds compared with eccentric hamstring angles which ranged from 31° to 38°. Therefore, within current literature, concentric and eccentric torques determined at different joint angles are used to represent the H/Q$_{func}$ ratio, rather than torque achieved at the same joint angle. This clearly does not help in elucidating the functional role that these muscles play in stabilizing the knee during childhood. Furthermore, the joint angle where non-contact ACL injury is mostly likely to occur is not at the point where peak torque is generated. Peak concentric and eccentric torque production is likely to occur in the mid–late range of the movement (around 30–80° of knee flexion), whereas it is well recognized that injury is likely to occur when the knee is closer to full extension (0–30°). Based on this knowledge, it would seem more appropriate to calculate the H/Q$_{func}$ ratio using angle-specific torque values close to full extension. It is clear that more data are required on the H/Q$_{func}$ ratio during childhood, especially using angle-specific data and in females. Whether this will change our understanding of the age- and sex-associated changes in dynamic knee stability and the susceptibility to knee injury during childhood remains to be established.

Neuromuscular functioning and knee stability

As the main role of the ACL is to resist anterior tibial translation in a landing task, co-contraction of the hamstring muscles could reduce ACL loading and injury potential. Although the amplitude of this neuromuscular response is important during dynamic tasks, it is probably the speed of this response that is most critical to ACL protection. There is a wealth of research from adults indicating that the timely activation of the hamstring muscles can assist in protecting the ACL from mechanical strain by stabilizing the tibia, thus reducing anterior and rotary tibial translation (e.g. Shultz and Perrin 1999). Most of the available studies examining the role of neuromuscular functioning as a relative risk factor for knee injury have focused on adults. For example, a recent study examining age-related changes in neuromuscular functioning and joint stability reported that functional knee stability in terms of anterior tibial translation appears to be unaffected in people between 20 and 70 years of age (Melnyk *et al.* 2008). However, the importance of exploring an activation-driven model of joint stability during childhood should not be overlooked as changes to a joint stabilization task, clinical pathology, experience and, importantly, growth and maturation, all influence activation patterns.

Age- and sex-associated development of co-contraction ratios and electromechanical delay

Chapter 3 in this book describes the age- and sex-associated development in neuromuscular functioning during childhood, primarily in relation to motor control. The focus of this section is to examine the age- and sex-associated changes that occur in the speed of neuromuscular functioning (specifically latency time, more commonly referred to as electromechanical delay (EMD)). There is still some debate as to whether reflective muscular activation and joint stiffening can occur quickly enough to protect the joint once a large force is applied to the ligament, and data during childhood are particularly sparse. There are few studies that have examined the age- and sex-associated changes in knee muscle activation during landing or pivoting tasks. Russell *et al.* (2007) determined co-contraction ratios (CCR) of the hamstrings and quadriceps during a landing task in prepubertal children compared with those of adults. They reported significant age differences in relation to prelanding CCR, indicating that adults preactivated their hamstring muscles prior to landing to a greater extent than children. This suggests that this feedforward mechanism is more mature in adults compared with children. However, there were no age- or sex-related differences in activation levels during the reflective or voluntary muscle activation phases. These data are supported by the work of Lazaridis *et al.* (2010) who also demonstrated higher and longer preactivation in adult males compared with prepubertal boys for the calf muscles. They provide an interesting insight into landing mechanisms during the prepubertal years. As the incidence rate of ACL injury is smaller in prepubertal children than in adults, they probably rely on different strategies to control the forces of landing. This may include relying on a proximal strategy (one that uses the large muscles in the hip and torso) as opposed to a knee and ankle strategy (Russell *et al.* 2007). As there appear to be no comparable data to support this view, it remains an interesting but unproven hypothesis.

EMD has been implicated as a risk factor for knee injury in adults (Troy Blackburn *et al.* 2009). Figure 11.2 demonstrates one method of calculation of EMD, assuming the onset of electrical activity as a ±15µV deviation from baseline measurement to the onset of force production, which is assumed as 9.6 Nm. Data from adults suggest that EMDs vary between 30 to 50 ms up to as much as a few hundred milliseconds, depending on the muscle examined and movement velocity (Shultz and Perrin 1999).

Considering this time lapse and the need to develop sufficient muscle tension rapidly enough to provide dynamic knee stability, EMD should be considered when evaluating muscular responses to an imposed perturbation or injurious stress. It is important to note that it has been suggested that EMD is influenced by a number of factors including the transmission of muscle force through the series of elastic component (SEC), changes in the ability of the action potential

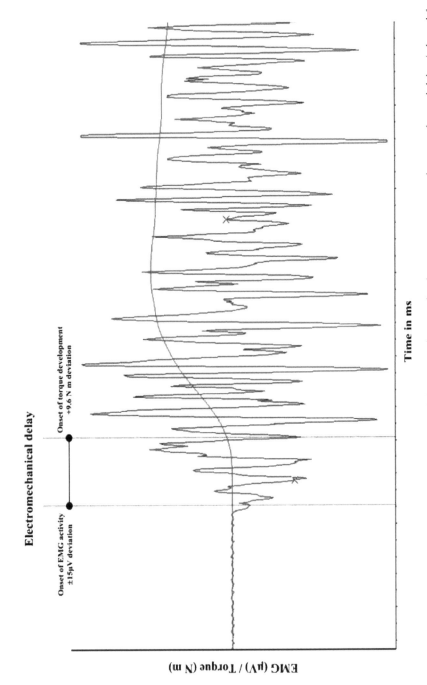

Figure 11.2 Determination of the onset of EMG activity and the onset of torque development to measure electromechanical delay (adapted from Zhou *et al.* (1995)).

to propogate and the properties of the excitation contraction coupling process (Howatson 2010). Recent adult data comparing sex differences in EMD of the hamstrings during eccentric muscle actions showed no significant sex difference (Troy Blackburn *et al.* 2009). However, conflicting data are available for other muscle groups (Bell and Jacobs 1986). Very few studies appear to have examined EMD during childhood (Cohen *et al.* 2010; Falk *et al.* 2009; Grosset *et al.* 2010; Zhou *et al.* 1995) and there appears to be only one longitudinal study.

The work of Grosset *et al.* (2010) focused on ankle stiffness and EMD of the triceps surae in normal and diseased children. They reported a greater EMD in children suffering with Legg-Calvé-Perthes disease (a hip disorder) compared to healthy controls, albeit only in six children. Cohen *et al.* (2010) also found significantly greater EMD in young children (9–12 years) compared to adults (65 ms vs. 57 ms) for knee and elbow extension and flexion during isometric actions. However, they did not show any significant difference between the endurance trained and untrained children, suggesting that level of training status did not have an effect on EMD in 9–12-year-olds.

Falk *et al.* (2009) reported a significantly longer EMD in prepubertal boys compared with adult males and Zhou *et al.* (1995) found significantly longer EMD values in 8–12-year-olds (61 ms for boys; 58 ms for girls) compared with 13–16-year-olds (44 ms for boys; 47 ms for girls) and adults (40 ms for males; 46 ms for females). This longer EMD in children may be as a result of differences in muscle composition. However, current limited evidence suggests that differences in muscle composition are not sufficient to account for the child–adult differences. Therefore, differences in muscle activation, such as excitation–contraction coupling and muscle-fibre conduction velocity have been implicated in this longer EMD. This lower rate of force development may therefore reduce muscle-tendinous stiffness and increase the potential for injury in children. However, Grosset *et al.* (2007) reported greater electrically stimulated EMD for the triceps surae in 7-year-old compared with 11-year-old prepubertal children, indicating a potential increase in muscle-tendinous stiffness with age independent of maturation. However, this hypothesis requires further investigation, employing longitudinal studies throughout childhood and including female participants.

A number of adult studies have suggested that males demonstrate a shorter EMD compared to females and have attributed this to greater musculotendinous stiffness in males (Troy Blackburn *et al.* 2009; Zhou *et al.* 1995). Conflicting data from our laboratory on adults have indicated no sex difference in EMD of the hamstrings during eccentric muscle actions over a range of velocities (De Ste Croix *et al.* 2011). Only one study appears to have explored sex differences in EMD of the knee extensors from isometric actions (Zhou *et al.* 1995) during childhood, and no sex differences were reported in either the 8–12-year-old or 13–16-year-old age groups. Whether EMD accounts for the greater relative risk of non-contact ACL injury in girls is unclear as there are no studies that have examined sex differences in EMD in children during eccentric muscle actions. It is surprising that there is only one study on the EMD of young girls given that they are the most at-risk group for non-contact ACL injury. It remains to be identified how EMD may change during childhood, particularly for the knee flexors and during eccentric actions. Further study is needed to explore the age- and sex-related changes in EMD, linked to the relative risk of injury.

Effects of fatigue on dynamic stability

After a fatiguing exercise bout, biomechanical and neuromuscular factors such as muscle activation patterns, coactivation, kinematics and kinetics and stiffness properties are altered (Padua *et al.* 2006). Adult studies have reported that submaximal fatigue not only increases anterior tibial translation but that this is accompanied by significantly longer latency of the hamstring muscles, subsequently decreasing joint stability (Melnyk and Gollhofer 2007). In a fatigued state, adults also use antagonist inhibition strategies by reducing hamstring activation (Padua *et al.* 2006). The work of Padua *et al.* (2006) demonstrated greater coactivation ratios in females compared to males in a fatigued state and also suggested that adults move to an ankle-dominant strategy compared to knee strategy to protect the knee on landing. It is well recognized in the available literature that injury to the ACL appears to be more prevalent in the latter stages of sporting performance and most likely when muscle fatigue is present (Small *et al.* 2010). For example a recent study has indicated that the H/Q_{func} ratio significantly decreases at the end of each half of a soccer match using a simulated soccer specific fatiguing task (Small *et al.* 2010). There appear to be no comparable data available during childhood but if fatigue has a similar effect on children, then the ability to resist fatigue and maintain joint stability should form a major part of prevention programmes. Work by Kawakami *et al.* (1993) suggested that at least for the elbow flexors, concentric and eccentric torque production decreases at a similar rate with advancing muscular fatigue in 13-year-old boys. These limited data would suggest that the H/Q_{func} ratio would remain similar in the fatigued and non-fatigued state in children. However, this suggestion requires further investigation, especially on the knee joint and using female participants across the age range.

Available data are reasonably consistent, indicating that children fatigue at a slower rate than adults during either a single or repeated bout of high-intensity exercise (De Ste Croix *et al.* 2009; Ratel *et al.* 2006). The ability of children to maintain force output during repeated bouts of high-intensity exercise may be related to lower levels of fatigue, which in turn may be reflective of different muscle characteristics compared to adults. Proposed mechanisms for the greater fatigue resistance in children compared with adults include: (a) children use oxidative pathways quicker than adults and therefore lead to a lower accumulation of by-products; (b) children's lower ability to activate their type II muscle fibres; and (c) possible faster phosphocreatine resynthesis, improved acid-base regulation and faster removal of metabolic by-products. Similar mechanisms have been proposed for a possible sex difference in fatigue resistance, with females showing less fatigue than males during repeated high-intensity tasks (Hicks *et al.* 2001). One of the few studies to examine this sex difference in young children and adults found no sex differences in either age group (De Ste Croix *et al.* 2009).

However, most of these data are from limited sources that do not adequately cover the period of maturation and have used varying methods to induce fatigue. Importantly, no available studies have examined the effect of eccentric fatigue on eccentric torque production during childhood or between the sexes.

Data on the effects of muscle action specific fatigue on either the H/Q_{func} ratio or EMD are sparse. It is well recognized in adults that fatigue affects eccentric and concentric actions differently and that generally, eccentric actions are more fatigue-resistant than concentric actions (Roig *et al.* 2009). Therefore, in a fatigued state the H/Q_{func} ratio should increase and the knee should be more stable. However, no studies appear to have reported angle-specific H/Q_{func} ratio after fatiguing tasks. Interestingly, a recent study on adults suggests that heavy-intensity aerobic training reduces the H/Q_{func} ratio but does not affect the H/Q_{con} ratio which points to greater fatigue during eccentric muscle actions (Oliveira *et al.* 2009).

Howatson (2010) examined the chronic effects of eccentric fatigue and muscle damage on the biceps brachii in male adults and reported significantly greater EMD up to 96 hours following the exercise bout. This finding was despite an apparent return of muscle function and suggests that caution should be taken if a task requiring a fast reaction time or fast generation of high forces is needed following this type of exercise. Zhou *et al.* (1996) also demonstrated a significant increase in EMD following four bouts of 30s all-out cycling exercise in adult males. All of the available adult studies seem to show a significant increase in EMD after fatiguing trails which would predispose the knee to greater injury risk. The mechanisms involved in the increase in EMD after fatigue could be due to the deterioration in muscle conductive, contractile or elastic properties and requires further study. Unfortunately, there are no current data available that permit us to explore the age- and sex-associated changes in EMD after fatiguing exercise in children. These data are urgently needed to elucidate the relative risk of non-contact knee injury during childhood when the child is in a fatigued state.

Clinical and practical applications

Eccentric hamstring training to reduce relative risk of injury

Biomechanical analyses using kinematic and ground-reaction force data in adults seem to confirm that hamstring injury is most likely to occur during the terminal swing phase as a consequence of an eccentric muscle action (Schache *et al.* 2009). These data reinforce the importance of hamstring training and conditioning in the reduction in the relative risk of injury. A number of studies have attempted to examine the effectiveness of hamstring training and plyometric training on reducing the relative risk of ACL injury (Myer *et al.* 2006a, 2006b, 2008). Evidence from adults has indicated that individuals who routinely perform sport activities under stressful conditions can protect against the rotational stresses at the knee joint by increasing eccentric hamstring actions, as the limb is striking the ground (Bonci 1999). For example, an injury-prevention programme by Lim *et al.* (2009) on young girls indicated positive changes in biomechanical factors associated with non-contact ACL injury (e.g. knee internal rotation angle; knee valgus moment; knee flexion angle). A systematic review by Roig *et al.* (2009) has pointed to effective improvements in eccentric strength following eccentric training at high intensities in adults. Data from children are limited and the large majority of studies have focused on high-school girls, probably due to the available incidence data that suggest that this is the most 'at-risk' group for non-contact ACL injury (Hewett *et al.* 1999; McLeod *et al.* 2009; Myer *et al.* 2006a, 2006b, 2008). These studies are relatively consistent, demonstrating a reduction in the relative risk of injury in young girls irrespective of the type of neuromuscular training employed.

It is somewhat surprising that, despite the larger absolute incidence rate in boys, there are very few studies that have looked at neuromuscular training of the hamstrings during eccentric exercise in boys (Emery and Meeuwisse 2010; Junge *et al.* 2002; Nelson and Bandy 2004). The proposed components of prevention programmes have been subject to much debate and surround developing strength ratios versus improving landing mechanics. However, most programmes currently aim to improve both movement strategies and neuromuscular/muscular performance, and routinely include strengthening, stretching, plyometric and balance components. It is beyond the scope of this chapter to explore these different neuromuscular training programmes. However, most importantly, they all contain some eccentric conditioning of the hamstring muscles, which emphasizes the importance of paying particular attention to eccentric hamstring training with the goal of injury prevention. For further information regarding training programmes the reader is directed towards the text of Hewett *et al.* (2007) on understanding and preventing non-contact ACL injuries.

Due to the individual timing and tempo of both muscle growth and neuromuscular development during childhood, it has been proposed that unequal development of these systems may result in neuromuscular imbalances that make the individual more susceptible to injury (McLeod *et al.* 2009). Therefore, identifying when neuromuscular training should start as an intervention to reduce possible neuromuscular imbalances is difficult due to the individual timing and tempo of growth and development of these systems. It would appear that training should be targeted to a period when rapid musculoskeletal growth is occurring and/or when limb lengths change decreasing balance and co-ordination. All of the available neuromuscular training studies during childhood have focused on older children (14–17-year-olds) except for the cluster-randomized controlled trial of Emery and Meeuwisse (2010), which looked at under 13- to under 18-year-old age groups. This current limitation with the extant literature makes the description of the effectiveness of neuromuscular training during childhood difficult.

Despite this limitation, most of the available data demonstrate the effectiveness of a range of neuromuscular training programmes (plyometrics, balance training, dynamic stabilization training, eccentric training, trunk and hip training, static stretching) on a range of biomechanical, muscular, neuromuscular and performance outcomes (e.g. hip adduction angle, maximum ankle eversion angle, isokinetic torque production, centre of pressure, vertical ground-reaction force, vertical jump height). Current data have provided us with a sense of which mechanisms may change after training and reduce the relative risk of injury during childhood. However, it is perhaps more important to explore if such programmes reduce injury rates. A systematic review by Hewett *et al.* (2005) indicated that all but one study that met the inclusion criteria have demonstrated significant reductions in injury rates following intervention programmes to reduce injury in females ranging from 36 to 87 per cent. Although these studies clearly indicate a role of neuromuscular training to reduce injury risk, it must be noted that these studies pertain to females and generally to the 14–16-year-old age group. Whether such programmes have the same protective effects in boys and throughout childhood remain to be identified.

There are a number of adult studies that have demonstrated a significant increase in the H/Q_{con} ratio after a prevention programme (Hewett *et al.* 1996; Wilkerson *et al.* 2004). One recent adult study also showed improvements in the H/Q_{func} ratio after six weeks of eccentric hamstring training in 20-year-old females (0.96 pre-training to 1.08 post-training).

There appear to be no studies that have investigated changes in the H/Q_{func} ratio in children after a training programme. However, one recent study has indicated that six weeks of neuromuscular training significantly increased balance and proprioceptive capabilities of 15–16-year-old female basketball players (McLeod *et al.* 2009). Myer *et al.* (2008) examined the effectiveness of a trunk- and hip-focused neuromuscular training programme on knee and hip isokinetic strength in 15-year-old girls. After a ten-week training programme, the experimental group improved their hip strength by 16 per cent, whereas the control group reported no significant increases in strength. The authors, however, do acknowledge the limitations of their findings as the strength measure was determined from concentric rather than eccentric actions and during open rather than closed chain actions. Further work by this group have also explored differences in the effects of plyometric versus dynamic stabilization and balance training on lower extremity biomechanics, power and balance (Myer *et al.* 2006a, 2006b). In the biomechanical 3D motion analysis, they reported that both types of training reduced lower extremity valgus measurements with plyometric exercise predominately influencing sagittal plane kinematics during a drop vertical jump and balance training during single-legged drop landing. Nelson and Bandy (2004) also reported improved hamstring flexibility in 16-year-old high-school boys after a six-week eccentric training programme, having linked the proposed benefits of enhanced flexibility to the reduction of injury. Emery and Meeuwisse (2010) recently reported that a soccer-specific neuromuscular training programme for 13–18-year-old males and females was protective of all injuries and reduced the injury rate to 2.08 per 1,000 player hours in the training group compared with 3.35 per 1,000 player hours in the control group. Therefore, the majority of available data clearly indicates that neuromuscular training appears to reduce the relative risk of injury and enhance physical performance in late adolescence in females.

However, little is known regarding the same protective effects and performance enhancement in boys and throughout the period of childhood. It is well recognized that selection into elite sport is biased towards early maturing children and it may be pertinent to explore these effects in early maturers and in younger age groups. Well-designed longitudinal studies are required to further elucidate answers to the effectiveness of such programmes throughout childhood.

Clinical implications for reduced stability of the knee

Neuromuscular disorders during childhood may predispose the clinically susceptible child to an increased injury risk. For example, Grosset *et al.* (2010) demonstrated a longer EMD and greater ankle stiffness in immobilized children due to hip osteochondritis. Developmental disorders such as cerebral palsy (CP) and developmental co-ordination disorder (DCD) are discussed in more detail in Chapters 12 and 13 of this book. Within the context of this chapter, it is worth mentioning one study, which demonstrated high H/Q_{func} ratios in 11-year-old children with CP (Damiano *et al.* 2001). The authors concluded that, despite a reduced capacity for both concentric and eccentric muscle torque, children with CP demonstrate a bias towards greater eccentric torque production than concentric torque production. Recent research has also examined the effectiveness of eccentric conditioning of the hamstring muscles in improving neuromuscular performance in children and adolescents with CP. Reid *et al.* (2010) reported that, after a six-week resistance training programme focusing on eccentric actions, children with CP exhibit reduced co-contraction of the knee, improving net torque production. This in turn will have significant effects on H/Q_{func} ratios and subsequent relative risk of injury to limb joints for children with clinical conditions.

Key points

- Relative risk of knee injury is greater in children than adults, and dynamic knee stability is reduced in girls compared with boys.
- Scientific results relating to age-related changes in the H/Q_{func} ratio are conflicting, and this may be prescribed in part to training status in children. Limited data suggest that there are no sex differences in the H/Q_{func} ratio during childhood.
- Young children appear to have a longer EMD than adults, but there are no sex differences in EMD during childhood. EMD would appear to decrease throughout puberty but longitudinal study is required to support this view.
- Children are more resistant to muscle fatigue than adults, but how this may affect the H/Q_{func} ratio remains to be investigated.
- Muscle fatigue negatively affects EMD in adults, but data on children are sparse and the influence of fatigue on neuromuscular performance in children remains to be investigated.
- Neuromuscular training programmes in children appear to improve a number

of muscular, neuromuscular and biomechanical parameters related to injury prevention and thus reduce the relative risk of injury in children.

References

Aagaard, P., Simonsen, E.B., Magnusson, S.P., Larsson, B. and Dyhre-Poulsen, P. (1998) 'A new concept for isokinetic hamstring:quadriceps muscle strength ratio', *American Journal of Sports Medicine*, 26: 231–7.

Aagaard, P., Simonsen, E.B., Trolle, M., Bangsbo, J. and Klausen, K. (1995) 'Isokinetic hamstring/quadriceps strength ratio: influence from joint angular velocity, gravity correction and contraction mode', *Acta Physiologica Scandinavia*, 154: 421–7.

Barber-Westin, S.D., Galloway, M., Noyes, F.R., Corbett, G., Walsh, C. (2005) 'Assessment of lower limb neuromuscular control in prepubescent athletes', *American Journal of Sports Medicine*, 33: 1853–60.

Bell, D.G. and Jacobs, I. (1986) 'Electro-mechanical response times and rate of force development in males and females', *Medicine and Science in Sports and Exercise*, 18: 31–6.

Blimkie, C.J.R. and Macauley, D. (2001) 'Muscle strength', in N. Armstrong and W. Van-Mechelen (eds) *Pediatric Exercise Science and Medicine*, Oxford: Oxford University Press, pp. 23–36.

Bonci, C. (1999) 'Assessment and evaluation of predisposing factors to anterior cruciate ligament injury', *Journal of Athletic Training*, 34: 155–64.

Brooks, G. and Fahey, F. (1987) *Fundamentals of Human Performance*, New York: Macmillan.

Brophy, R., Silvers, H., Gonzales, T. and Mandelbaum, B. (2010) 'Gender influences: the role of leg dominance in ACL injury among soccer players', *British Journal of Sports Medicine*, 44: 694–7.

Cohen, R., Mitchell, C., Dotan, R., Gabriel, D., Klentrou, P. and Falk, B. (2010) 'Do neuromuscular adaptations occur in endurance-trained boys and men?', *Applied Physiology, Nutrition and Metabolism*, 35: 471–9.

Collard, D., Chinapaw, M., van Mechelen, W. and Verhagen, E. (2009) 'Design of the iPlay study: systematic development of a physical activity injury prevention programme for primary school children', *Sports Medicine*, 39: 889–902.

Coombs, R. and Garbutt, G. (2002) 'Developments in the use of the hamstring/quadriceps ratio for the assessment of muscle balance', *Journal of Sports Science and Medicine*, 1: 56–62.

Damiano, D.L., Martellotta, T.L., Quinlivan, J.M. and Abel, M.F. (2001) 'Deficits in eccentric versus concentric torque in children with spastic cerebral palsy', *Medicine and Science in Sports and Exercise*, 33: 117–22.

Dauty, M., Potiron-Josse, M. and Rochcongar, P. (2003) 'Identification of previous hamstring injury by isokinetic concentric and eccentric torque measurement in elite soccer players', *Isokinetics and Exercise Science*, 11: 134–44.

Decker, M.J., Torry, M.R., Wyland, D.J., Sterett, W.I. and Steadman, J.R. (2003) 'Gender differences in lower extremity kinematics, kinetics and energy absorption during landing', *Clinical Biomechanics*, 18: 662–9.

Deighan, M.A., De Ste Croix, M.B.A. and Armstrong, N. (2003) 'Reliability of isokinetic concentric and eccentric knee and elbow extension and flexion in 9/10 year old boys', *Isokinetics and Exercise Science*, 11: 109–15.

de Loës, M., Dahlstedt, L.J. and Thomée, R.A. (2000) '7-year study on risks and costs of knee injuries in male and female youth participants in 12 sports', *Scandanavian Journal of Medicine and Science in Sports*, 10: 90–7.

De Ste Croix, M.B.A., Deighan, M.A. and Armstrong, N. (2003) 'Assessment and interpretation of isokinetic muscle strength during growth and maturation', *Sports Medicine*, 33: 727–43.

De Ste Croix, M.B.A., Deighan, M.A. and Armstrong, N. (2007) 'Functional eccentric-concentric ratio of knee extensors and flexors in pre-pubertal children, teenagers and adults', *International Journal of Sports Medicine*, 28: 768–72.

De Ste Croix, M.B.A., Deighan, M.A., Ratel, S. and Armstrong, N. (2009) 'Age- and sex-associated differences in isokinetic knee muscle endurance between young children and adults', *Applied Physiology, Nutrition and Metabolism*, 34: 725–31.

De Ste Croix, M.B.A., Elnagar, Y.O., James, D.V.B. and Iga, J. (2011) 'Sex differences in electromechanical delay of the hamstrings during muscle actions' (under review).

Emery, C. and Meeuwisse, W. (2010) 'The effectiveness of a neuromuscular prevention strategy to reduce injuries in youth soccer: a cluster-randomised controlled trial', *British Journal of Sports Medicine*, 44: 555–62.

Enoka, R.M. (2002) *Neuromechanics of Human Movement*, Champaign, IL: Human Kinetics.

Falk, B., Usselman, C., Dotan, R., Brunton, L., Klentrou, P., Shaw, J. and Gabriel, D. (2009) 'Child-adult differences in muscle strength and activation pattern during isometric elbow flexion and extension', *Applied Physiology, Nutrition and Metabolism*, 34: 609–15.

Fecteau, D., Gravel, J., D'Angelo, A., Martin, E. and Amre, D. (2008) 'The effect of concentrating periods of physical activity on the risk of injury in organized sports in a pediatric population', *Clinical Journal of Sport Medicine*, 18: 410–15.

Forbes, H., Sutcliffe, S., Lovell, A., McNaughton, L. and Siegler, J. (2009a) 'Isokinetic thigh muscle ratios in youth football: effects of age and dominance', *International Journal of Sports Medicine*, 30: 602–6.

Forbes, H., Bullers, A., Lovell, A., McNaughton, L., Polman, R. and Siegler, J. (2009b) 'Relative torque profiles of elite male youth footballers: effects of age and pubertal development', *International Journal of Sports Medicine*, 30: 592–7.

Ford, K.R., Myer, G.D. and Hewett, T.E. (2003) 'Valgus knee motion during landing in high school female and male basketball players', *Medicine and Science in Sports and Exercise*, 35: 1745–50.

Funasaki, H. (2009) 'Issues related to sports injuries during growth periods in children and teens', *Advances in Exercise and Sports Physiology*, 15: 43.

Gallagher, S.S., Finison, K., Guyer, B. and Goodenough, S. (1984) 'The incidence of injuries among 87,000 Massachusetts children and adolescents: results of the 1980–81 Statewide Childhood Injury Prevention Program Surveillance System', *American Journal of Public Health*, 74: 1340–7.

Gerodimos, V., Mandou, V., Zafeiridis, A., Ioakimidis, P., Stavropoulos, N. and Kellis, S. (2003) 'Isokinetic peak torque and hamstring/quadriceps ratios in young basketball players', *Journal of Sports Medicine and Physical Fitness*, 43: 444–52.

Goldblatt, J. and Richmond, J. (2003) 'Anatomy and biomechanics of the knee', *Operative Techniques in Sports Medicine*, 11: 172–86.

Griffin, L.Y., Albohm, M.J., Arendt, E.A., Bahr, R., Beynnon, B.D., DeMaio, M. *et al.* (2006) 'Understanding and preventing noncontact anterior cruciate ligament injuries: a review of the Hunt Valley II meeting, January 2005', *American Journal of Sports Medicine*, 34: 1513–32.

Grosset, J-F., Mora, I., Lambertz, D. and Perot, C. (2007) 'Changes in stretch reflexes and muscle stiffness with age in prepubescent children', *Journal of Applied Physiology*, 102: 2352–60.

Grosset, J-F., Lapole, T., Mora, I., Verhaeghe, M., Doutrellot, P-L. and Perot, C. (2010) 'Follow-up of ankle stiffness and electromechanical delay in immobilized children: three case studies', *Journal of Electromyography and Kinesiology*, 20: 642–7.

Hamstra-Wright, K.L., Swanik, C.B., Sitler, M.R., Swanik, K.A., Ferber, R., Ridenour, M. and Huzel, K.C. (2006) 'Gender comparisons of dynamic restraint and motor skill in children', *Clinical Journal of Sport Medicine*, 16: 56–62.

Hewett, T.E., Myer, G.D. and Ford, K.R. (2001) 'Prevention of anterior cruciate ligament injuries', *Current Women's Health Report*, 1: 218–24.

Hewett, T.E., Shultz, S.J. and Griffin, L.Y. (2007) *Understanding and Preventing Noncontact ACL Injuries*, Champaign, IL: Human Kinetics.

Hewett, T.E., Lindenfeld, T.N., Riccobene, J.V. and Noyes, F.R. (1999) 'The effect of neuromuscular training on the incidence of knee injury in female athletes: a prospective study', *American Journal of Sports Medicine*, 27: 699–706.

Hewett, T.E., Stroupe, A.L., Nance, T.A. and Noyes, F.R. (1996) 'Plyometric training in female athletes: decreased impact forces and increased hamstring torques', *American Journal of Sports Medicine*, 24: 765–73.

Hewett, T.E., Zazulak, B.T., Myer, G.D. and Ford, K.R. (2005) 'Review of electromyographic activation levels, timing differences, and increased anterior cruciate ligament injury incidence in female athletes', *British Journal of Sports Medicine*, 39: 347–50.

Hicks, A.L., Kent-Braun, J. and Ditor, D.S. (2001) 'Sex differences in human skeletal muscle fatigue', *Exercise and Sport Sciences Reviews*, 29: 109–12.

Holm, I. and Vøllestad, N. (2008) 'Significant effect of gender on hamstring-to-quadriceps strength ratio and static balance in prepubescent children from 7 to 12 years of age', *American Journal of Sports Medicine*, 36: 2007–14.

Howatson, G. (2010) 'The impact of damaging exercise on electromechanical delay in biceps brachii', *Journal of Electromyography and Kinesiology*, 20: 477–82.

Huston, L.J. (2007) 'Clinical biomechanical studies in ACL injury risk factors', in T.E. Hewitt, S.J. Schultz and L.Y. Griffin (eds) *Understanding and Preventing Noncontact ACL Injuries*, Champaign, IL: Human Kinetics, pp. 141–53.

Huston, L.J. and Wojtys, E.M. (1996) 'Neuromuscular performance characteristics in elite female athletes', *American Journal of Sports Medicine*, 24: 427–36.

Huston, L.J., Vibert, B., Ashton-Miller, J.A. and Wojtys, E.M. (2001) 'Gender differences in knee angle when landing from a drop-jump', *American Journal of Knee Surgery*, 14: 215–19.

Iga, J., George, K., Lees, A. and Reilly, T. (2006) 'Reliability of assessing indices of isokinetic leg strength in pubertal soccer players', *Pediatric Exercise Science*, 18: 436–46.

Iga, J., George, K., Lees, A. and Reilly, T. (2009) 'Cross-sectional investigation of indices of isokinetic leg strength in youth soccer players and untrained individuals', *Scandanavian Journal of Medicine and Science in Sports*, 19: 714–20.

Junge, A., Roesch, D., Peterson, L., Graf-Baumann, T. and Dvorak, J. (2002) 'Prevention of soccer injuries: a prospective intervention study in youth amateur players', *American Journal of Sports Medicine*, 30: 652–9.

Kaminski, T.W., Wabbersen, C.V. and Murphy, R.M. (1998) 'Concentric versus enhanced eccentric hamstring strength training: clinical implications', *Journal of Athletic Training*, 33: 216–21.

Kawakami, Y., Kanehisa, H., Ikegawa, S. and Fukunaga, T. (1993) 'Concentric and eccentric muscle strength before, during and after fatigue in 13 year-old boys', *European Journal of Applied Physiology and Occupational Physiology*, 67: 121–4.

Kellis, E. and Katis, A. (2007) 'Quantification of functional knee flexor to extensor moment ratio using isokinetics and electromyography', *Journal of Athletic Training*, 42: 477–86.

Komi, P. and Bosco, C. (1978) 'Utilization of stored elastic energy in leg extensor muscles by men and women', *Medicine and Science in Sports*, 10: 261–5.

Lazaridis, S., Bassa, E., Patikas, D., Giakas, G., Gollhofer, A. and Kotzamanidis, C. (2010) 'Neuromuscular differences between prepubescent boys and adult men during drop jump', *European Journal of Applied Physiology*, 110: 67–75.

Lim, B-O., Lee, Y., Kim, J., An, K., Yoo, J. and Kwon, Y. (2009) 'Effects of sports injury prevention training on the biomechanical risk factors of anterior cruciate ligament injury in high school female basketball players', *American Journal of Sports Medicine*, 37: 1728–35.

Lloyd, D., Buchanan, T. and Besier, T. (2005) 'Neuromuscular biomechanical modelling to understand knee ligament loading', *Medicine and Science in Sports and Exercise*, 37: 1939–47.

McLeod, T., Armstrong, T., Miller, M. and Sauers, J. (2009) 'Balance improvements in female high school basketball players after a 6-week neuromuscular training program', *Journal of Sport Rehabilitation*, 18: 465–81.

Melnyk, M. and Gollhofer, A. (2007) 'Submaximal fatigue of the hamstrings impairs specific reflex components and knee stability', *Knee Surgery Sports Traumatology Arthroscopy*, 15: 525–32.

Melnyk, M., Luebken, F.V., Hartmann, J., Claes, L., Gollhofer, A. and Friemert, B. (2008) 'Effects of age on neuromuscular knee joint control', *European Journal of Applied Physiology*, 103: 523–8.

Myer, G., Ford, K. and Hewett, T. (2004) 'Rationale and clinical techniques for anterior cruciate ligament injury prevention among female athletes', *Journal of Athletic Training*, 39: 352–64.

Myer, G., Brent, J., Ford, K. and Hewett, T. (2008) 'A pilot study to determine the effect of trunk and hip focused neuromuscular training on hip and knee isokinetic strength', *British Journal of Sports Medicine*, 42: 614–19.

Myer, G., Ford, K., Brent, J. and Hewett, T. (2006a) 'The effects of plyometric versus dynamic stabilization and balance training on power, balance, and landing force in female athletes', *Journal of Strength and Conditioning Research*, 20: 345–53.

Myer, G., Ford, K., McLean, S. and Hewett, T. (2006b) 'The effects of plyometric versus dynamic stabilization and balance training on lower extremity biomechanics', *American Journal of Sports Medicine*, 34: 445–55.

Myer, G., Ford, K., Khoury, J., Succop, P. and Hewett, T. (2010) 'Clinical correlates to laboratory measures for use in non-contact anterior cruciate ligament injury risk prediction algorithm', *Clinical Biomechanics*, 25: 693–9.

Myers, C. and Hawkins, D. (2010) 'Alterations to movement mechanics can greatly reduce anterior cruciate ligament loading without reducing performance', *Journal of Biomechanics*, 43: 2657–65.

Nelson, R. and Bandy, W. (2004) 'Eccentric training and static stretching improve hamstring flexibility of high school males', *Journal of Athletic Training*, 39: 254–8.

Oliveira, A. de S.C., Caputo, F., Gonçalves, M. and Denadai, B.S. (2009) 'Heavy-intensity aerobic exercise affects the isokinetic torque and functional but not conventional hamstrings:quadriceps ratios', *Journal of Electromyography and Kinesiology*, 19: 1079–85.

Padua, D., Arnold, B., Perrin, D., Gansneder, B., Carcia, C. and Granata, K. (2006) 'Fatigue, vertical leg stiffness and stiffness control strategies in males and females', *Journal of Athletic Training*, 41: 294–304.

Palmieri-Smith, R.M. and Thomas, A.C. (2009) 'A neuromuscular mechanism of post-traumatic osteoarthritis associated with ACL injury', *Sport and Exercise Sciences Reviews*, 37: 147–54.

Ratel, S., Duché, P. and Williams, C.A. (2006) 'Muscle fatigue during high intensity exercise in children', *Sports Medicine*, 36: 1031–66.

Reid, S., Hamer, P., Alderson, J. and Lloyd D. (2010) 'Neuromuscular adaptations to eccentric strength training in children and adolescents with cerebral palsy', *Developmental Medicine and Child Neurology*, 52: 358–63.

Riemann, B. and Lephart, S. (2002) 'The sensorimotor system. Part I. The physiologic basis of functional joint stability', *Journal of Athletic Training*, 37: 71–9.

Roig, M., O'Brien, K., Kirk, G., Murray, R., McKinnon, P., Shadgan, B. and Reid, W.D. (2009) 'The effects of eccentric versus concentric resistance training on muscle strength and mass in healthy adults: a systematic review with meta-analysis', *British Journal of Sports Medicine*, 43: 556–69.

Russell, P.J., Croce, R.V., Swartz, E.E., Decoster, L.C. (2007) 'Knee-muscle activation during landings: developmental and gender comparisons', *Medicine and Science in Sports and Exercise*, 39: 159–69.

Schache, A., Wrigley, T., Baker, R. and Pandy, M. (2009) 'Biomechanical response to hamstring muscle strain injury', *Gait & Posture*, 29: 332–8.

Seger, J. and Thorstenson, A. (2000) 'Muscle strength and electromyogram in boys and girls followed through puberty', *European Journal of Applied Physiology*, 81: 54–61.

Senter, C. and Hame, S.L. (2006) 'Biomechanical analysis of tibial torque and knee flexion angle: implications for understanding knee injury', *Sports Medicine*, 36: 635–42.

Shea, K.G., Pfeiffer, R., Wang, J.H., Curtin, M. and Apel, P.J. (2004) 'Anterior cruciate ligament injury in pediatric and adolescent soccer players: an analysis of insurance data', *Journal of Pediatric Orthopedics*, 24: 623–8.

Shultz, S.J. (2007) 'Hormonal influences on ligament biology', in T.E. Hewett, S.J. Shultz and L.Y. Griffin (eds) *Understanding and Preventing Noncontact ACL Injuries*, Champaign, IL: Human Kinetics, pp. 219–38.

Shultz, S. and Perrin, D. (1999) 'Using surface electromyography to assess sex differences in neuromuscular response characteristics', *Journal of Athletic Training*, 34: 165–76.

Small, K., McNaughton, L., Greig, M. and Lovell, R. (2010) 'The effects of multidirectional soccer-specific fatigue on markers of hamstring injury risk', *Journal of Science & Medicine in Sport*, 13: 120–6.

Troy Blackburn, J., Bell, D., Norcross, M., Hudson, J. and Engstrom, L. (2009) 'Comparison of hamstring neuromechanical properties between healthy males and females and the influence of musculotendinous stiffness', *Journal of Electromyography and Kinesiology*, 19: e362–9.

Veigel, J. and Pleacher, M. (2008) 'Injury prevention in youth sports', *Current Sports Medicine Reports*, 7: 348–53.

Weltman, A., Janney, C., Rians, C.B., Strand, K., Berg, B., Tippitt, S., Wise, J., Cahill, B.R. and Katch, F.I. (1986) 'The effects of hydraulic resistance strength training in pre-pubertal males', *Medicine and Science in Sports and Exercise*, 18: 629–38.

Westing, S.H. and Seger, J.Y. (1989) 'Eccentric and concentric torque-velocity characteristics, torque output comparisons, and gravity effect torque corrections for the quadriceps and hamstring muscles in females', *International Journal of Sports Medicine*, 10: 175–80.

Wikstrom, E.A., Tillman, M.D., Chmielewski, T.L. and Borsa, P.A. (2006) 'Measurement and evaluation of dynamic joint stability of the knee and ankle after injury', *Sports Medicine*, 36: 393–411.

Wilkerson, G.B., Colston, M.A., Short, N.I., Neal, K.L., Hoewischer, P.E. and Pixley, J.J. (2004) 'Neuromuscular changes in female collegiate athletes resulting from a plyometric jump-training program', *Journal of Athletic Training*, 39: 17–23.

Yeung, S.S., Suen, A.M.Y. and Yeung, E.W. (2009) 'A prospective cohort study of hamstring injuries in competitive sprinters: preseason muscle imbalance as a possible risk factor', *British Journal of Sports Medicine*, 43: 589–95.

Yu, B., Kirkendall, D., Taft, T. and Garrett, W. (2002) 'Lower extremity motor control-related and other risk factors for non-contact anterior cruciate ligament injuries', in J. Beaty (ed.) *Instructional Course Lectures*, Rosemont, IL: AAOS, pp. 315–24.

Yu, B., McClure, S.B., Onate, J.A., Guskiewicz, K.M., Kirkendall, D.T. and Garrett, W.E. (2005) 'Age and gender effects on lower extremity kinematics of youth soccer players in a stop-jump task', *American Journal of Sports Medicine*, 33: 1356–64.

Zhou, S., Lawson, D.L., Morrison, W.E. and Fairweather, I. (1995) 'Electromechanical delay of knee extensors: the normal range and the effects of age and gender', *Journal of Human Movement Studies*, 28: 127–46.

Zhou, S., McKenna, M.J., Lawson, D.L., Morrison, W.E. and Fairweather, I. (1996) 'Effects of fatigue and sprint training on electromechanical delay of knee extensor muscles', *European Journal of Applied Physiology and Occupational Physiology*, 72: 410–16.

12 Developmental Coordination Disorder
Biomechanical and neuromuscular considerations

Jill Whitall and Jane Clark

Introduction

Running, jumping, kicking and throwing are the hallmarks of childhood. Typically developing children engage in these gross motor activities early in their lives and many go on to acquire the complex motor skills required for sports and dance that are rooted in these early behaviours. However, there is a small portion of children for whom the coordination and control required for these gross motor activities is nearly impossible to achieve. For decades, these children were labelled 'clumsy', but today many of these children are diagnosed as having Developmental Coordination Disorder (DCD). While some children with DCD will have difficulties only with fine motor (primarily handwriting) skills, the vast majority will also have motor difficulties in locomotor and postural tasks. In this chapter, we examine the biomechanical and neuromuscular aspects of motor skill development in children with DCD. We also refer the reader to Chapter 2, which includes a discussion of this disorder relative specifically to sensorimotor integration and development.

Basic theoretical concepts

What is Developmental Coordination Disorder?

These criteria come from the *Diagnostic Statistical Manual IV-TR*, often referred to by its initials as the DSM-IV-TR (APA 2000). In the DSM-IV-TR, the criteria for DCD include:

A   Performance in daily activities that require motor coordination to be substantially below that expected given the person's chronological age and intelligence.
B   The disturbance in performance significantly interferes with academic achievement or activities of daily living.

C   The disturbance is not due to a general medical condition (e.g. cerebral palsy or muscular dystrophy) and does not meet criteria for pervasive developmental disorder.

D   If mental retardation is present, the motor difficulties are in excess of those usually associated with it.

The DSM-IV-TR puts the prevalence of DCD at 6 per cent of the population, but other estimates vary widely: from 2 per cent in the UK (Lingam *et al.* 2009) to as high as 19 per cent in Greece (Tsiotra *et al.* 2006), making it one of the most common developmental disorders of childhood.

Importance of DCD

The significance of having DCD is high because these children are at risk for low academic performance (Dewey *et al.* 2002), low social esteem and popularity (Piek *et al.* 2000), anxiety/depression (Piek *et al.* 2007), low levels of physical fitness and activity (Cairney *et al.* in press; Faught *et al.* 2005; Schott *et al.* 2007) and increased long-term health risk factors (Chirico *et al.* 2011). The fact that many, if not the majority of these children, do not outgrow the movement impairments (Losse *et al.* 1991), means that their quality of daily living (Cousins and Smyth 2003; Missiuna *et al.* 2008) and the cardiovascular risk factors for long-term health consequences are also compromised for adolescents and young adults (Cairney *et al.* 2007; Faught *et al.* 2005). The consequences of having DCD for older adults have thus far not been investigated. Children with DCD often have comorbidity with other learning disorders, in particular Attention Deficit Hyperactive Disorder (ADHD) with which there can be as much as 30–55 per cent overlap (Kopp *et al.* 2010; Watemberg *et al.* 2007). Finally, DCD is a very heterogeneous disorder with many presumed subtypes (Macnab *et al.* 2001; Visser 2003). It is not uncommon to find children with a diagnosis of DCD whose major problems are quite different, for example, difficulties in handwriting but not balance and vice versa. Consequently, understanding the underlying impairments of DCD is difficult.

Assessment and diagnosis of children with DCD

According to the DSM-IV-TR (APA 2000), the first criterion (A) for identifying a child with DCD is that their motor performance in those tasks that require motor coordination should be substantially below that of their age-matched peers. Parents, teachers and caregivers are often the first to notice the children's movement difficulties in the typical activities of their lives such as dressing, colouring or gross motor play activities. However, it is through formal evaluation with assessment instruments designed to identify children with movement difficulties, that the children's motor performance is determined to be age-appropriate or not. In addition to the children's motor competence, some verification of criterion B such as the Developmental Coordination Disorder Questionnaire (Wilson *et al.* 2000), teachers filling out the Movement Assessment Battery for Children (MABC) questionnaire or family history is required. Criteria C and D require physician assessment (preferably) or parental self-report.

Diagnosis of this disorder to satisfy criterion A is commonly pinned upon the performance of children on one of two tests: Movement Assessment Battery for Children, 2nd edition (MABC-2) (Henderson and Sugden 2007), or the Bruininks-Oseretsky Test of Motor Proficiency, 2nd edition (BOT-2) (Bruininks and Bruininks 2005). In Australia, the McCarron Assessment of Neuromuscular Development (MAND) is often used (McCarron 1982). However, most researchers in the field use the MABC. In 2001 and 2006, researchers agreed on test performance cut points that classified those at or below the 5th percentile as having DCD and those that are between the 6th and 15th percentile to be 'at risk' for DCD (Geuze *et al.* 2001). These standards were subsequently embraced in a 2006 consensus report by a group of researchers studying DCD (Sugden 2006).

In general, the motor performance assessed falls into the four broad categories:

A   Manual dexterity or fine motor coordination.
B   Static balance.
C   Dynamic balance or mobility (walking and jumping).
D   Hand–eye coordination with balls (catching and throwing).

These categories provide a framework for the behaviours we will examine in this chapter. However, in the interests of space, we will omit manual dexterity here because there are so many studies on fine motor skills, and particularly hand-writing and finger tapping, that we could not do them justice.

Biomechanical and neuromotor principles and variables relevant to DCD

The human body is comprised of many parts or segments that must move together in a coordinated fashion to move itself from one place to another, as well as to interact and act upon other objects in its environment. As the child in Figure 12.1 illustrates, lifting the body up requires that the segments of the leg (foot, shank and thigh) act in such a way as to lift the upper segments (HAT – or **h**ead, **a**rms and **t**runk). Moving the body can be measured in terms of *interlimb* coordination, defined as the manner in which pairs of limbs interact with each other. *Intralimb* coordination is defined as the manner in which the joints or segments of a single limb interact with each other (e.g. the thigh and shank). Coordination variables are often reported as either cross-correlations (particularly for continuous data) or phasing relationships that define whether the limbs or segments act simultaneously (0° or 0 per cent) in alternation (180° or 50 per cent) or something in between.

Figure 12.1 Child in the act of hopping.

Even a task that does not require the body to move, but rather to stand still, for example in a one-legged stand, can be very challenging. The complexity of aligning the body's multiple body segments over a small base of support (e.g. one foot positioned on a small wooden rocker plate) presents a significant challenge to the human motor system. Biomechanically, the problem in this task is to keep the centre of mass (COM) – the point in the body where the mass of the system is located over the base of support – in this task, one foot. Many motor batteries test this very simple skill by asking children to stand on one or two feet with their eyes open or closed. In field tests, the children are timed on how long they can hold the position without falling over. In the laboratory, there are many variables that can be used to capture the neuromotor control strategies for these balance tasks. For a good review of these variables, the reader is referred to a paper by Prieto *et al.* (1996).

When our movements involve another object (e.g. a ball, pen or another person) the movement problem becomes even more complex. Hand–eye or foot–eye coordinations are critical to the play-game skills of childhood and are foundational to the games and sports of our culture at all ages. Not only are the coordering of body segments and balance part of the challenge, but also now the actor has to position the segments such that they are coordered with the object that has become part of the task. For example, when catching a ball (Figure 12.2), all these components are involved as the child 'tracks' with his eyes the moving ball and times the hands to close at just the right moment to secure the ball. For this child, the ball was successfully caught, but the body is not in a well-balanced position for the catch. In addition to biomechanical variables, measures of spatial accuracy, timing (accuracy) and optimal force (accuracy) now become important to define the skilfulness of an individual.

Figure 12.2 Child in the act of catching a ball.

The examples in Figures 12.1 and 12.2 demonstrate the fundamental biomechanical and neuromotor principle of all human movement – that is, all actions take place in time and space and involve force. To understand the process of motor control of this multi-segmented body, we need to understand the challenge the neuromotor system faces. Indeed, the CNS must solve the problems of redundancies, time delays, sequencing and redundant degrees of freedom (Rosenbaum 2009; Turvey *et al.* 1982). There are many proposed conceptualizations about how these problems are solved, and we refer the reader to the excellent textbook by Schmidt and Lee (2011) for an introduction to these alternate theories and models. Here, we offer a control theoretic perspective on how the central nervous system (CNS) solves these problems. This approach includes aspects of both the information processing and the dynamical systems approaches as described in Chapter 2.

The control theoretic approach models the behaviour of the neuromuscular system as a set of processes underlying a goal-directed movement. The first step is the selection of the goal. Based on its previous experience, the system will select a 'motor command' – a muscle activation plan that should achieve the intended goal. Of course, the system must 'know' where the body is at all times; referred to as, *state estimation*. The body has millions of 'sensors' to provide this information. Particularly well suited to determine where the body is at any one time are the sensors of the vestibular and somatosensory systems that, along with the visual system, provide a good estimation of where our bodies are in space (see Chapter 2 for details). As we move to a desired goal, however, we may deviate or be perturbed in our movement. The nervous system therefore must be able to detect from the ongoing feedback any errors, deviations or perturbations (*error detection*). However, from the time the error is noticed and evaluated until the time another response can be generated (reaction time) is considerable. To shorten this time delay, the CNS uses a forward estimation to condition the sensors for a specific outcome. If that outcome does not occur, the sensors can respond more quickly. This is the process of *feedforward* or anticipation. Together, motor command, state estimation, error detection and forward estimation comprise what is referred to as the *internal model*, a representation of the environmental and internal dynamics. This model is not a structure in the CNS, but rather a way of describing or understanding the processes that take place in the CNS to use our sensors in the service of motor control and coordination.

DCD and selected motor skills

Mobility: getting from here to there

At birth, the human body is an awkward top-heavy segmented object that has neither the motor control nor strength to move about the world. As we have seen in Chapter 9, upright bipedal locomotion is not usually achieved for almost a year and, for some infants, almost two years. Interestingly, as difficult as hands-free walking is, no reports have ever documented that children with DCD are late walkers unless they also are born at a low birth weight (Holsti *et al.* 2002). However, as the gross motor mobility tasks increase in complexity, such as increased postural and navigational demands, children with DCD have more difficulties compared with their typically developing peers.

The bipedal gait pattern itself has been well studied in typically developing infants and young children (see Chapter 9) primarily as a reference for those infants and children with severe motor impairments such as cerebral palsy (see Chapter 13). Descriptions of children with DCD have been mostly limited to qualitative observations in books. Using a classification system based on spatial-temporal parameters (e.g. time in single stance, step length), Woodruff *et al.* (2002) found that their small sample of children with DCD exhibited 'abnormal' walking patterns, although there was no systematic departure from the typically developing children's gait patterns that would characterize the gait of children with DCD.

In 2006, Deconinck *et al.* (2006a) conducted a study that is the only one currently available that provides a rigorous kinematic analysis of the walking patterns of children with DCD. This study has its limitations, namely due to its small sample size (ten children), narrow age range (mean 7.4 ± 0.86) and few clearly classified children having DCD (only four of the ten children had scores below the 5th percentile on the MABC test and the other six were between the 5th and 15th percentile). In addition, the data were collected on a motorized treadmill that put the children in an unfamiliar and challenging environment as well as one that provides a 'structure' to the action (i.e. a driven belt that gives the motor system information that constrains it). Notwithstanding these limitations, the study by Deconinck *et al.* (2006a) affords several insights that suggest follow-up studies. Like Woodruff *et al.* (2002), this study also found that the children had abnormal patterns and differences in their absolute time and distance gait measures. Children with DCD took shorter strides (in time and space) and higher step frequencies. However, when their gait cycles were normalized in duration, the relative phases of the gait cycles were similar to their typically developing (TD) peers. This would suggest that their interlimb gait coordination was normal. Analysis of their limb and trunk configurations in space, however, indicated that the children with DCD had greater forward trunk inclination and correspondingly decreased hip angle and increased knee angles, indicating a somewhat crouched walking configuration. Speculatively, it is possible that the children made these postural adjustments due to the instability of their locomotion on the treadmill. Without data from over-ground walking, however, conclusions about the postural configurations are only conjecture.

The same group of researchers (Rosengren *et al.* 2009) used elliptical Fourier analysis (EFA) to explore the possible differences and similarities in walking pattern complexity and variability of children with DCD and TD children. In this study, EFA was used to characterize the closed-loop contours of the segmental and joint angle phase portraits of both leg segments by calculating the number of harmonics required to fit the data to the children's phase portraits. Again, with a small sample that was limited in its age range and walking on a treadmill (possibly the same data set as Deconinck *et al.* (2006a)), the authors found 'pattern' differences. The children with DCD walked with more variability and complexity than the TD children. In addition, they found that the children with DCD had asymmetric leg patterns.

So, while there seem to be no *striking* differences in the walking patterns of children with DCD compared to their TD peers, the evidence suggests that there are differences, albeit subtle, in the walking gait of children with DCD. Another way to attack the problem was offered by Chia *et al.* (2010). They reasoned that if children with DCD were less coordinated (i.e. awkward or clumsy) in their locomotor gait, then they would be expected to expend more energy walking and running because their gaits were more inefficient. Surprisingly, their findings were to the contrary. Children with DCD did not differ from their age-matched TD peers in the oxygen cost of their locomotion. However, for running, the children with DCD reported that they were working harder compared to the reports from the TD children. Indeed, a higher number of children with DCD did not complete the full testing protocol. A possible explanation for this failure to complete the testing might be that children with DCD withdraw from performing motor tasks when the exertion demands become too high.

Walking is a well-practised locomotor skill. Perhaps the difficulties children with DCD might have with such gross motor skills will only be revealed when the child is 'challenged' during walking. Several studies support this argument. When children performed a concurrent task while walking, Cherng *et al.* (2009) found no differences in kinematic walking variables (e.g. cadence, stride length, speed and time in double support) between the children with DCD and their TD counterparts. Indeed, when the concurrent task was cognitively challenging (repeating digit series), both the children with DCD and TD children were affected, but not differentially. However, when the concurrent task was a motor task (carrying objects), the children with DCD slowed their cadence significantly more than their TD peers when the motor task was difficult (e.g. carrying seven marbles on a tray). Again, the sample size in this study was small, and the children with DCD were defined as those with MABC scores below the 15th not 5th percentile. Thus, the cognitive tasks appeared equally attention-demanding between the two groups, while the motor task was more demanding for the group with DCD, suggesting that motor performance is more demanding for the children with DCD even when they do not show overt performance differences.

Another challenge to the children's walking skill is to test their ability to navigate over an obstacle put in front of them. Seven- to nine-year-old children with DCD were compared to their TD peers as they walked over obstacles of various heights (Deconinck *et al.* 2010). The children with DCD had no more difficulty with the task than did their TD peers in anticipating and adjusting to the obstacle in their pathway. However, the children with DCD exhibited less ability to control their COM medio-laterally, indicating a postural control problem. Postural control in relation to DCD is discussed later in this chapter.

Finally, two studies investigated the ability of children to simultaneously time a clap and march on the spot. In contrast with walking, where the arms swing naturally without intention to couple, in this task, intentional timing of two independently well-practised actions requires the children to be proficient at multi-limb coordination. In one study of ten children with DCD (<15th percentile on the MABC), the task was to clap and march to an auditory cue set at four different frequencies (Whitall *et al.* 2006). Thus, on top of coordinating the two sets of limbs, the task required timing to a cue, although the accuracy of the timing itself was not measured. Children with and without DCD were less accurate at coordinating their limb girdles together (using mean absolute deviation of relative phasing between the clap and a footfall) than adults, but the children with DCD were also more variable in their phasing than both TD children and adults. Individually, children with DCD were less likely than TD children to adopt the absolute (simultaneous) coupling required of the task. This difference became more emphasized with faster frequencies. These results indicate that children with DCD have difficulty with managing large moments of inertia but do not provide information about whether these children are inherently more variable when they try to intentionally coordinate their limbs or whether the additional challenge of keeping to a beat causes the variability. This question was resolved in a follow-up study of this task where children adopted a preferred rate of performing the task (no beat) under conditions of full vision and hearing, vision occluded by blindfold, hearing occluded by headphones with white noise and no vision or hearing of the task (Mackenzie *et al.* 2008). Contrary to the researchers' hypotheses, there was no effect of sensory information on task performance for any group, suggesting that vision and audition are not salient sources of information for this particular task. However, even without the auditory cue, children with DCD were inherently more variable than TD children who, in turn, were more variable than adults. This variability was seen in both the relative phasing between the limb girdles (inter-girdle coordination) and also the within-limb girdle variability of the arms and legs separately as they perform each clap or footfall. Taken together, these studies provide additional support for the idea that walking needs to be challenged in order to show differences between children with and without DCD and that children with DCD allocate considerable 'resources' to seemingly automatic motor performance, namely, walking.

Postural control: orientation and equilibrium

The human body is comprised of multiple segments (e.g. shank, thigh, torso, etc.) that are 'deformable' at every joint. Maintaining posture, upright or otherwise, requires these joints to be 'controlled'. Postural control, therefore, comprises two tasks: postural equilibrium and orientation (Horak *et al.* 1996). In Chapter 7, we have seen how postural control develops in TD children. Not surprisingly, most tests of motor development include a measure of quiet (or static) balance. Children with DCD often have difficulties with these items, but usually only when the task is a challenging posture (e.g. one-footed) or when one or more sources of sensory information are removed (e.g. no vision). In the early studies of postural control in children with DCD, investigators reported that they were not able to balance as long as TD children (Hoare 1994). Not all children with

DCD have balance problems, but when investigators look at the children who perform poorly on the timed balance tests, they find these children have larger centre of pressure (COP) excursions in both anterior–posterior (AP) and medio-lateral (ML) directions (Geuze 2003; Tsai *et al.* 2008). Furthermore, as the task demands increase (e.g. one-footed standing, closing eyes), children with DCD have more difficulties controlling their balance as measured by COP excursions or the time duration of unperturbed standing.

When we control our posture, we need to manage the pull of gravity and the forces that are caused by segmental actions. Imagine, for example, that you have to point at the target or reach for and drink from a cup. These actions of the upper extremity will create torques that the postural system must accommodate or postural control will be lost. Not surprisingly, children with DCD show difficulties with managing these intentional as well as unintended torques. In TD children and adults, postural adjustments are made in anticipation of the disturbance. So, just before an arm raise, the CNS activates the trunk and leg muscles prior to the movement in *anticipation* of the arm's action to counteract the mechanical effects of the planned action (Belen'kii *et al.* 1967). These anticipatory postural adjustments (APAs) have been demonstrated in adults (Cordo and Nashner 1982) and young children (Schmitz and Assaiante 2002). But in children with DCD, the evidence suggests that APAs are qualitatively different from those of their TD cohort (Johnston *et al.* 2002; Przysucha *et al.* 2008). In addition, they are slower to make their postural adjustments (Jucaite *et al.* 2003). Even in the simplest task, namely, adjusting to a moving platform under the children's feet, children with movement problems activate their muscles in a less well-organized fashion than their TD peers (Williams and Woollacott 1997).

In addition to controlling the multi-segmented body in different orientations, the problem of postural control is complicated in that this control must be accomplished in a constantly changing environment. So, in quiet standing on two feet fixating on a target to focus on or reaching for a fixed target, the sensory information to interpret is minimized. But in the 'real' world, our postural control must be achieved in a complex, ever-changing sensory-rich environment. Not surprisingly, researchers studying posture have investigated whether sensory information processing is a source of difficulty for children with DCD. An early study by Wann *et al.* (1998) suggested that some children with DCD relied more on visual input in postural control than their TD peers. A similar conclusion was reached by Deconinck and his colleagues (2008), but not by Przysucha and Taylor (2004) or Grove and Lazarus (2007). To date, it is unclear if children with DCD rely on one sense more than another in their postural control. However, it is clear that they have difficulties when sensory information is unavailable, unreliable or in conflict.

Upper extremity skills: interactions with objects

When we interact with objects, we often do so with our upper extremities. Many of the play-sport skills of childhood are captured in this set of actions: for example, throwing, striking and catching. In Chapter 9, we have seen how throwing and striking skills develop in TD children. The MABC and BOT tests use throwing and/or striking a ball as test items, which discriminate between typically coordinated and poorly coordinated children and many studies refer to 'inaccurate throwing' or 'uncoordinated throwing movements' in this population, yet there are virtually no biomechanical studies of throwing or striking in these children. In this section, then, we will concentrate on the upper extremities as they 'interact', 'intercept' and/or 'receive' objects while sitting or standing. We start with the simplest (aiming and pointing), then take up reach and grasp, and finally, we examine the evidence about how children with DCD catch objects.

Aiming/pointing to a target

Simple aiming movements, where a child must move from a start location to a pre-specified target, require three fundamental actions: (1) control of the trunk (in a seated position); (2) visual regard of the target; and (3) transport of the whole arm to the target through shoulder flexion and elbow extension (typically movement of the wrist/hand/fingers is not required because either a stylus is held or a finger is pointed through the movement). In this section, we will not consider the postural control requirements because most studies of upper extremity skills do not report the relevant variables. Furthermore, there is evidence that the fundamental visual pathway is not impaired in these children (Henderson *et al.* 1994; Mon-Williams *et al.* 1994, 1996). Therefore, we will concentrate on the visuo-motor coordination to control the task. As described earlier, the underlying neuromotor processes for this visuo-motor coordination include estimating the spatial location of the arm in relation to the target (state estimation), planning the movement to the target and detecting any errors that occur along the way through use of feedforward and/or feedback processes. Simple aiming movements have two phases: (1) the initial acceleration phase that is assumed to be open-loop in nature and reflect feedforward control; and (2) the deceleration phase which is under feedback control.

In simple aiming tasks, when the instruction is to point as fast and accurately as possible to targets in different spatial positions, children with DCD demonstrate slower movement time (MT), a longer trajectory path illustrating an inaccurate path to the target and greater end-point errors when they reach the target when compared to their TD peers (Ameratunga *et al.* 2004). Since these characteristics are not independent of one another, as a package they can be taken to suggest that more spatial errors are made during the transport phase, but it is not possible to say which underlying processes are involved or anything about the initial planning of the movement from these data. Many studies have illustrated that children with DCD take longer to plan a movement before actually moving as reflected in increased reaction times (RT). In simple aiming tasks these children can decrease their RT if they are given advanced pre-cueing about target selection (Pettit *et al.* 2008). However, they are still more inaccurate than TD children who do not substantively change their RT with pre-cueing, suggesting that the TD children are able to plan optimally.

By introducing perturbations to the aiming task and portioning MT into temporal phases, a better understanding of underlying principles can be attained. For example, Plumb *et al.* (2008) used a step-perturbation paradigm where a simple aiming movement to a target was sometimes perturbed (e.g. target moving after the start), requiring a correction. The authors tested whether children with DCD would show an additional impairment in the perturbation conditions (relative to unperturbed conditions) compared to the impairment shown by TD children – that is, they were looking for a group by condition interaction in addition to the usual group effect of children with DCD being slower and less accurate. They did not find a meaningful interaction, suggesting that the children with DCD performed similarly to the TD children in adapting to the perturbation. They also did not find differences in the initial feedforward component of the reach as reflected in peak speed and time to peak speed. They concluded that children with DCD have fundamental difficulties in generating the basic movement that could account for the noted difficulties in correcting on-going movements. This study is interesting not just for its conclusions about the underlying processes of aiming, but also for the fact that the study design had to be altered for the children with DCD – that is, the original design called for the children to stand and use a thin pencil-like stylus for the task, but children with DCD were allowed to sit and use a thicker stylus. And yet, even with the easier task conditions, the children with DCD still performed worse than TD children. These unexpected problems with postural control and grasping illustrate how pervasive the motor problems are for these children.

Other studies do not necessarily support the conclusion by Plumb *et al.* (2008) that online control by itself is not a primary deficit. Using a double-step reaching paradigm, children were asked to start at a central target, then press a peripheral target and return to a central target (Hyde and Wilson 2010). On some trials the central target jumped at movement onset to one of two peripheral locations (jump trials). As would be expected, children with DCD were slower to initiate a movement (longer RT) regardless of condition, but in this case an interaction with condition on MT occurred as these children took a proportionally far longer time in the jump trials. Essentially, the children were slow in inhibiting the response to go back to the centre, suggesting impairment in the ability to make rapid online adjustments.

Finally, another set of experiments using a computer set-up with peripheral targets displayed in a circle explored whether children with DCD have problems when learning a new skill. The initial study in this vein was conducted much earlier (Missiuna 1994) and the results were somewhat unexpected. Children with DCD demonstrated slower RTs and MTs, but their rate and pattern of learning were the same as matched controls. Furthermore, they transferred this learning to new target distances and to targets with a different visual stimulus, in exactly the same way as the TD children. Limitations of this study included the lack of any spatial error information. Recent studies have used a more sophisticated measurement approach with a motor adaptation paradigm (Kagerer *et al.* 2004, 2006). The motor-adaptation approach tests motor learning by altering the relationship between the movements made by the child and the resulting visual image of that movement. After baseline trials, children were exposed to a 45° visual feedback rotation which initially took them off-target, after which they gradually adapted (learned) the new relationship and reduced the spatial error (Kagerer *et al.* 2004). Children with DCD were actually less affected by the visual distortion and were able to adapt similarly to TD children. However, after the distortion was removed the TD children showed an initial negative after-effect of moving in the opposite direction until they de-adapted. Children with DCD showed less effect, suggesting that they had not learned the new relationship (i.e. an internal model) very well. Children with DCD were slower but equally accurate in the baseline condition and adapted at a similar rate. However, they were more variable in the adaptation phase, suggesting different motor control (as opposed to motor learning) abilities. In a follow-up study, Kagerer *et al.* (2006) demonstrated that if the distortion was large (60°) and the adaptation period long, then children with DCD adapted as well as TD children, including demonstrating after-effects. However, if the distortion was gradually introduced (under perceptual threshold), only the TD children responded. These results clearly indicate that children with DCD do not detect small error signals and lend support to the idea that they do have difficulty with online corrections as well as planning the initial movement.

In summary, the literature on aiming and particularly the use of computer displays and tablets provides a description of slow and inaccurate arm movement with some insight into decrements of feedback and feedforward mechanisms. Further, findings from Huh *et al.* (1998) demonstrating prolonged agonist activity in children with DCD suggest that these children have difficulty with time-appropriate muscle activation and de-activation. Such findings highlight the need for investigation into the biomechanical mechanisms underlying DCD-related reductions in motor performance.

Reaching to grasp an object

From a seated position, the biomechanical tasks for reaching to grasp are similar to pointing but with the addition of thumb–finger movements (opening and closing grip apertures). Smyth *et al.* 2001 report that, as with aiming, children with DCD are less accurate in transporting the hand. Additionally, they spend less time in the deceleration phase of hand transport, suggesting that they do not rely on feedback. When vision was removed as movement began, TD children spent more time decelerating and reached peak grip aperture earlier. These findings led the authors to conclude that children with DCD were less dependent on vision than TD children. Similar findings were observed by Zoia *et al.* (2005) who also concluded that visual feedback was used differently in children with DCD. In general, these findings support the idea that online control is not utilized as much for these children, which may be a reason why they are clumsy. Possibly, this idea is related to the aforementioned fact that they can only detect a large error signal and therefore ignore other information. A recent experiment by Viswanathan *et al.* (2010) adds more information about differences in the underlying processes involved in reaching between children with DCD and their TD peers. These authors investigated anticipatory modifications made in the transport section of the reach, comparing the three tasks of reach to grasp a dowel, reach to grasp and lift the dowel, and reach to grasp, lift and place the dowel at the starting position. The rationale was that anticipatory modifications of the reach component as the tasks of lift, and lift and place were added would reflect feedforward processing. A key aspect of the approach was to use a method of partitioning MT into four components instead of the usual two, based on the time of peak acceleration of the first movement unit (in positive acceleration), the time of peak velocity and the time of peak deceleration of the first movement unit (in negative deceleration phase) respectively.

Figure 12.3 illustrates the four phases in a TD child and a child with DCD. As can be seen in this figure, the velocity and acceleration trajectories for the DCD children are less smooth, showing a greater number and earlier corrections. The authors defined only the T1 component as reflecting feedforward planning and found that children with DCD showed increased time for feedback reliance (T2–T4) compared to controls. This difference increased with the more complex tasks. More importantly, these children did not show task-related anticipatory feedback modifications in T2 (representing corrections based on a forward model) or T4 (representing corrections based on external stimuli), which were seen in children at both 8 and 10 years of age. Thus, although children with DCD appeared to rely more on feedback control, supporting some of the earlier experiments, they do not appear to use this information well and produce jerkier movements as well as show increased latency between the grasp and lift. In addition to demonstrating that both feedforward and feedback processes seem to be affected in DCD, the results of this study also indicated that 10-year-old children with DCD perform similarly to 6–8-year-old TD children, suggesting developmental delay rather than an atypical developmental trajectory.

In summary, the reach to grasp data support the aiming data in finding slow, jerky, inaccurate movements with additional findings that (1) visual information is not utilized in the same way and (2) subsequent movements involve longer transition times and a lack of anticipatory modifications found in typically developing children. In addition, a breakdown of the MT components seems to implicate both feedforward and feedback processes as problematic in children with DCD.

Catching

In addition to the biomechanical constraints of reaching and grasping, catching involves timing the closure of the fingers to a moving object, either in a standing or moving position that requires increased postural control and sometimes using two hands rather than one. Only one research group has focused on describing catching in this population over a series of papers. Astill and Utley (Astill and Utley 2006; Utley *et al.* 2007) began by videoing children (7–8 years of age) with and without DCD performing 30 two-handed catches from a ball-machine. The

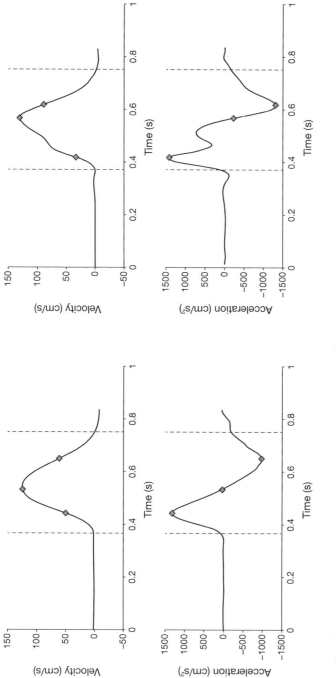

Figure 12.3 Representative velocity (above) and acceleration (below) of hand path in a reach to grasp task (R_G) when using preferred hand for a 10-year-old TD child (left) and age-matched child with DCD (right). The dotted lines represent the movement onset and offset respectively. The grey squares represent portioning into T1 (feedforward), and T2, T3 and T4 (feedback) components.

age-matched TD children were approximately twice as successful in catching the balls (~24 to 12). A kinematic analysis showed reduced range of motion and tighter coupling both between and within limbs for the DCD group and more intra-individual variability of coupling in the TD group. The authors concluded that children with DCD showed less capacity to vary their limb movements and thus had a less adaptable movement system that limits their success. When the ball was thrown towards the left or right shoulder, as compared to the midline, however, the children with DCD exhibited a lower degree of limb coupling than the TD group, thus supporting an argument that TD children can adapt to different constraints better than those with DCD and that the latter tend to 'freeze' their degrees of freedom to perform the task in the midline (Utley *et al.* 2007).

Using an alternative approach Utley and Astill (2007) investigated the qualitative components of catching. First, they undertook a cross-sectional study on 18 DCD and 18 age-matched TD children (7–10 years of age) to investigate the three component actions of two-handed catching (hand, arm and body) hypothesized by Haywood and Getchell (2008). They determined that the developmental sequences observed in these children met the pre-longitudinal screening criteria proposed by (Roberton *et al.* 1980) and thus were able to use these sequences to evaluate a smaller group of children with and without DCD. The results were predictable in finding developmental delay. Children with DCD had less advanced modal components of arm and body separately as well as developmental profile as a whole. They also had a greater range of developmental profiles as well as an unstable modal profile. Thus, children with DCD had not acquired a stable coordination pattern for the task and, in terms of the internal model concept that we have referred to earlier, they do not have a good inverse model upon which to initially start their movement (Plumb *et al.* 2008). Taken together, these reports by Utley and Astill imply that children with DCD were most likely to stand with their arms outstretched and joints fixed, leaving them little adaptability if the ball flight was not straight to them.

One limitation of the above analyses, aside from the small sample size and the use of the 15th percentile, is the lack of differentiation between transport and grasp phase and specifically an assessment of the grip opening and closing function in relation to a moving target. Using a different paradigm of a pendulum ball machine, Astill and Utley (2008) identified 10 children below the 5th percentile on the Movement ABC and analysed both the transport and grasp phases of 15 two-handed catches with a very complete kinematic analysis. For the transport phase, children with DCD had a longer RT and MT and an earlier deacceleration phase, suggesting that they required more time to plan but exhibited a shorter preplanned aspect of the reach at the same time as they took longer to reach by intercepting the ball later in the swing. Given that the ball flight was relatively predictable, the success rate for both sets of children was higher than using a ball machine (but still higher in TD children). Notably, children with DCD initiated the grasp phase earlier and reached maximum aperture earlier and more variably than TD children who were more like adults. Finally, the maximum aperture was larger in children with DCD, indicating a compensatory strategy. The authors conclude that children with DCD have problems in regulating the temporal relationship between the transport and grasp phases of the catch, leading to a decomposition of the movement similar to that seen in patients with cerebellar lesions (Zackowski *et al.* 2002). Furthermore, there is clear evidence again that problems exist in both the initial inverse internal model that plans the movement, the ability to use feedback during the movement and, likely, the forward internal model that allows for early corrections.

Other studies of catching do not exactly mirror the findings of the Astill and Utley (2008) study, but this may be owing to different experimental paradigms or severity of DCD. For example, Deconinck *et al.* (2006b), in a study of boys who were not all below the 5th percentile and who were catching one-handed from a ball machine, did not find a difference in movement onset or moment of hand closure and found a decreased, not increased maximal aperture. The authors still concluded that these boys had a problem with task execution even though they could broadly adapt to a change of ball speed. Another study, also on boys (n=12 per group; age 9–10 years; <15th percentile) used underhand throwing from a person (albeit relatively reliable in tossing) and tested 60 catches in central and lateral locations, under blocked and randomized conditions (Przysucha and Maraj 2010). No effect of randomization was found for the number of balls caught, but TD boys had nearly perfect scores while those with DCD caught more balls in central (73 per cent) than lateral trials (47 per cent). Using a qualitative biomechanical analysis (similar to Utley and Astill (2007)), they determined that major problems for the lateral trials occurred with the timing of the unilateral grasp and also the positioning of the trunk and arms (but not hands, as per Utley and Astill (2007)). The authors concluded that the primary problems are at the intersegmental coordination at the behavioural level, especially when the ball is not thrown directly to them.

In summary, the catching data highlight the problems of inefficient, decomposed, intersegmental coordination as well as posture deficiencies. Timing of the finger movements to grasp a moving target is also a major problem and this ability, known as coincident anticipation timing, needs further study. From a neuromotor control perspective, the issues of movement planning and fast, slow feedback corrections are all likely impaired, but it is not clear that each child has the exact same problems.

Recommendations for future research

Considering the breadth of data provided in this review, we will restrict ourselves to suggesting key biomechanical and neuromotor-based studies within one review area. For upper extremity skills there are two avenues of future research. The first is to continue to add to the broad database of descriptive research by including an analysis of the skills of throwing and striking in a similar vein to Utley and Astill's investigation of catching. This analysis might begin by determining impairments in quantitative parameters such as ball velocity and distance thrown, as well as qualitative impairments of the throwing form itself (see Chapter 9 for discussion of developmental sequence analysis). The importance of these data is in understanding exactly how limited these children are in complex tasks in both closed and open skill situations. From a long-term perspective, this knowledge may help parents, teachers or clinicians make informed decisions about the usefulness of trying to ameliorate vs. compensate for poor throwing/striking skills. Subsequent analysis of kinematic, kinetic and EMG parameters might also be useful, but only if it contributes to understanding how the impairments are instantiated and therefore how the impairments might be reduced. This leads to a second avenue for research since much progress has been made on investigating whether feedforward or feedback or both sets of processes are implicated. If we assume that the specific upper extremity skill is less important than the mechanistic underpinnings of poor performance, future studies could determine an empirical method of allocating an individual's problems to either feedforward or feedback processes (or both) in order to begin optimizing interventions that can target these processes for an individual. Both sets of studies need follow-up short- and long-term interventions to test whether the developmentally delayed appearance of the activities can, in fact, at least be partially remedied. If not, then alternative compensatory activities may need to be promoted for this population (Missiuna et al. 2006).

Such studies should include manipulation of different sensory sources of information as well as ideally be combined with a neurophysiological investigation. Spatial imaging techniques such as structural magnetic resonance imaging and resting state functional connectivity, and temporal techniques such as electro-encephalography, positron emission tomography or magnetoencephalography and transcranial magnetic stimulation techniques will parallel behavioural data with knowledge of the architecture of these children's brains compared to TD children. Although the definition of children with DCD does not include hard neurological signs, it is plausible that soft neurological signs are present and can be detected by newer neurophysiological techniques, as seen in very recent studies.

Clinical and practical applications

Children with DCD typically perform balance, locomotion and upper extremity skills at a performance level that is delayed for approximately two to four years on average. Major characteristics of the skills are slowness, variability and inaccuracy. In addition, the underlying neuromuscular mechanisms that are implicated in 'causing' the delay are assumed to involve sensorimotor integration in both feedforward and feedback processes. We have not discussed potential neuro-anatomical correlates in this chapter, but the cerebellum is one plausible site of structural/functional delay or atypicality (Bo et al. 2008). These findings have implications for parents, teachers and clinicians.

For parents of children who appear markedly behind their peers in motor skills, there are several implications. First, unless they believe that their child has not had a 'normal' motoric environment to grow up in, they should realize that motor coordination problems are not likely to disappear or be 'grown out of' and that there are potentially serious sequelae to this condition, as outlined earlier. Second, they should strongly consider getting their child tested so that extra services might be available to them. Third, they should take some responsibility in providing extra motor practice conditions for their children. Although we did not review the intervention data here, there is consensus that task-related practice is beneficial for these children (Pless and Carlsson 2000; Sigmundsson *et al.* 1998; Sugden and Chambers 1998).

For classroom and physical education teachers with pupils who appear substantively more clumsy and uncoordinated than their peers, there are options. First, they too should advocate for formal testing if the child has not been given a diagnosis. Regardless of diagnosis, however, there are two options for helping these children. First, they can be encouraged to repetitively and progressively practise basic skills in a non-threatening environment. Second, there are possibilities that compensatory activities can be used to help them function in a way that keeps them more closely in line with their peers without increasing the social stigmatization that is often a by-product. In the classroom, this might include aids for writing such as computers (with poor aiming skills, these children's writing is also impacted). In the physical education class this might include allowing these children to concentrate on activities that do not require high hand–eye coordination or complex interactions such as individual activities like jogging or walking. Such activities include the health benefits that sports provide, but not the high pressure to perform well. Involving these children as scorers or in other peripheral roles can also help keep them involved in the complex sports.

For clinicians who are treating an already diagnosed child, the implications are first to determine whether one area of motor skill performance demands particular attention over another, and subsequently to determine a course of treatment. Given the major findings, across the skills reviewed here, of slowness, variability and inaccuracy, the therapist should use repetitive and progressive task-related practice of skills that the individual child wants to learn. There is evidence that involving the child in this decision is useful for providing motivation to practice (Polatajko *et al.* 2001). In terms of working on specific sensorimotor integration in feedforward or feedback situations, one way to ensure practice of both is to have some practice performed in a predictable environment (to promote feedforward processes) and other practice performed in a less predictable environment (to promote feedback processes).

Key points

- DCD is a prevalent developmental disorder that involves a substantially poorer performance of motor skills than age-matched peers. It is not caused by mental retardation or known neurological deficits. It is heterogeneous, has consequential sequelae and persists for most into adulthood.
- DCD-related performance deficits in upper extremity skills such as aiming, reaching and catching result in slow, variable and inaccurate movements.
- DCD-related performance deficits in locomotor skills are more obvious when

there is an additional challenge such as a secondary motor task, an obstacle or a complex timing or coordination task.

- DCD-related performance deficits in posture involve greater sway in quiet standing and are also more evident with challenging postures and sensory conditions.
- Across all three skills reviewed, there appear to be underlying deficits in both feedforward and feedback neuromuscular mechanisms.
- Comprehensive sophisticated biomechanical analyses of these children need to be undertaken as well as simultaneous assessments of neuro-anatomical correlates. Ultimately, well-motivated intervention studies are needed.

References

Ameratunga, D., Johnston, L. and Burns, Y. (2004) 'Goal-directed upper limb movements by children with and without DCD: a window into perceptuo-motor dysfunction?', *Physiotherapy Research International*, 9: 1–12.

APA (2000) *Diagnostic and Statistical Manual of Mental Disorders IV-TR (DSM-IV-TR)*, Washington, DC: American Psychiatric Association.

Astill, S. and Utley, A. (2006) 'Two-handed catching in children with developmental coordination disorder', *Motor Control*, 10: 109–24.

Astill, S. and Utley, A. (2008) 'Coupling of the reach and grasp phase during catching in children with Developmental Coordination Disorder', *Journal of Motor Behavior*, 40: 315–24.

Belen'kii, V.Y., Gurfinkel, V.S. and Pal'tsev, Y.I. (1967) 'Control elements of voluntary movements', *Biofizika*, 12: 135–41.

Bo, J., Bastian, A.J., Kagerer, F.A., Contreras-Vidal, J.L. and Clark, J.E. (2008) 'Temporal variability in continuous versus disconcontinuous drawing for children with Developmental Coordination Disorder', *Neuroscience Letters*, 431: 215–30.

Bruininks, R.H. and Bruininks, B.D. (2005) *Bruininks–Oseretsky Test of Motor Proficiency*, 2nd edn, Minneapolis, MN: Pearson Assessment.

Cairney, J., Hay, J., Veldhuizen, S. and Faught, B.E. (in press) 'Trajectories of cardio-respiratory fitness in children with and without developmental coordination disorder: a longitudinal analysis', *British Journal of Sports Medicine*.

Cairney, J., Hay, J.A., Faught, B.E., Flouris, A. and Klentrou, P. (2007) 'Developmental coordination disorder and cardiorespiratory fitness in children', *Pediatric Exercise Science*, 19: 20–8.

Cherng, R.J., Liang, L.Y., Chen, Y.J. and Chen, J.Y. (2009) 'The effects of a motor and a cognitive concurrent task on walking in children with developmental coordination disorder', *Gait & Posture*, 29: 204–7.

Chia, L.C., Guelfi, K.J. and Licari, M.K. (2010) 'A comparison of the oxygen cost of locomotion in children with and without developmental coordination disorder', *Developmental Medicine and Child Neurology*, 52: 251–5.

Chirico, D., O'Leary, D., Cairney, J., Klentrou, P., Haluka, K., Hay, J. and Faught, B. (2011) 'Left ventricular structure and function in children with and without developmental coordination disorder', *Research in Developmental Disabilities*, 32: 115–23.

Cordo, P.J. and Nashner, L.M. (1982) 'Properties of postural adjustments associated with rapid arm movements', *Journal of Neurophysiology*, 47: 287–302.

Cousins, M. and Smyth, M.M. (2003) 'Developmental coordination impairments in adulthood', *Human Movement Science*, 22: 433–59.

Deconinck, F.J., Savelsbergh, G.J., De Clercq, D. and Lenoir, M. (2010) 'Balance problems during obstacle crossing in children with Developmental Coordination Disorder', *Gait & Posture*, 32: 327–31.

Deconinck, F.J., De Clercq, D., Savelsbergh, G.J., van Coster, .R., Oostra, A., Dewitte, G. and Lenoir, M. (2006a) 'Differences in gait between children with and without developmental coordination disorder', *Motor Control*, 10: 125–42.

Deconinck, F.J., De Clercq, D., Savelsbergh, G.J.P., van Coster, R., Oostra, A., Dewitte, G. and Lenoir, M. (2006b) 'Adaptations to task constraints in catching by boys with DCD', *Adapted Physical Activity Quarterly*, 23: 14–30.

Deconinck, F.J., De Clercq, D., van Coster, R., Oostra, A., Dewitte, G., Savelsbergh, G.J.P., Cambler, D. and Lenoir, M. (2008) 'Sensory contributions to balance in boys with Developmental Coordination Disorder', *Adapted Physical Activity Quarterly*, 25: 17–35.

Dewey, D., Kaplan, B.J., Crawford, S.G. and Wilson, B.N. (2002) 'Developmental coordination disorder: associated problems in attention, learning, and psychosocial adjustment', *Human Movement Science*, 21: 905–18.

Faught, B.E., Hay, J.A., Cairney, J. and Flouris, A. (2005) 'Increased risk for coronary vascular disease in children with developmental coordination disorder 1', *Journal of Adolescent Health*, 37: 376–80.

Geuze, R.H. (2003) 'Static balance and developmental coordination disorder', *Human Movement Science*, 22: 527–48.

Geuze, R.H., Jongmans, M.J., Schoemaker, M.M. and Smits-Engelsman, B.C. (2001) 'Clinical and research diagnostic criteria for developmental coordination disorder: a review and discussion', *Human Movement Science*, 20: 7–47.

Grove, C.R. and Lazarus, J-A. (2007) 'Impaired re-weighting of sensory feedback for maintenance of postural control in children with developmental coordination disorder', *Human Movement Science*, 26: 457–76.

Haywood, K. and Getchell, N. (2008) *Life Span Motor Development*, Champaign, IL: Human Kinetics.

Henderson, S. and Sugden, D.A. (2007) *Movement Assessment Battery for Children 2*, San Antonio, TX: Psychological Corporation.

Henderson, S.E., Barnett, A.L. and Henderson, L. (1994) 'Visuospatial difficulties and clumsiness: on the interpretation of conjoined deficits', *Journal of Child Psychology and Psychiatry*, 35: 961–9.

Hoare, D. (1994) 'Subtypes of developmental coordination disorder', *Adapted Physical Activity Quarterly*, 11: 158–69.

Holsti, L., Grunau, R.V. and Whitfield, M.F. (2002) 'Developmental coordination disorder in extremely low birth weight children at nine years', *Journal of Developmental & Behavioral Pediatrics*, 23: 9–15.

Horak, F.B., Macpherson, J.M., Rowell, L.B. and Shepard, J.T. (1996) 'Postural orientation and equilibrium', in L.B. Rowell and J.T. Shepars (eds) *Handbook of Physiology. Section 12: Exercise: Regulation and Integration of Multiple Systems*, Oxford: Oxford University Press, pp. 255–92.

Huh, J., Williams, H.G. and Burke, J.R. (1998) 'Development of bilateral motor control in children with developmental coordination disorders', *Developmental Medicine and Child Neurology*, 40: 474–84.

Hyde, C. and Wilson, P. (2010) 'Online motor control in children with developmental coordination disorder: chronometric analysis of double-step reaching performance', *Child: Care, Health and Development*, 37: 111–22.

Johnston, L.M., Burns, Y.R., Brauer, S.G. and Richardson, C.A. (2002) 'Differences in postural control and movement performance during goal directed reaching in children with developmental coordination disorder', *Human Movement Science*, 21: 583–601.

Jucaite, A., Fernell, E., Forssberg, H. and Hadders-Algra, M. (2003) 'Deficient coordination of associated postural adjustments during a lifting task in children with neurodevelopmental disorders', *Developmental Medicine and Child Neurology*, 45: 731–42.

Kagerer, F.A., Bo, J., Contreras-Vidal, J.L. and Clark, J.E. (2004) 'Visuomotor adaptation in children with developmental coordination disorder', *Motor Control*, 8: 450–60.

Kagerer, F.A., Contreras-Vidal, J.L., Bo, J. and Clark, J.E. (2006) 'Abrupt, but not gradual visuomotor distortion facilitates adaptation in children with developmental coordination disorder', *Human Movement Science*, 25: 622–33.

Kopp, S., Beckung, E. and Gillberg, C. (2010) 'Developmental coordination disorder and other motor control problems in girls with autism spectrum disorder and/or attention-deficit/hyperactivity disorder', *Research in Developmental Disabilities*, 31: 350–61.

Lingam, R., Hunt, L., Golding, J., Jongmans, M. and Emond, A. (2009) 'Prevalence of developmental coordination disorder using the DSM-IV at 7 years of age: a UK population-based study', *Pediatrics*, 123: e693–700.

Losse, A., Henderson, S.E., Elliman, D., Hall, D., Knight, E. and Jongmans, M. (1991) 'Clumsiness in children – do they grow out of it? A 10-year follow-up study', *Developmental Medicine and Child Neurology*, 33: 55–68.

Mackenzie, S.J., Getchell, N., Deutsch, K., Wilms-Floet, A., Clark, J.E. and Whitall, J. (2008) 'Multi-limb coordination and rhythmic variability under varying sensory availability conditions in children with DCD', *Human Movement Science*, 27: 256–69.

Macnab, J.J., Miller, L.T. and Polatajko, H.J. (2001) 'The search for subtypes of DCD: is cluster analysis the answer?', *Human Movement Science*, 20: 49–72.

McCarron, L.T. (1982) *McCarron Assessment of Neuromuscular Development*, Dallas, TX: McCarron-Dial Systems.

Missiuna, C. (1994) 'Motor skill acquisition in children with developmental coordination disorder', *Adapted Physical Activity Quarterly*, 11: 214–35.

Missiuna, C., Rivard, L. and Bartlett, D. (2006) 'Exploring assessment tools and the target of intervention for children with Developmental Coordination Disorder', *Physical & Occupational Therapy in Pediatrics*, 26: 71–89.

Missiuna, C., Moll, S., King, G., Stewart, D. and Macdonald, K. (2008) 'Life experiences of young adults who have coordination difficulties', *Canadian Journal of Occupational Therapy*, 75: 157–66.

Mon-Williams, M.A., Wann, J.P. and Pascal, E. (1994) 'Ophthalmic factors in developmental coordination disorder', *Adapted Physical Activity Quarterly*, 11: 170–8.

Mon-Williams, M.A., Mackie, R.T., McCulloch, D.L. and Pascal, E. (1996) 'Visual evoked potentials in children with developmental coordination disorder', *Ophthalmic and Physiological Optics*, 16: 178–83.

Pettit, L., Charles, J., Wilson, A.D., Plumb, M.S., Brockman, A., Williams, J.H.G. and Mon-Williams, M. (2008) 'Constrained action selection in children with developmental coordination disorder', *Human Movement Science*, 27: 286–95.

Piek, J.P., Dworcan, M., Barrett, N.C. and Coleman, R. (2000) 'Determinants of self-worth in children with and without developmental coordination disorder', *International Journal of Disability, Development and Education*, 47: 259–72.

Piek, J.P., Rigoli, D., Pearsall-Jones, J.G., Martin, N.C., Hay, D.A., Bennett, K.S. and Levy, F. (2007) 'Depressive symptomatology in child and adolescent twins with attention-deficit hyperactivity disorder and/or developmental coordination disorder', *Twin Research and Human Genetics*, 10: 587–96.

Pless, M. and Carlsson, M. (2000) 'Effects of motor skill intervention on developmental coordination disorder: a meta-analysis', *Adapted Physical Activity Quarterly*, 17: 381–401.

Plumb, M.S., Wilson, A.D., Mulroue, A., Brockman, A., Williams, J.H.G. and Mon-Williams, M. (2008) 'Online corrections in children with and without DCD', *Human Movement Science*, 27: 695–704.

Polatajko, H.J., Mandich, A.D., Miller, L.T. and Macnab, J.J. (2001) 'Cognitive orientation to daily occupational performance (CO-OP): part II – the evidence', *Physical & Occupational Therapy in Pediatrics*, 20: 83–106.

Prieto, T.E., Myklebust, J.B., Hoffmann, R.G., Lovett, E.G. and Myklebust, B.M. (1996) 'Measures of postural steadiness: differences between healthy young and elderly adults', *IEEE Transactions on Biomedical Engineering*, 43: 956–66.

Przysucha, E.P. and Maraj, B.K. (2010) 'Movement coordination in ball catching: comparison between boys with and without developmental coordination disorder', *Research Quarterly for Exercise and Sport*, 81: 152–61.

Przysucha, E.P. and Taylor, M.J. (2004) 'Control of stance and developmental coordination disorder: the role of visual information', *Adapted Physical Activity Quarterly*, 21: 19–33.

Przysucha, E.P., Taylor, M.J. and Weber, D. (2008) 'The nature and control of postural adaptations of boys with and without developmental coordination disorder', *Adapted Physical Activity Quarterly*, 25: 1–16.

Roberton, M.A., Williams, K. and Langendorfer, S. (1980) 'Pre-longitudinal screening of motor development sequences', *Research Quarterly for Exercise and Sport*, 51: 724–31.

Rosenbaum, D.A. (2009) *Human Motor Control*, San Diego, CA: Academic Press.

Rosengren, K.S., Deconinck, F.J., Diberardino, L.A., 3rd, Polk, J.D., Spencer-Smith, J., De Clercq, D. and Lenoir, M. (2009) 'Differences in gait complexity and variability between children with and without developmental coordination disorder', *Gait & Posture*, 29: 225–9.

Schmidt, R.A. and Lee, T. (2011) *Motor Control and Learning: A Behavioral Emphasis*, Champaign, IL: Human Kinetics.

Schmitz, C. and Assaiante, C. (2002) 'Developmental sequence in the acquisition of anticipation during a new co-ordination in a bimanual load-lifting task in children', *Neuroscience Letters*, 330: 215–18.

Schott, N., Alof, V., Hultsch, D. and Meermann, D. (2007) 'Physical fitness in children with developmental coordination disorder', *Research Quarterly for Exercise and Sport*, 78: 438–50.

Sigmundsson, H., Pedersen, A.V., Whiting, H.T.A. and Ingvaldsen, R.P. (1998) 'We can cure your child's clumsiness! A review of the intervention methods', *Scandinavian Journal of Rehabilitation Medicine*, 30: 101–6.

Smyth, M.M., Anderson, H.I. and Churchill, A.C. (2001) 'Visual information and the control of reaching in children: a comparison between children with and without developmental coordination disorder', *Journal of Motor Behavior*, 33: 306–20.

Sugden, D. (2006) *Leeds Consensus Statement: Developmental Coordination Disorder as a Specific Learning Difficulty*, Leeds: Economic & Social Research Council.

Sugden, D. and Chambers, M.E. (1998) 'Intervention approaches and children with developmental coordination disorder', *Pediatric Rehabilitation*, 2: 139–47.

Tsai, C.L., Wu, S.K. and Huang, C.H. (2008) 'Static balance in children with developmental coordination disorder', *Human Movement Science*, 27: 142–53.

Tsiotra, G.D., Flouris, A.D., Koutedakis, Y., Faught, B.E., Nevill, A.M., Lane, A.M. and Skenteris, N. (2006) 'A comparison of developmental coordination disorder prevalence rates in Canadian and Greek children', *Journal of Adolescent Health*, 39: 125–7.

Turvey, M.T., Fitch, H.L., Tuller, B. and Kelso, J.A.S. (1982) 'The Bernstein perspective: I. The problems of degrees of freedom and context-conditioned variability', in J.A.S. Kelso (ed.) *Human Motor Behavior: An Introduction*, Hillsdale, NJ: Lawrence Erlbaum.

Utley, A. and Astill, S.L. (2007) 'Developmental sequences of two-handed catching: how do children with and without developmental coordination disorder differ?', *Physiotherapy Theory and Practice*, 23: 65–82.

Utley, A., Steenbergen, B. and Astill, S.L. (2007) 'Ball catching in children with developmental coordination disorder: control of degrees of freedom', *Developmental Medicine and Child Neurology*, 49: 34–8.

Visser, J. (2003) 'Developmental coordination disorder: a review of research on subtypes and comorbidities', *Human Movement Science*, 22: 479–93.

Viswanathan, P., Kagerer, F.A. and Whitall, J. (2010) 'Sequential reaching and grasping: predictive control in children with Developmental Coordination Disorder', paper presented at the Society for Neuroscience, San Diego, November.

Wann, J.P., Mon-Williams, M. and Rushton, K. (1998) 'Postural control and co-ordination disorders: the swinging room revisited', *Human Movement Science*, 17: 491–513.

Watemberg, N., Waiserberg, N., Zuk, L. and Lerman-Sagie, T. (2007) 'Developmental coordination disorder in children with attention-deficit-hyperactivity disorder and physical therapy intervention', *Developmental Medicine and Child Neurology*, 49: 920–5.

Whitall, J., Getchell, N., McMenamin, S., Horn, C., Wilms-Floet, A. and Clark, J.E. (2006) 'Perception-action coupling in children with and without DCD: frequency locking between task-relevant auditory signals and motor responses in a dual-motor task', *Child: Care, Health and Development*, 32: 679–92.

Williams, H.G. and Woollacott, M.H. (1997) 'Characteristics of neuromuscular responses underlying posture control in clumsy children', in J.E. Clark and J.H. Humphrey (eds) *Motor Development: Research & Reviews*, Reston, VA: NASPE, pp. 8–23.

Wilson, B.N., Kaplan, B.J., Crawford, S.G., Campbell, A. and Dewey, D. (2000) 'Reliability and validity of a parent questionnaire on childhood motor skills', *The American Journal of Occupational Therapy*, 54: 484–93.

Woodruff, S.J., Bothwell-Myers, C., Tingley, M. and Albert, W.J. (2002) 'Gait pattern classification of children with Developmental Coordination Disorder', *Adapted Physical Activity Quarterly*, 19: 378–91.

Zackowski, K.M., Thach, W.T., Jr. and Bastian, A.J. (2002) 'Cerebellar subjects show impaired coupling of reach and grasp movements', *Experimental Brain Research*, 146: 511–22.

Zoia, S., Castiello, U., Blason, L. and Scabar, A. (2005) 'Reaching in children with and without developmental coordination disorder under normal and perturbed vision 1', *Developmental Neuropsychology*, 27: 257–73.

13 Biomechnical and neuromuscular aspects of motor development in children with cerebral palsy

Laura Prosser and Diane Damiano

Introduction

In this chapter we will discuss how the process and outcome of motor development can be altered substantially by a brain injury early in life, specifically one that affects the sensory and motor areas as seen in children with cerebral palsy (CP). Current knowledge of developmental biomechanics in CP will be primarily reviewed; however, some inferences will be made from the study of older children with CP when our current understanding of atypical motor development is limited. After a brief overview of CP, the impact of growth and sensory impairments on motor development biomechanics will be briefly discussed, followed by the development of force production and neuromuscular mechanics, motor development, postural control and fundamental motor skills.

Definition, prevalence and clinical presentation of cerebral palsy

Cerebral palsy is defined as 'a group of permanent disorders of the development of movement and posture, causing activity limitation, that are attributed to non-progressive disturbances that occurred in the developing fetal or infant brain' (Rosenbaum *et al.* 2007, p. 9). CP is the most common neuromuscular disorder in children, with high economic and emotional costs. The reported prevalence of CP in the United States is 3.6 cases per 1,000 live births (Yeargin-Allsopp *et al.* 2008), and in northwest Europe, it is approximately 2.5 per 1,000, which has increased in the past several decades (Odding *et al.* 2006). Children, adolescents and adults with CP report lower quality of life or lower self-esteem than their typically developing (TD) peers (Livingston *et al.* 2007; Majnemer *et al.* 2007; Minter *et al.* 2007). The estimated net lifetime cost for a person with CP is over $900,000, and the estimated total lifetime costs for all persons born with CP in 2000 in the United States is $11.5 billion dollars (Honeycutt *et al.* 2004). Given the high individual and societal costs, improving existing methods of prevention, assessment and treatment for CP are important clinical and research objectives.

The motor disorder in CP is 'often accompanied by disturbances of sensation, perception, cognition, communication and behaviour, by epilepsy and by secondary musculoskeletal problems' (Rosenbaum *et al.* 2007, p. 9). Impairments caused by the primary neural insult include poor coordination, altered muscle activation, spasticity and a primary inability to produce force. These limit the amount and type of physical activity, leading to further muscle weakness, altered muscle physiology and decreased range of motion, all of which contribute to compensatory movement patterns that limit function. While the neurological lesion is non-progressive, musculoskeletal consequences often worsen over time and with growth.

The severity and anatomic extent of CP vary and are influenced by the type, size and location of the underlying central nervous system (CNS) lesion. The brain injury can be the result of periventricular leukomalacia in infants born preterm (damage to the white matter tracts transmitting information to and from the body), focal haemorrhagic infarcts (bleeding in one area of the brain), neuronal loss secondary to prolonged anoxia (loss of oxygen) or brain malformations (Bax *et al.* 2006). Mobility limitations in children with CP range from mild difficulties with high-level skills such as running to complete dependence on others for care.

The heterogeneity of CP has led to wide variability across studies in the types of patients studied and results, many of which appear to be conflicting, making it difficult to draw conclusions or generalize results. Methods to classify the motor presentation are critical in the clinical management of CP. Traditional classification schemes included the anatomical distribution of the involvement (e.g. diplegia, hemiplegia or quadriplegia) combined with the physiological phenotype of the tone or movement disorder (e.g. spasticity or dystonia). However, the terms 'diplegia' and 'quadriplegia', which have long been used in clinical practice, are imprecise, and their use is now discouraged, with preference for the terms 'unilateral' and 'bilateral'. As the field transitioned in recent years from an impairment to activity-based model of rehabilitation, functional classification scales have emerged. The Gross Motor Function Classification System (GMFCS) groups children based on functional mobility (Palisano *et al.* 1997). Similarly, the Manual Ability Classification Scale (MACS) classifies upper extremity function (Eliasson *et al.* 2006b). A combination of functional and anatomical information is now recommended over the classic subtypes described above (Gorter *et al.* 2004; Rosenbaum *et al.* 2007).

Challenges of studying the biomechanics of motor development in cerebral palsy

While some children may demonstrate clearly abnormal motor patterns or neurological signs in infancy, the diagnosis of CP is often not suspected until the achievement of significant motor milestones, such as walking, are delayed. The prediction of CP is improving using tests that will be described later in this chapter. However, a definitive diagnosis is often still not made until a child is 18–24 months of age. Given this limitation, studies of motor development in infants who will later be diagnosed with CP must use other factors, such as degree of prematurity (comparison of full-term to preterm infants), presence of brain lesion (comparison of infants with a specific brain injury to those without) or gross motor test scores (comparison of infants with motor delays to those without), in an effort to identify children who are 'at risk' prior to the definitive determination of a CP diagnosis. While not all of these children will be diagnosed with CP, these strategies are useful in gaining information about motor development in those likely to develop CP.

Basic theoretical concepts

Growth and maturation

A child's size, body proportions and rate of skeletal growth affect his or her developmental trajectory. For information relating to growth and maturation in TD children, we refer the reader to Chapter 1. In children with developmental disabilities, variations in these factors may have an even greater impact. Abnormal growth and maturation can affect health and survival as well as social, physical and psychological functioning (Kuperminc and Stevenson 2008). Growth reference standards created from 360 children with quadriplegic CP in the United States were below their national counterparts in stature and weight, with CP-related difference in stature increasing with age (Krick *et al.* 1996). Poor growth in children with CP can often be attributed to malnutrition, endocrinopathy or poor skeletal development and should be treated when indicated (Kuperminc and Stevenson 2008).

Skeletal maturation demonstrates greater variability in children with CP, with evidence of both accelerated and delayed maturation, the patterns of which are inconsistent with the degree of motor involvement (Gollapudi *et al.* 2007; Henderson *et al.* 2005b; van Eck *et al.* 2008). Perhaps more importantly, low bone mineral density is common in children with CP. With the more impaired children at greater risk, physical limitations appear to dramatically affect bone health (Henderson *et al.* 2005a).

Sensory system and biomechanical implications

In Chapter 2, it is described how the sensory systems develop in TD children. In children with CP, there is a high prevalence of sensory abnormalities, including visual or auditory impairments, vestibular problems or tactile and proprioceptive deficits, which could have an adverse effect on motor function.

There is evidence of structural damage in sensory pathways throughout the nervous system. Fukuhara *et al.* (2010) demonstrated that some children with CP show evidence of sensory changes on the myelin sheath of peripheral nerves, postulated to be the result of chronic overactivity of muscle spindles due to spasticity, which is a heightened response to muscle stretch. Hoon Jr *et al.* (2009) showed that damage to sensory white matter pathways in the brain were more extensive in children with CP than in motor pathways. Reduced cortical activity in somatosensory areas has also been reported in children with CP compared to their peers during tactile discrimination tasks (Wingert *et al.* 2010). It is unknown if sensory damage in the CNS is a direct result of the primary perinatal injury or a secondary result of altered sensorimotor experiences.

The development of sensory deficits and how they relate to motor development is poorly understood. A recent study suggested that sensory stimulation, such as proprioceptive input or muscle activation through stretch, electrical stimulation or voluntary contraction may be an effective avenue to drive motor recovery (Behrman *et al.* 2006). Clinical trials are needed that investigate the effect of sensory training on motor performance.

Motor development in children with CP

Neuromuscular activation

Abnormal neuromuscular activation is the primary impairment in children with CP. Despite the static nature of the cerebral lesion, the resulting musculoskeletal impairments progress with growth and the repetition of unrefined, abnormal or compensatory movements during development. Such findings underscore the need to directly address activation deficits early to mitigate these secondary sequelae. Figure 13.1 illustrates a conceptual framework of the development and impact of motor impairment in CP.

Abnormal neuromuscular activation is present both at rest and during movement in CP. An altered state of resting muscle tone is one element of the clinical diagnosis of CP. Spasticity, an involuntary heightened velocity-dependent response to passive stretch and dystonia, involuntary abnormal movements or postures that may occur even when the child is attempting to rest, are the most common types of altered muscle tone. These result from damage to central neural mechanisms (Sanger *et al.* 2003). Abnormalities are present in spinal pathways in CP as well (Achache *et al.* 2010). In typical voluntary movement, reciprocal inhibition facilitates coordination through relaxation of antagonist muscles during activation of agonists. Individuals with CP may demonstrate concurrent muscle activation in the agonist and antagonist, termed 'reciprocal excitation'. This phenomenon was not observed in healthy adults, or in those who had a stroke, but is a characteristic of infant reflexes (Myklebust *et al.* 1982). This suggests that the neurological damage in CP causes retention of immature muscle activation patterns.

The ability of children with CP to voluntarily activate their muscles can be as low as 25 per cent of that of TD children during maximal effort tasks (Stackhouse *et al.* 2005) and displays less precise modulation of force output during submaximal tasks (Bandholm *et al.* 2009). In an apparent paradox, children with CP also demonstrate significantly greater levels and durations of muscle activity during functional movement compared to their peers (Lauer *et al.* 2007), even at a young age (Prosser *et al.* 2010b). This excessive and inefficient muscle activity contributes to greater co-contraction, which further inhibits coordinated and energetically efficient movement (Unnithan *et al.* 1996).

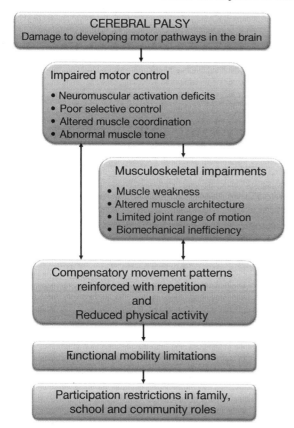

Figure 13.1 Schematic summary of the progression from an injury to the brain to primary and secondary impairments and consequences on activity, mobility and participation.

Sutherland *et al.* (1988) documented the development of phasic muscle activity in lower extremity muscles during walking in young children with TD. The gastrocsoleus demonstrated two different patterns, with the pattern that the authors labelled as 'mature' becoming increasingly prevalent with age. Interestingly, the 'immature' pattern of premature activation in late swing is commonly seen in children with CP (Romkes and Brunner 2007; Sutherland *et al.* 1988). In another example of immature muscle activation patterns, the mean frequency of quadriceps EMG signals (a measure of motor unit activation synchrony) was higher in a younger age group (2–6 years) of children with CP compared to an older group (7–14 years) with CP (Lauer *et al.* 2010). In contrast to the TD comparison groups, this pattern may indicate a motor strategy in the younger group with CP of continuous motor unit activation with little rate or time modulation, which improves somewhat over time, but remains elevated compared to TD peers.

The end result of these impairments in neuromuscular activation, including involuntary motor responses, poor force generation and inappropriate muscle activity, is impaired precise motor control of the trunk and limbs. In fact, selective voluntary motor control, the ability to freely move individual joints as desired, is impaired at all functional levels in CP (Fowler *et al.* 2010) and has been shown to explain more of the variance in gross motor function than spasticity, range of motion, or gestational age (Ostensjo *et al.* 2004).

Strength and muscle size

Compared to TD children (see Chapter 4), children with CP demonstrate reduced maximum voluntary muscle force production (Poon and Hui-Chan 2009; Wiley and Damiano 1998). Such reductions in strength are related to poor motor function (Damiano and Abel 1998). Several muscle factors compound the neurological factors and further contribute to weakness in CP. Muscle tissue is highly plastic, and the amount and intensity of physical activity is a major factor in muscle development. A number of investigators have reported reduced muscle size in individuals with CP relative to their TD peers (Malaiya *et al.* 2007; Moreau *et al.* 2009a; Stackhouse *et al.* 2007).

Furthermore, muscle length growth often does not keep up with bone growth in children with CP. Gastrocnemius-soleus muscle-tendon unit length was shorter in preterm infants compared to full-term infants at term age, 6 weeks and 12 weeks of age (Grant-Beuttler *et al.* 2009). This shortening of the muscle-tendon unit relative to bone growth (as measured by joint range of motion over time) worsens during early childhood and into adolescence in those with CP (Nordmark *et al.* 2009).

Muscle architecture

In typical skeletal muscle there is a relative balance of Type I (slow-twitch) and Type II (fast-twitch) fibres, and this distribution remains constant throughout development (Brooke and Engel 1969). In children with CP, there is a predominance of Type I at the expense of Type II fibres (Marbini *et al.* 2002). A predominance of slow-twitch fibres, used to maintain a muscle's endurance, may result from prolonged muscle activation (Prosser *et al.* 2010b) over time and may contribute to the counter-intuitive phenomenon of fatigue resistance (Moreau *et al.* 2009b; Stackhouse *et al.* 2005) in children with CP. Muscle architecture is also altered at the cellular level and at the neuromuscular junctions in individuals with CP (Friden and Lieber 2003; Theroux *et al.* 2005). Further reducing the amount of true contractile muscle tissue in children with CP is the greater proportion of adipose (fatty) tissue throughout their muscles (Johnson *et al.* 2009) and selective muscle atrophy, which creates greater variation in muscle fibre diameter in this population (Rose *et al.* 1994). Since CP is a not a primary muscle disorder, longitudinal studies beginning in infancy or early childhood are needed to better understand how and why alterations both within the muscle fibre itself and in the supporting connective tissue develop.

General movements

Infant movements have been studied in depth to determine how early CP can be identified. If the diagnosis of CP can be made sooner, more intensive and focused treatment for the child could begin during infancy. Einspieler and Prechtl (2005) classified early infant movements starting from preterm birth to the first several months of post-term life and examined their relationship to motor outcome in childhood. Abnormalities in these movement patterns are predictive of CP (Einspieler and Prechtl 2005) with unique predictors for spastic and dystonic types (Einspieler *et al.* 2002). Direct relationships between the degree of white matter damage in the brain and general movement quality have been identified as early as 1 month of age in infants born prior to 30 weeks of gestation (Spittle *et al.* 2008b).

Writhing movements are present from birth to 6–9 weeks of age and involve the whole body, including head, trunk, arms and legs. Movements occur in variable sequences, with variable force and speed, but always begin and end gradually. In infants with nervous system damage, abnormal general movements are observed during the first two months of life in addition to or instead of writhing movements. In typical infants, fidgety movements gradually emerge from writhing movements at 6–9 weeks of age and are present until 15–20 weeks of age. These are continuous, circular movements of small amplitude and moderate speed. In infants with nervous system damage, fidgety movements may be abnormal or absent. Fidgety movements gradually disappear with the emergence of intentional antigravity movements.

Gross motor development

The Test of Infant Motor Performance (TIMP) has been developed to predict motor function outcome in the first few months of life (Campbell *et al.* 1995). Scoring is based on the infant's ability or inability to perform specific elicited motor tasks and on behavioural observation. It is validated for use from 34 weeks postconceptional age to 4 months post-term, has good predictive ability and is sensitive to change in motor function (Lekskulchai and Cole 2001; Spittle *et al.* 2008a).

Infants who were identified as high-risk for CP at birth but did not develop CP demonstrated motor delays at 3 months but recovered to attain age-appropriate mean scores on the TIMP by 6 months of age (Campbell and Wilhelm 1985). These findings of motor 'recovery' in some children at high risk for CP have been supported by Pin *et al.* (2009). Alternately, some infants with neurological damage may demonstrate extensor hypertonicity that can be misjudged by their caregivers as representing advanced motor skills, such as rolling from prone to supine without trunk rotation, or an early ability to fully bear weight in supported standing (Pin *et al.* 2009). These studies demonstrate the need for accurate prediction of functional outcome to ensure that families receive reliable information about the needs of their infants and to facilitate the delivery of the most effective treatment regimens.

The Gross Motor Function Measure (GMFM) is a widely used tool to document motor delay in children with CP and to measure change in motor ability (Russell *et al.* 1989). The aforementioned GMFCS was developed from the GMFM database of a large cohort of children with CP to classify differing levels of functional mobility in CP. The classification ranges from I to V, with level I representing mild impairment primarily affecting only more advanced motor skills, such as running and jumping, and level V representing severe impairment with dependence on others for mobility and self-care. Levels II through IV represent decreasing levels of ambulatory ability. In individuals with CP, GMFCS level has been shown to remain relatively stable over the lifespan (McCormick *et al.* 2007; Palisano *et al.* 2006), except during the first two years of life (Gorter *et al.* 2009). Forty-two per cent of children classified before 2 years of age were reclassified to a different level after the age of 2. Most of these children were re-classified from levels II and III, which are difficult to distinguish at a young age. Such findings suggest that there may be a considerable window of opportunity to influence the trajectory of motor development prior to 2 years of age in children with CP.

Gross motor development peaks at the mean age of 6–7 years in children with CP, with a subsequent decline in motor ability observed particularly at GMFCS levels II, III and IV (Beckung *et al.* 2007; Harries *et al.* 2004). Even more revealing is the age by which children with CP reach 90 per cent of their motor development potential (termed 'age-90'). With the creation of gross motor development curves in children with CP, the age-90 was calculated for each level of the GMFCS in a total of 657 children (Rosenbaum *et al.* 2002). The age-90 for levels I through V, are 4.8, 4.4, 3.7, 3.5 and 2.7 years, respectively. This further emphasizes the need for intensive intervention *prior* to nearing the developmental peak in order to effectively alter the trajectory of motor development. The greater the level of severity, the shorter this therapeutic window may be.

Postural control

Postural control is the foundation for functional mobility. Chapter 8 describes the development of balance control in TD children. Differences between infants with CP and their TD peers are seen early in the process of developing postural control and are often the first signs that an infant is not developing as expected.

Head control is the first skill acquired in the development of upright postural control. Lacey *et al.* (1985) evaluated 104 infants born preterm at 25–33 weeks of gestation for protective head lateral rotation in prone and head lag during pull to sit; 11 of the 104 children were later diagnosed with CP. For protective head turning, infants with later abnormal development showed persistence of the earlier, less mature pattern of movement (slow, gravity-assisted rotation with movement at the lower trunk and pelvis rather than progression to isolated head extension and rotation without spinal extension). Similarly, for head position during pull to sit, the infants with later abnormal outcome demonstrated persistence of head lag rather than progression to eventual head righting. These results imply that more primitive movement patterns are retained in CP instead of developmental progression to more advanced movement patterns.

Sitting is one of the most important functional postures for humans. Children are required to sit for extended amounts of time when they attend school and often assume various sitting positions during play. Impairment in sitting can significantly limit participation in family, social and academic activities. One method of measuring postural stability is to track the centre of pressure (COP) movement in various positions. During seated postural sway analysis, Deffeyes *et al.* (2009) reported that infants who had been diagnosed or were at risk for CP demonstrated more repeated, or stereotypical, COP patterns and less variability in their movement coordination compared to infants with TD. These findings suggest that impairments in seated postural control are present early in development in children with CP.

Different detrimental effects of preterm birth and periventricular leukomalacia on the ability to generate seated postural adjustments have been described in young children (Hadders-Algra *et al.* 1999). Preterm children were able to generate various postural responses, but had a limited capacity to modulate the amplitude of muscle activity when balance is perturbed. Conversely, a generally limited repertoire of postural responses was a consistent observation in the children born preterm who also had periventricular leukomalacia. Postural control limitations contribute to functional limitations, with a more stable head, a more mobile trunk and a more stable pelvis being related to better functional performance and/or a better quality of reaching in children with CP (van der Heide *et al.* 2005a).

Trunk rotation is also limited in children born preterm. De Groot *et al.* (1995) studied postural control in 37 preterm infants and 20 full-term infants during sitting. The authors demonstrated that at 52 weeks of age, a majority of the preterm infants could sit without using their hands for support, but were unable to rotate their trunk during play in this position. The preterm infants demonstrated greater trunk extensor muscle activity and shoulder retraction during various motor tasks, including sitting, when compared to the full-term infants. This increased muscle activity may be required for these children to maintain sitting. However, the muscle activity in preterm infants was significantly different from the typical pattern of smooth, coordinated and phasic activity of the trunk muscles in the full-term infants that allows constant adaptation to the environment through small changes in movement direction and body position.

Another example that children with CP retain immature movement patterns is provided by Woollacott *et al.* (1998). These authors investigated postural responses to perturbations during static standing in young children with CP and TD with varying levels of walking experience. Regardless of walking experience, the children with CP demonstrated disorganized muscle responses, a high level of coactivation and a proximal to distal pattern of muscle activation. This pattern was most similar to the pattern observed in the 10-month-old infants with TD who were not yet walking independently. The timing of muscle activation was in contrast to the distinct, phasic bursts of distal to proximal muscle activity in the experienced walkers with TD.

Reaching

Movements of the proximal limb segments position the hand for object manipulation. Successful reaching is important for self-care and functional independence. The development of vision contributes significantly to the development of reaching and grasping (O'Connor *et al.* 2009). Retinopathy of prematurity is a common sequela of preterm birth; therefore, children with CP often have visual impairments. The influence of visual deficits on fine motor delay should be differentially distinguished from neuromotor factors.

Reaching movements in children with CP tend to be slower compared to their peers with TD, trajectories are less direct and may consist of multiple rather than a single movement unit (van der Heide *et al.* 2005b). The quality of reaching is related to the extent and location of the lesion, the severity of the motor disorder, the degree of biceps spasticity and functional performance (van der Heide *et al.* 2005b). CP-related deficits in the development of anticipatory reaching control have also been reported at a young age (van der Meer *et al.* 1995), suggesting impairments in the planning as well as execution of upper extremity movements.

Postural control during the development of reaching skill is also atypical in children with CP. During reaching in supine, preterm infants demonstrated slower and less trunk movement than full-term infants (Fallang *et al.* 2003). Counterintuitively, the more 'still' the postural behaviours in the preterm infants, the better their reaching quality, indicating that reduced trunk movement in the early development of reaching may be a compensatory strategy to provide stability for limb movement. Conversely, kinematic analysis of the trunk during seated reaching in children with CP reveals that a 'stiffer' trunk is related to poorer functional reaching performance (van der Heide *et al.* 2005a). A more mobile trunk, with a stable pelvis and head, was observed in children with TD, supporting the suggestion that a stable pelvic base, with controlled flexibility and postural adjustments in the trunk, allows for head stability and functional hand–eye reaching coordination. Considering these studies together, it appears that while children with CP may develop early postural compensations that appear to improve function in the beginning stages of skill development, the persistence of those compensatory strategies leads to long-term functional limitations in the ability to skilfully position the trunk for more complex arm and hand use (Figure 13.2).

The development of hand function and precision grip control can continue to improve through adolescence in individuals with CP (Eliasson *et al.* 2006a). Even more encouraging are the results of a randomized controlled trial of an 8-week upper limb movement training programme in 2-month-old infants born at less than 33 weeks gestational age. After training, the treatment group had more toy–hand contacts than both the full-term and preterm groups who received social training only. The treatment group was not different from the full-term group in

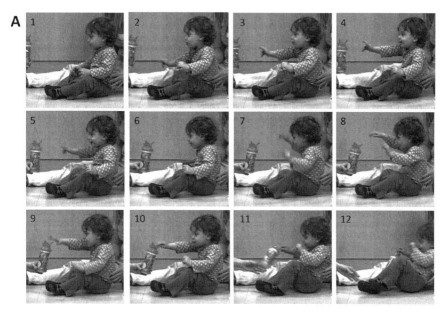

Figure 13.2 Sequential images of two children reaching forwards for a bottle. (A) This set of images shows an 18.5 months old (corrected for premature birth) boy with spastic diplegic cerebral palsy. He was unable to sit on his own on the floor for more than several seconds at a time. Note his flexed trunk and knees, which contributed to his sitting balance difficulties and were partially a result of tight hamstring muscles. He was not able to reach the bottle on his first attempt (frame 4). He was successful on the second attempt (frame 10), but fell when bringing the bottle towards his body (frames 11–12).

(B) This set of images shows a girl of similar age with typical motor development. She was stable in floor sitting, reaching beyond her base of support, and demonstrated a more upright head and trunk posture.

total toy–hand contact time, or in the number of contacts with an open hand. However, the preterm social group demonstrated poorer performance in both areas (Heathcock *et al.* 2008). This study demonstrates that early and focused training can reduce or eliminate the deficits in early reaching behaviour observed in preterm infants at risk for CP.

Kicking

Purposeful lower limb movements gradually emerge from the spontaneous general limb movements observed in the first few months of life. Unlike full-term infants, infants born preterm did not increase their leg-kicking frequency in response to feedback (Heathcock *et al.* 2005). However, purposeful leg movements in preterm infants can be trained to match the levels of infants born full-term (Heathcock and Galloway 2009).

Geerdink *et al.* (1996) documented kicking patterns in full-term and preterm infants born between 27 and 34 weeks of gestation at 6, 12 and 18 weeks of age. Differences in kick frequency, intra-kick pauses and joint coupling between preterm and full-term infants were observed at 6 and 12 weeks, but few differences remained at 18 weeks. This group of preterm infants was specifically selected for having no evidence of periventricular leukomalacia and likely represents the portion of preterm infants who 'recover' or develop minor neurological dysfunction, but not CP. In contrast, infants with periventricular leukomalacia demonstrated different kicking patterns through at least 26 weeks of age compared to full-term infants, including an inability to disassociate the joints of the lower extremity (Vaal *et al.* 2000), meaning that infants at risk for CP extended or flexed their whole leg as a unit and were less able to move their hips, knees and ankles separately from one another. Leg movements are therefore more stereotypical with less variability and flexibility within and across tasks. This tendency is more prevalent in those born more prematurely and may be predictive of later delays in walking (Jeng *et al.* 2002, 2004).

These observations on the biomechanical differences in kicking between children at risk for CP and their typically developing peers have provided useful insights into the motor control mechanisms underlying CP-related motor deficits. However, detailed kinematic analysis of leg movements may not be as clinically useful in the initial few weeks or months after birth as other assessment tools. The observation of general movements was more related to the presence of brain injury and neurological outcome than detailed kinematic analysis of kicking patterns in preterm infants at 31–9 weeks post-conceptional age (Droit *et al.* 1996). Van der Heide *et al.* (1999) concurred that *qualitative* assessments of leg and general movements were more appropriate and sensitive measures than kinematic analysis at 1 and 3 months of age. This distinction is important for clinicians who are observing young infants at risk for CP and predicting future motor ability.

Transitioning to standing

Transitional movements such as moving from a sitting to standing position are critical for functional independence, and they are common as therapeutic targets. Unfortunately, transitional movements have not been studied in depth in children with CP. In one of the few studies, Dan *et al.* (2000) demonstrated that children with CP hyperextend their neck during transitions from squatting to standing and excessively flex their neck during the reverse movement. Even when children with CP use similar movement patterns to those described in children with TD, they may have significantly less intra- and inter-individual variation in those movement strategies (Mewasingh *et al.* 2002). Such results provide further evidence of the limited and stereotyped movement repertoire in children with CP. Perhaps intervention should be focused much earlier in development when the abnormal movements are first observed, so that alternate movement strategies can be introduced before these less flexible and variable movements become too well established.

Walking

Independent ambulation is one of the most important goals for families and care-givers of children with CP. Consequently, ambulatory prognosis is an essential aspect of family education and counselling during the early years of the child's life. Age at achievement of independent sitting is one of the best predictors of later ambulation and may explain up to 91 per cent of the variance in ambulatory ability (Watt *et al.* 1989). Most studies agree that the ability to sit independently by the age of 2 years is related to the eventual achievement of independent walking (with or without an assistive device) (Badell-Ribera 1985; da Paz Jr *et al.* 1994; Fedrizzi *et al.* 2000; Watt *et al.* 1989).

Other milestones that have a strong association with the attainment of independent walking include head control by 9 months, crawling by 30–36 months and the ability to bear weight through the hands in prone and to roll from supine to prone by 18 months of age (Bottos *et al.* 1995; da Paz Jr *et al.* 1994; Fedrizzi *et al.* 2000). Factors related to more impaired ambulation are the achievement of head control after the age of 20 months and the persistence of infant reflexes at 2 years of age (da Paz Jr *et al.* 1994; Watt *et al.* 1989).

Badell-Ribera (1985) described several skills at 1.5 to 2.5 years of age that predicted the functional level of later ambulation in 50 children with CP. The children who could never maintain sitting with hand support did not develop the ability to ambulate, even with assistance. Those who required considerable assis-tance to walk in the future were able to maintain sitting with hand support by the age of 2 years, but were not able to assume a sitting position or crawl by this age. Children who became household ambulators with an assistive device by the age of 5.5 years could transition to sitting from prone by age 2 years and could crawl symmetrically by 2.5 years, but never learned to reciprocally crawl. Children who would later ambulate limited distances outdoors with an assistive device could also transition to sitting from prone and crawl symmetrically by age 2.5 years, but developed reciprocal crawling between 3–3.5 years of age. All children who developed community ambulation without assistive devices by the age of 6 years crawled reciprocally by the age of 2.5 years and never used a symmetrical crawling pattern.

As with kicking movements, infant stepping patterns may provide fundamental insights into how impaired walking patterns develop. Davis *et al.* (1994) demonstrated that low-risk preterm infants performed coordinated and alternating stepping at 1, 6 and 9 months of age on a treadmill. Stepping frequency peaked at 6 months for most infants and was responsive to changes in treadmill speed. These patterns were similar to those observed from an earlier study of full-term infants (Thelen and Ulrich 1991), which may suggest a better prognosis for them.

In contrast, results from a longitudinal study by Angulo-Barroso *et al.* (2010) show that infants who were later diagnosed with CP demonstrated differences in stepping patterns. Fifteen infants at risk for neuromotor delay were assessed every two months from when they took at least six steps in one minute when supported over a treadmill until the onset of walking or 24 months corrected age. As a group, the infants increased the number of alternating steps, flat foot contacts versus toe contacts and high-level physical activity. However, those children who were eventually identified as having CP demonstrated lower values on all three measures compared to those who were not later diagnosed with CP. Higher rates of alternating stepping and greater levels of high-level physical activity were related to earlier onset of walking.

As with other motor skills, focused training for stepping is feasible at a young age and may improve later functional outcomes. Bodkin *et al.* (2003) showed that after six months of step training initiated at 5 months of age, a preterm infant with an intraventricular haemorrhage increased the percentage of alternating steps while stepping asymmetries decreased. The child remained delayed in motor skills through 18 months of age, but was discharged from physical therapy by 2 years of age due to age-appropriate gross motor skills. No gait deviations were observed at that time.

Walking patterns in children with CP often resemble the immature patterns that are observed in children with TD during supported walking or in the first few months of walking ability (Leonard *et al.* 1991). They do not develop to the level of refined coordination or advanced skill that is typical, except perhaps in children at GMFCS level I. Prosser *et al.* (2010a) demonstrated that young children with CP showed slower walking speed and cadence, shorter step length and reduced single support time, compared with a TD group, even when matched for months of walking experience rather than age. These same children had greater total muscle activation and differences in activation timing of trunk and hip muscles during walking compared to the TD group (Prosser *et al.* 2010b). The authors suggest that this excessive, non-reciprocal trunk muscle activity may be a compensation for poor postural control and may be a strategy to limit postural sway in an effort to maintain upright posture. However, this compensation strategy may limit the ability to precisely control changes in the body's centre of mass during dynamic movements (Hsue *et al.* 2009).

Similar to gross motor function, walking ability in children with CP typically plateaus during school age and then deteriorates by adolescence and young adulthood (Day *et al.* 2007). CP-related signs of gait deterioration include increased double support time and reduced joint excursion at the pelvis, hip and knee (Johnson *et al.* 1997). Gait decline may be a consequence of muscle weakness and postural instability that cannot keep up with growth and increasing musculoskeletal impairments.

Compounding the issues related to growth and musculoskeletal impairments is the relative inactivity of children with CP compared to their peers. Physical activity of school-age children can be 20 to 75 per cent less in ambulatory children with CP compared to controls and is related to functional level (Bjornson *et al.* 2007; van den Berg-Emons *et al.* 1995). Reduced activity likely contributes to motor disability in children with CP and increasing activity levels has been the focus of recent advocacy (Damiano 2006).

Clinical and practical applications

Therapeutic interventions are the only avenue currently available to clinically address the primary impairment of poor neuromuscular control in children with CP. However, research in this area has only recently begun to emerge. Most of the available scientific evidence focuses instead on interventions that target secondary impairments such as decreased flexibility, weakness or poor endurance.

The majority of medical and surgical interventions similarly target secondary musculoskeletal deformities, and they are typically reactive to bone or muscle deformities rather than preventative. Botulinum toxin injections or medication may temporarily decrease spasticity. Soft tissue lengthenings increase joint range of motion. Osteotomies reduce bony abnormalities that result from years of abnormal muscle and positional forces, but do not correct the originating abnormal forces. In fact, while kinematics may improve, patterns of muscular activation rarely change after orthopaedic procedures (Brunt and Scarborough 1988; Lee *et al.* 1992).

While these interventions remain an important part of managing the course of CP throughout the lifespan, they do little to address the primary issue of poor neuromuscular activation. Pharmaceutical or stem-cell interventions may be available in the near future to stimulate brain recovery, but are currently not developed sufficiently for widespread clinical use and will likely need to be combined with motor training that directs recovery in specific pathways.

Current efforts focus on the design of treatment strategies in accordance with recent findings in neuroscience about cortical motor plasticity and how specific practice paradigms can promote learning and lasting improvements in posture and movement control (Kleim *et al.* 2004; Nudo 2003). Three basic themes regarding practice have emerged from research in various neurological patient populations. For optimal outcomes, practice should be intensive, correct and initiated early in the course of the disorder.

Intensive and challenging motor practice can induce training-dependent changes in the cortical structures of adults of varying disabilities (Hlustik *et al.* 2004; Kleim *et al.* 2004; Nudo 2003; Pascual-Leone *et al.* 1995; Winchester *et al.* 2005) and promote learning of skilled movement in adults with neurological injury (Behrman and Harkema 2000; Kunkel *et al.* 2003). The most effective rehabilitation programmes for patients with neurological injury include both intensive and early intervention (Dobkin *et al.* 2007; Horn *et al.* 2005; Wolf *et al.* 2006).

Children with CP develop compensatory movement strategies as a result of poor postural control, muscle weakness, poor coordination and spasticity. If motor patterns are reinforced with repetition, these abnormal patterns are reinforced daily with each functional movement, over time and throughout development. Changing the well-established abnormal motor patterns after years of reinforcement through conservative treatment is challenging at best and may not be possible. Rehabilitation clinicians are in an optimal position to intervene in this process. If effective strategies are identified, there may be greater potential to improve function in younger children, who have less reinforced abnormal movement patterns. Such practice may prevent some of the compensatory movement patterns from developing.

Applying neuro-rehabilitation strategies used for adults, who acquired CNS damage later in life, to individuals with CP may not result in similar outcomes. First, adults who sustain an injury typically have a functioning nervous system until the injury. Children with CP have never functioned with a 'typical' CNS. Second, if intensive treatment is more beneficial the sooner it is provided after injury, interventions for CP would necessitate application during the first few years of life. The nervous systems of infants and young children may be more 'plastic' and more responsive to training than in adults or even older children, and the efficacy of early training should be explored (Johnston *et al.* 2009). In fact, early and intensive treatment programmes can be feasible in infants and young children, and several have demonstrated encouraging scientific results (Cope *et al.* 2008; Girolami and Campbell 1994; Goodman *et al.* 1985; Heathcock and Galloway 2009; Heathcock *et al.* 2008; Lekskulchai and Cole 2001; Richards *et al.* 1997; Ulrich *et al.* 2001).

Certainly, the age of the child impacts the type of activities that are appropriate to address in therapy. It also impacts the relative meaning of 'intense' practice. Overstimulation of infants must be avoided, and the line between intense and perhaps too intense needs to be deciphered.

Given the examples of immature movement patterns in children with CP that are similar to much earlier stages of typical development, one might conclude that additional practice is a simple solution. Yet, more practice alone does not alter movement patterns. More importantly, 'correct' practice must be achieved in order to train the nervous system to control the body more effectively and efficiently. Treatments that may facilitate the learning of 'correct' movement patterns are those that allow the child to initially produce the desired movement pattern with assistance, which is gradually reduced as the child learns to produce the movement on his or her own without compensations. Specific interventions may include functional electrical stimulation and robotic-assisted movement that can facilitate the learning of new movement patterns. Other treatments that may train more efficient movement are the use of external exercise devices, aquatic programmes and therapist-guided functional training, which may allow more effective movement patterns to develop in safe environments despite weakness and poor postural control.

Key points

- The primary impairment of children with CP is abnormal neuromuscular activation, caused by damage to cerebral motor pathways. Deficits during volitional movement are observed in selective voluntary control and co-ordination of muscle activation.
- While the neurological lesion is non-progressive, musculoskeletal consequences often worsen with growth and the accumulating effects of altered physical activity.
- Infant movements and early postural control in children later diagnosed with CP are different from their TD peers.
- The development of fundamental motor skills in children with CP often does not progress beyond the 'immature' patterns demonstrated during the early stages of typical skill development, preventing the achievement of more advanced and efficient motor patterns.
- Gross motor function peaks at a mean age of 6–7 years in children with CP. While neural plasticity is possible throughout the lifespan, the best opportunity to change the trajectory of motor function may be during the early years of life, prior to the reinforcement of compensatory movement strategies.
- Current treatment strategies should incorporate neuroscientific evidence about the potential for neural plasticity in response to correct practice, intensive training and the early initiation of habilitation programmes.

References

Achache, V., Roche, N., Lamy, J.C., Boakye, M., Lackmy, A., Gastal, A., Quentin, V. and Katz, R. (2010) 'Transmission within several spinal pathways in adults with cerebral palsy', *Brain*, 133: 1470–83.

Angulo-Barroso, R.M., Tiernan, C.W., Chen, L.C., Ulrich, D. and Neary, H. (2010) 'Treadmill responses and physical activity levels of infants at risk for neuromotor delay', *Pediatric Physical Therapy*, 22: 61–8.

Badell-Ribera, A. (1985) 'Cerebral palsy: postural-locomotor prognosis in spastic diplegia', *Archives of Physical Medicine and Rehabilitation*, 66: 614–19.

Bandholm, T., Rose, M.H., Sløk, R., Sonne-Holm, S. and Jensen, B.R. (2009) 'Ankle torque steadiness is related to muscle activation variability and coactivation in children with cerebral palsy', *Muscle & Nerve*, 40: 402–10.

Bax, M., Tydeman, C. and Flodmark, O. (2006) 'Clinical and MRI correlates of cerebral palsy: The European Cerebral Palsy Study', *Journal of the American Medical Association*, 296: 1602–8.

Beckung, E., Carlsson, G., Carlsdotter, S. and Uvebrant, P. (2007) 'The natural history of gross motor development in children with cerebral palsy aged 1 to 15 years', *Developmental Medicine and Child Neurology*, 49: 751–6.

Behrman, A.L. and Harkema, S.J. (2000) 'Locomotor training after human spinal cord injury: a series of case studies', *Physical Therapy*, 80: 688–700.

Behrman, A.L., Bowden, M.G. and Nair, P.M. (2006) 'Neuroplasticity after spinal cord injury and training: an emerging paradigm shift in rehabilitation and walking recovery', *Physical Therapy*, 86: 1406–25.

Bjornson, K.F., Belza, B., Kartin, D., Logsdon, R. and McLaughlin, J.F. (2007) 'Ambulatory physical activity performance in youth with cerebral palsy and youth who are developing typically', *Physical Therapy*, 87: 248–57.

Bodkin, A.W., Baxter, R.S. and Heriza, C.B. (2003) 'Treadmill training for an infant born preterm with a grade III intraventricular hemorrhage', *Physical Therapy*, 83: 1107–18.

Bottos, M., Puato, M.L., Vianello, A. and Facchin, P. (1995) 'Locomotion patterns in cerebral palsy syndromes', *Developmental Medicine and Child Neurology*, 37: 883–99.

Brooke, M.H. and Engel, W.K. (1969) 'The histographic analysis of human muscle biopsies with regard to fiber types. 4. Children's biopsies', *Neurology*, 19: 591–605.

Brunt, D. and Scarborough, N. (1988) 'Ankle muscle activity during gait in children with cerebral palsy and equinovarus deformity', *Archives of Physical Medicine and Rehabilitation*, 69: 115–17.

Campbell, S.K. and Wilhelm, I.J. (1985) 'Development from birth to 3 years of age of 15 children at high risk for central nervous system dysfunction. Interim report', *Physical Therapy*, 65: 463–9.

Campbell, S.K., Kolobe, T.H., Osten, E.T., Lenke, M. and Girolami, G.L. (1995) 'Construct validity of the Test of Infant Motor Performance', *Physical Therapy*, 75: 585–96.

Cope, S.M., Forst, H.C., Bibis, D. and Liu, X.C. (2008) 'Modified constraint-induced movement therapy for a 12-month-old child with hemiplegia: a case report', *American Journal of Occupational Therapy*, 62: 430–7.

da Paz Jr, A.C., Burnett, S.M. and Braga, L.W. (1994) 'Walking prognosis in cerebral palsy: a 22-year retrospective analysis', *Developmental Medicine and Child Neurology*, 36: 130–4.

Damiano, D.L. (2006) 'Activity, activity, activity: rethinking our physical therapy approach to cerebral palsy', *Physical Therapy*, 86: 1534–40.

Damiano, D.L. and Abel, M.F. (1998) 'Functional outcomes of strength training in spastic cerebral palsy', *Archives of Physical Medicine and Rehabilitation*, 79: 119–25.

Dan, B., Bouillot, E., Bengoetxea, A., Noel, P., Kahn, A. and Cheron, G. (2000) 'Head stability during whole body movements in spastic diplegia', *Brain Development*, 22: 99–101.

Davis, D.W., Thelen, E. and Keck, J. (1994) 'Treadmill stepping in infants born prematurely', *Early Human Development*, 39: 211–23.

Day, S.M., Wu, Y.W., Strauss, D.J., Shavelle, R.M. and Reynolds, R.J. (2007) 'Change in ambulatory ability of adolescents and young adults with cerebral palsy', *Developmental Medicine and Child Neurology*, 49: 647–53.

de Groot, L., Hopkins, B. and Touwen, B. (1995) 'Muscle power, sitting unsupported and trunk rotation in pre-term infants', *Early Human Development*, 43: 37–46.

Deffeyes, J.E., Harbourne, R.T., Kyvelidou, A., Stuberg, W.A. and Stergiou, N. (2009) 'Nonlinear analysis of sitting postural sway indicates developmental delay in infants', *Clinical Biomechanics*, 24: 564–70.

Dobkin, B., Barbeau, H., Deforge, D., Ditunno, J., Elashoff, R., Apple, D., Basso, M., Behrman, A., Harkema, S., Saulino, M. and Scott, M. (2007) 'The evolution of walking-related outcomes over the first 12 weeks of rehabilitation for incomplete traumatic spinal cord injury: the multicenter randomized spinal cord injury locomotor trial', *Neurorehabilitation and Neural Repair*, 21: 25–35.

Droit, S., Boldrini, A. and Cioni, G. (1996) 'Rhythmical leg movements in low-risk and brain-damaged preterm infants', *Early Human Development*, 44: 201–13.

Einspieler, C. and Prechtl, H.F. (2005) 'Prechtl's assessment of general movements: a diagnostic tool for the functional assessment of the young nervous system', *Mental Retardation and Developmental Disabilities Research Reviews*, 11: 61–7.

Einspieler, C., Cioni, G., Paolicelli, P.B., Bos, A.F., Dressler, A., Ferrari, F., Roversi, M.F. and Prechtl, H.F. (2002) 'The early markers for later dyskinetic cerebral palsy are different from those for spastic cerebral palsy', *Neuropediatrics*, 33: 73–8.

Eliasson, A.C., Forssberg, H., Hung, Y.C. and Gordon, A.M. (2006a) 'Development of hand function and precision grip control in individuals with cerebral palsy: a 13-year follow-up study', *Pediatrics*, 118: e1226–36.

Eliasson, A.C., Krumlinde-Sundholm, L., Rosblad, B., Beckung, E., Arner, M., Ohrvall, A.M. and Rosenbaum, P. (2006b) 'The Manual Ability Classification System (MACS) for children with cerebral palsy: scale development and evidence of validity and reliability', *Developmental Medicine and Child Neurology*, 48: 549–54.

Fallang, B., Saugstad, O.D. and Hadders-Algra, M. (2003) 'Postural adjustments in preterm infants at 4 and 6 months post-term during voluntary reaching in supine position', *Pediatric Research*, 54: 826–33.

Fedrizzi, E., Facchin, P., Marzaroli, M., Pagliano, E., Botteon, G., Percivalle, L. and Fazzi, E. (2000) 'Predictors of independent walking in children with spastic diplegia', *Journal of Child Neurology*, 15: 228–34.

Fowler, E.G., Staudt, L.A. and Greenberg, M.B. (2010) 'Lower-extremity selective voluntary motor control in patients with spastic cerebral palsy: increased distal motor impairment', *Developmental Medicine and Child Neurology*, 52: 264–9.

Friden, J. and Lieber, R.L. (2003) 'Spastic muscle cells are shorter and stiffer than normal cells', *Muscle & Nerve*, 27: 157–64.

Fukuhara, T., Namba, Y. and Yamadori, I. (2010) 'Peripheral sensory neuropathy observed in children with cerebral palsy: is chronic afferent excitation from muscle spindles a possible cause?', *Child's Nervous System*, 26: 751–4.

Geerdink, J.J., Hopkins, B., Beek, W.J. and Heriza, C.B. (1996) 'The organization of leg movements in preterm and full-term infants after term age', *Developmental Psychobiology*, 29: 335–51.

Girolami, G.L. and Campbell, S.K. (1994) 'Efficacy of a neuro-developmental treatment program to improve motor control of preterm infants', *Pediatric Physical Therapy*, 6: 175–84.

Gollapudi, K., Feeley, B.T. and Otsuka, N.Y. (2007) 'Advanced skeletal maturity in ambulatory cerebral palsy patients', *Journal of Pediatric Orthopedics*, 27: 295–8.

Goodman, M., Rothberg, A.D., Houston-McMillan, J.E., Cooper, P.A., Cartwright, J.D. and van der Velde, M.A. (1985) 'Effect of early neurodevelopmental therapy in normal and at-risk survivors of neonatal intensive care', *Lancet*, 2: 1327–30.

Gorter, J.W., Ketelaar, M., Rosenbaum, P., Helders, P.J. and Palisano, R. (2009) 'Use of the GMFCS in infants with CP: the need for reclassification at age 2 years or older', *Developmental Medicine and Child Neurology*, 51: 46–52.

Gorter, J.W., Rosenbaum, P.L., Hanna, S.E., Palisano, R.J., Bartlett, D.J., Russell, D.J.,

Walter, S.D., Raina, P., Galuppi, B.E. and Wood, E. (2004) 'Limb distribution, motor impairment, and functional classification of cerebral palsy', *Developmental Medicine and Child Neurology*, 46: 461–7.

Grant-Beuttler, M., Palisano, R.J., Miller, D.P., Reddien Wagner, B., Heriza, C.B. and Shewokis, P.A. (2009) 'Gastrocnemius-soleus muscle tendon unit changes over the first 12 weeks of adjusted age in infants born preterm', *Physical Therapy*, 89: 136–48.

Hadders-Algra, M., Brogren, E., Katz-Salamon, M. and Forssberg, H. (1999) 'Periventricular leuckomalacia and preterm birth have different detrimental effects on postural adjustments', *Brain*, 122 (Pt 4): 727–40.

Harries, N., Kassirer, M., Amichai, T. and Lahat, E. (2004) 'Changes over years in gross motor function of 3–8 year old children with cerebral palsy: using the Gross Motor Function Measure (GMFM-88)', *The Israel Medical Association Journal*, 6: 408–11.

Heathcock, J.C. and Galloway, J.C. (2009) 'Exploring objects with feet advances movement in infants born preterm: a randomized controlled trial', *Physical Therapy*, 89: 1027–38.

Heathcock, J.C., Lobo, M. and Galloway, J.C. (2008) 'Movement training advances the emergence of reaching in infants born at less than 33 weeks of gestational age: a randomized clinical trial', *Physical Therapy*, 88: 310–22.

Heathcock, J.C., Bhat, A.N., Lobo, M.A. and Galloway, J.C. (2005) 'The relative kicking frequency of infants born full-term and preterm during learning and short-term and long-term memory periods of the mobile paradigm', *Physical Therapy*, 85: 8–18.

Henderson, R.C., Kairalla, J.A., Barrington, J.W., Abbas, A. and Stevenson, R.D. (2005a) 'Longitudinal changes in bone density in children and adolescents with moderate to severe cerebral palsy', *Journal of Pediatrics*, 146: 769–75.

Henderson, R.C., Gilbert, S.R., Clement, M.E., Abbas, A., Worley, G. and Stevenson, R.D. (2005b) 'Altered skeletal maturation in moderate to severe cerebral palsy', *Developmental Medicine and Child Neurology*, 47: 229–36.

Hlustik, P., Solodkin, A., Noll, D.C. and Small, S.L. (2004) 'Cortical plasticity during three-week motor skill learning', *Journal of Clinical Neurophysiology*, 21: 180–91.

Honeycutt, A., Dunlap, L., Chen, H. and Al Homsi, G. (2004) 'Economic costs associated with mental retardation, cerebral palsy, hearing loss, and vision impairment – United States, 2003', *MMWR Weekly*, 53: 57–9.

Hoon Jr, A.H., Stashinko, E.E., Nagae, L.M., Lin, D.D., Keller, J., Bastian, A., Campbell, M.L., Levey, E., Mori, S. and Johnston, M.V. (2009) 'Sensory and motor deficits in children with cerebral palsy born preterm correlate with diffusion tensor imaging abnormalities in thalamocortical pathways', *Developmental Medicine and Child Neurology*, 51: 697–704.

Horn, S.D., DeJong, G., Smout, R.J., Gassaway, J., James, R. and Conroy, B. (2005) 'Stroke rehabilitation patients, practice, and outcomes: is earlier and more aggressive therapy better?', *Archives of Physical Medicine and Rehabilitation*, 86: S101–14.

Hsue, B.J., Miller, F. and Su, F.C. (2009) 'The dynamic balance of the children with cerebral palsy and typical developing during gait. Part I. Spatial relationship between com and cop trajectories', *Gait & Posture*, 29: 465–70.

Jeng, S.F., Chen, L.C. and Yau, K.I. (2002) 'Kinematic analysis of kicking movements in preterm infants with very low birth weight and full-term infants', *Physical Therapy*, 82: 148–59.

Jeng, S.F., Chen, L.C., Tsou, K.I., Chen, W.J. and Luo, H.J. (2004) 'Relationship between spontaneous kicking and age of walking attainment in preterm infants with very low birth weight and full-term infants', *Physical Therapy*, 84: 159–72.

Johnson, D.C., Damiano, D.L. and Abel, M.F. (1997) 'The evolution of gait in childhood and adolescent cerebral palsy', *Journal of Pediatric Orthopedics*, 17: 392–6.

Johnson, D.L., Miller, F., Subramanian, P. and Modlesky, C.M. (2009) 'Adipose tissue infiltration of skeletal muscle in children with cerebral palsy', *Journal of Pediatrics*, 154: 715–20.

Johnston, M.V., Ishida, A., Ishida, W.N., Matsushita, H.B., Nishimura, A. and Tsuji, M. (2009) 'Plasticity and injury in the developing brain', *Brain Development*, 31: 1–10.

Kleim, J.A., Hogg, T.M., VandenBerg, P.M., Cooper, N.R., Bruneau, R. and Remple, M. (2004) 'Cortical synaptogenesis and motor map reorganization occur during late, but not early, phase of motor skill learning', *Journal of Neuroscience*, 24: 628–33.

Krick, J., Murphy-Miller, P., Zeger, S. and Wright, E. (1996) 'Pattern of growth in children with cerebral palsy', *Journal of the American Dietetic Association*, 96: 680–5.

Kunkel, A., Kopp, B., Muller, G., Villringer, K., Villringer, A., Taub, E. and Flor, H. (2003) 'Constraint-induced movement therapy for motor recovery in chronic stroke patients', *Archives of Physical Medicine and Rehabilitation*, 80: 624–8.

Kuperminc, M.N. and Stevenson, R.D. (2008) 'Growth and nutrition disorders in children with cerebral palsy', *Developmental Disabilities Research Reviews*, 14: 137–46.

Lacey, J.L., Henderson-Smart, D.J., Edwards, D.A. and Storey, B. (1985) 'The early development of head control in preterm infants', *Early Human Development*, 11: 199–212.

Lauer, R.T., Pierce, S.R., Tucker, C.A., Barbe, M.F. and Prosser, L.A. (2010) 'Age and electromyographic frequency alterations during walking in children with cerebral palsy', *Gait & Posture*, 31: 136–9.

Lauer, R.T., Stackhouse, C.A., Shewokis, P.A., Smith, B.T., Tucker, C.A. and McCarthy, J. (2007) 'A time-frequency based electromyographic analysis technique for use in cerebral palsy', *Gait & Posture*, 26: 420–7.

Lee, E.H., Goh, J.C.H. and Bose, K. (1992) 'Value of gait analysis in the assessment of surgery in cerebral palsy', *Archives of Physical Medicine and Rehabilitation*, 73: 642–6.

Lekskulchai, R. and Cole, J. (2001) 'Effect of a developmental program on motor performance in infants born preterm', *Australian Journal of Physiotherapy*, 47: 169–76.

Leonard, C.T., Hirschfeld, H. and Forssberg, H. (1991) 'The development of independent walking in children with cerebral palsy', *Developmental Medicine and Child Neurology*, 33: 567–77.

Livingston, M.H., Rosenbaum, P.L., Russell, D.J. and Palisano, R.J. (2007) 'Quality of life among adolescents with cerebral palsy: what does the literature tell us?', *Developmental Medicine and Child Neurology*, 49: 225–31.

Majnemer, A., Shevell, M., Rosenbaum, P., Law, M. and Poulin, C. (2007) 'Determinants of life quality in school-age children with cerebral palsy', *Journal of Pediatrics*, 151: 470–5.

Malaiya, R., McNee, A.E., Fry, N.R., Eve, L.C., Gough, M. and Shortland, A.P. (2007) 'The morphology of the medial gastrocnemius in typically developing children and children with spastic hemiplegic cerebral palsy', *Journal of Electromyography and Kinesiology*, 17: 657–63.

Marbini, A., Ferrari, A., Cioni, G., Bellanova, M.F., Fusco, C. and Gemignani, F. (2002) 'Immunohistochemical study of muscle biopsy in children with cerebral palsy', *Brain Development*, 24: 63–6.

McCormick, A., Brien, M., Plourde, J., Wood, E., Rosenbaum, P. and McLean, J. (2007) 'Stability of the gross motor function classification system in adults with cerebral palsy', *Developmental Medicine and Child Neurology*, 49: 265–9.

Mewasingh, L.D., Demil, A., Christiaens, F.J., Missa, A.M., Cheron, G. and Dan, B. (2002) 'Motor strategies in standing up in leukomalacic spastic diplegia', *Brain Development*, 24: 291–5.

Minter, C., Tylkowski, C., Bylica-Perryman, K. and Knapp, D. (2007) 'A comprehensive review of function, satisfaction, and physical and psychological health in adults with cerebral palsy', *Developmental Medicine and Child Neurology*, 49 (Suppl.): 22–3.

Moreau, N.G., Teefey, S.A. and Damiano, D.L. (2009a) 'In vivo muscle architecture and size of the rectus femoris and vastus lateralis in children and adolescents with cerebral palsy', *Developmental Medicine and Child Neurology*, 51: 800–6.

Moreau, N.G., Li, L., Geaghan, J.P. and Damiano, D.L. (2009b) 'Contributors to fatigue resistance of the hamstrings and quadriceps in cerebral palsy', *Clinical Biomechanics*, 24: 355–60.

Myklebust, B.M., Gottlieb, G.L., Penn, R.D. and Agarwal, G.C. (1982) 'Reciprocal excitation of antagonistic muscles as a differentiating feature in spasticity', *Annals of Neurology*, 12: 367–74.

Nordmark, E., Hagglund, G., Lauge-Pedersen, H., Wagner, P. and Westbom, L. (2009) 'Development of lower limb range of motion from early childhood to adolescence in cerebral palsy: a population-based study', *BMC Medicine*, 7: 65.

Nudo, R.J. (2003) 'Adaptive plasticity in motor cortex: implications for rehabilitation after brain injury', *Journal of Rehabilitation Medicine*, 41 (Suppl.): 7–10.

O'Connor, A.R., Birch, E.E. and Spencer, R. (2009) 'Factors affecting development of motor skills in extremely low birth weight children', *Strabismus*, 17: 20–3.

Odding, E., Roebroeck, M.E. and Stam, H.J. (2006) 'The epidemiology of cerebral palsy: incidence, impairments and risk factors', *Disability and Rehabilitation*, 28: 183–91.

Ostensjo, S., Carlberg, E.B. and Vollestad, N.K. (2004) 'Motor impairments in young children with cerebral palsy: relationship to gross motor function and everyday activities', *Developmental Medicine and Child Neurology*, 46: 580–9.

Palisano, R.J., Cameron, D., Rosenbaum, P.L., Walter, S.D. and Russell, D. (2006) 'Stability of the gross motor function classification system', *Developmental Medicine and Child Neurology*, 48: 424–8.

Palisano, R., Rosenbaum, P., Walter, S., Russell, D., Wood, E. and Galuppi, B. (1997) 'Development and reliability of a system to classify gross motor function in children with cerebral palsy', *Developmental Medicine and Child Neurology*, 39: 214–23.

Pascual-Leone, A., Wassermann, E.M., Sadato, N. and Hallett, M. (1995) 'The role of reading activity on the modulation of motor cortical outputs to the reading hand in braille readers', *Annals of Neurology*, 38: 910–15.

Pin, T.W., Darrer, T., Eldridge, B. and Galea, M.P. (2009) 'Motor development from 4 to 8 months corrected age in infants born at or less than 29 weeks' gestation', *Developmental Medicine and Child Neurology*, 51: 739–45.

Poon, D.M. and Hui-Chan, C.W. (2009) 'Hyperactive stretch reflexes, co-contraction, and muscle weakness in children with cerebral palsy', *Developmental Medicine and Child Neurology*, 51: 128–35.

Prosser, L.A., Lauer, R.T., VanSant, A.F., Barbe, M.F. and Lee, S.C. (2010a) 'Variability and symmetry of gait in early walkers with and without bilateral cerebral palsy', *Gait & Posture*, 31: 522–6.

Prosser, L.A., Lee, S.C., VanSant, A.F., Barbe, M.F. and Lauer, R.T. (2010b) 'Trunk and hip muscle activation patterns are different during walking in young children with and without cerebral palsy', *Physical Therapy*, 90: 986–97.

Richards, C.L., Malouin, F., Dumas, F., Marcoux, S., Lepage, C. and Menier, C. (1997)

'Early and intensive treadmill locomotor training for young children with cerebral palsy: a feasibility study', *Pediatric Physical Therapy*, 9: 158–65.

Romkes, J. and Brunner, R. (2007) 'An electromyographic analysis of obligatory (hemiplegic cerebral palsy) and voluntary (normal) unilateral toe-walking', *Gait & Posture*, 26: 577–86.

Rose, J., Haskell, W.L., Gamble, J.G., Hamilton, R.L., Brown, D.A. and Rinsky, L. (1994) 'Muscle pathology and clinical measures of disability in children with cerebral palsy', *Journal of Orthopedics Research*, 12: 758–68.

Rosenbaum, P., Paneth, N., Leviton, A., Goldstein, M., Bax, M., Damiano, D., Dan, B. and Jacobsson, B. (2007) 'A report: the definition and classification of cerebral palsy April 2006', *Developmental Medicine and Child Neurology Supplement*, 109: 8–14.

Rosenbaum, P.L., Walter, S.D., Hanna, S.E., Palisano, R.J., Russell, D.J., Raina, P., Wood, E., Bartlett, D.J. and Galuppi, B.E. (2002) 'Prognosis for gross motor function in cerebral palsy: creation of motor development curves', *Journal of the American Medical Association*, 288: 1357–63.

Russell, D.J., Rosenbaum, P.L., Cadman, D.T., Gowland, C., Hardy, S. and Jarvis, S. (1989) 'The gross motor function measure: a means to evaluate the effects of physical therapy', *Developmental Medicine and Child Neurology*, 31: 341–52.

Sanger, T.D., Delgado, M.R., Gaebler-Spira, D., Hallett, M. and Mink, J.W. (2003) 'Classification and definition of disorders causing hypertonia in childhood', *Pediatrics*, 111: e89–97.

Spittle, A.J., Doyle, L.W. and Boyd, R.N. (2008a) 'A systematic review of the clinimetric properties of neuromotor assessments for preterm infants during the first year of life', *Developmental Medicine and Child Neurology*, 50: 254–66.

Spittle, A.J., Brown, N.C., Doyle, L.W., Boyd, R.N., Hunt, R.W., Bear, M. and Inder, T.E. (2008b) 'Quality of general movements is related to white matter pathology in very preterm infants', *Pediatrics*, 121: e1184–9.

Stackhouse, S.K., Binder-Macleod, S.A. and Lee, S.C. (2005) 'Voluntary muscle activation, contractile properties, and fatigability in children with and without cerebral palsy', *Muscle & Nerve*, 31: 594–601.

Stackhouse, S.K., Binder-Macleod, S.A., Stackhouse, C.A., McCarthy, J.J., Prosser, L.A. and Lee, S.C. (2007) 'Neuromuscular electrical stimulation versus volitional isometric strength training in children with spastic diplegic cerebral palsy: a preliminary study', *Neurorehabilitation and Neural Repair*, 21: 475–85.

Sutherland, D., Olshen, R., Biden, E. and Wyatt, M. (1988) *The Development of Mature Walking*, Cambridge: Cambridge University Press.

Thelen, E. and Ulrich, B.D. (1991) 'Hidden skills: a dynamic systems analysis of treadmill stepping during the first year', *Monographs of the Society for Research in Child Development*, 56: 1–98; discussion 99–104.

Theroux, M.C., Oberman, K.G., Lahaye, J., Boyce, B.A., Duhadaway, D., Miller, F. and Akins, R.E. (2005) 'Dysmorphic neuromuscular junctions associated with motor ability in cerebral palsy', *Muscle & Nerve*, 32: 626–32.

Ulrich, D.A., Ulrich, B.D., Angulo-Kinzler, R.M. and Yun, J. (2001) 'Treadmill training of infants with Down syndrome: evidence-based developmental outcomes', *Pediatrics*, 108: e84.

Unnithan, V.B., Dowling, J.J., Frost, G. and Bar-Or, O. (1996) 'Role of cocontraction in the O_2 cost of walking in children with cerebral palsy', *Medicine and Science in Sports and Exercise*, 28: 1498–504.

Vaal, J., van Soest, A.J., Hopkins, B., Sie, L.T. and van der Knaap, M.S. (2000)

'Development of spontaneous leg movements in infants with and without periventricular leukomalacia', *Experimental Brain Research*, 135: 94–105.

van den Berg-Emons, H.J., Saris, W.H., de Barbanson, D.C., Westerterp, K.R., Huson, A. and van Baak, M.A. (1995) 'Daily physical activity of schoolchildren with spastic diplegia and of healthy control subjects', *Journal of Pediatrics*, 127: 578–84.

van der Heide, J., Paolicelli, P.B., Boldrini, A. and Cioni, G. (1999) 'Kinematic and qualitative analysis of lower-extremity movements in preterm infants with brain lesions', *Physical Therapy*, 79: 546–57.

van der Heide, J.C., Fock, J.M., Otten, B., Stremmelaar, E. and Hadders-Algra, M. (2005a) 'Kinematic characteristics of postural control during reaching in preterm children with cerebral palsy', *Pediatric Research*, 58: 586–93.

van der Heide, J.C., Fock, J.M., Otten, B., Stremmelaar, E. and Hadders-Algra, M. (2005b) 'Kinematic characteristics of reaching movements in preterm children with cerebral palsy', *Pediatric Research*, 57: 883–9.

van der Meer, A.L., van der Weel, F.R., Lee, D.N., Laing, I.A. and Lin, J.P. (1995) 'Development of prospective control of catching moving objects in preterm at-risk infants', *Developmental Medicine and Child Neurology*, 37: 145–58.

van Eck, M., Dallmeijer, A.J., Voorman, J.M. and Becher, J.G. (2008) 'Skeletal maturation in children with cerebral palsy and its relationship with motor functioning', *Developmental Medicine and Child Neurology*, 50: 515–19.

Watt, J.M., Robertson, C.M. and Grace, M.G. (1989) 'Early prognosis for ambulation of neonatal intensive care survivors with cerebral palsy', *Developmental Medicine and Child Neurology*, 31: 766–73.

Wiley, M.E. and Damiano, D.L. (1998) 'Lower-extremity strength profiles in spastic cerebral palsy', *Developmental Medicine and Child Neurology*, 40: 100–17.

Winchester, P., McColl, R., Querry, R., Foreman, N., Mosby, J., Tansey, K. and Williamson, J. (2005) 'Changes in supraspinal activation patterns following robotic locomotor therapy in motor-incomplete spinal cord injury', *Neurorehabilitation and Neural Repair*, 19: 313–24.

Wingert, J.R., Sinclair, R.J., Dixit, S., Damiano, D.L. and Burton, H. (2010) 'Somatosensory-evoked cortical activity in spastic diplegic cerebral palsy', *Human Brain Mapping*, 31: 1772–85.

Wolf, S.L., Winstein, C.J., Miller, J.P., Taub, E., Uswatte, G., Morris, D., Giuliani, C., Light, K.E. and Nichols-Larsen, D. (2006) 'Effect of constraint-induced movement therapy on upper extremity function 3 to 9 months after stroke: the EXCITE randomized clinical trial', *Journal of the American Medical Association*, 296: 2095–104.

Woollacott, M.H., Burtner, P., Jensen, J., Jasiewicz, J., Roncesvalles, N. and Sveistrup, H. (1998) 'Development of postural responses during standing in healthy children and children with spastic diplegia', *Neuroscience and Biobehavioral Reviews*, 22: 583–9.

Yeargin-Allsopp, M., Van Naarden Braun, K., Doernberg, N.S., Benedict, R.E., Kirby, R.S. and Durkin, M.S. (2008) 'Prevalence of cerebral palsy in 8-year-old children in three areas of the United States in 2002: a multisite collaboration', *Pediatrics*, 121: 547–54.

Index

Note: Page numbers followed by 'f' refer to figures and followed by 't' refer to tables.